Microbial Bioprocessing of Agri-Food Wastes

This book gives a comprehensive overview of recent advances in the valorization of agri-food waste and discusses the main process conditions needed to overcome the difficulties of using waste as alternative raw materials. It also discusses specific methodologies, opportunistic microbes for biomass valorization, the sustainable production of agri-food waste, as well as examines the assessment and management of bioactive molecules production from microbial-valorization of agri-food waste. The authors provide technical concepts on the production of various bio-products and their commercial interest including agri-food waste utilization in the microbial synthesis of proteins, the valorization of horticulture waste, the sustainable production of pectin via microbial fermentation, as well as other food and pharmacological applications.

This book is intended for bioengineers, biologists, biochemists, biotechnologists, microbiologists, food technologists, enzymologists, and related professionals and researchers.

- Explores recent advances in the valorization of agri-food waste
- Provides technical concepts on the production of various bio-products of commercial interest
- Discusses the main process conditions to overcome the difficulties of using waste as alternative raw materials
- Introduces technical-economic details on the advantages and disadvantages of exploring the waste recovery chain
- Explores the main technological advances in the recovery of residues in functional products.

Advances and Applications in Biotechnology
Series Editors
Gustavo Molina, Institute of Science and Technology – UFVJM, Diamantina – Brazil

Vijai Kumar Gupta, Biorefining and Advanced Materials Research Center; Center for Safe and Improved Food, SRUC, UK

Microbial Bioprocessing of Agri-food Wastes: Bioactive Molecules Volume 1
Gustavo Molina, Minaxi Sharma, Rachid Benhida, Vijai Kumar Gupta, and Ramesh Chander Kuhad

Microbial Bioprocessing of Agri-food Wastes: Industrial Applications Volume 2
Gustavo Molina, Minaxi Sharma, Rachid Benhida, Vijai Kumar Gupta, and Ramesh Chander Kuhad

Microbial Bioprocessing of Agri-food Wastes: Industrial Enzymes Volume 3
Gustavo Molina, Minaxi Sharma, Vipin Chandra Kalia, Franciele Maria Pelissari, and Vijai Kumar Gupta

Microbial Bioprocessing of Agri-food Wastes: Food Ingredients Volume 4
Gustavo Molina, Minaxi Sharma, Vipin Chandra Kalia, Franciele Maria Pelissari, and Vijai Kumar Gupta

Microbial Bioprocessing of Agri-Food Wastes

Bioactive Molecules

Edited by
Gustavo Molina
Minaxi Sharma
Rachid Benhida
Vijai Kumar Gupta
Ramesh Chander Kuhad

CRC Press
Taylor & Francis Group
Boca Raton London New York

CRC Press is an imprint of the
Taylor & Francis Group, an **informa** business

Designed cover image: © Shutterstock

First edition published 2023
by CRC Press
2385 NW Executive Center Drive, Suite 320, Boca Raton FL 33431

and by CRC Press
4 Park Square, Milton Park, Abingdon, Oxon, OX14 4RN

CRC Press is an imprint of Taylor & Francis Group, LLC

ISBN: 978-0-367-62518-4 (hbk)
ISBN: 978-0-367-65336-1 (pbk)
ISBN: 978-1-003-12897-7 (ebk)

DOI: 10.1201/9781003128977

Typeset in Times
by MPS Limited, Dehradun

Contents

Editor Biographies

Prof. Gustavo Molina graduated in Food Engineering, received his Master's (2010) and PhD (2014) degree at the University of Campinas – Unicamp (Campinas – Brazil), and part of his doctoral research was developed at the Laboratoire de Génie Chimique et Biochimique at the Université Blaise Pascal (Clermont-Ferrand – France). He is an associate professor at UFVJM (Diamantina – Brazil) in Food Engineering and a supervisor of students and researchers, as the head of the Laboratory of Food Biotechnology. In 2016–2017, he developed his Postdoctoral research at the Institut Polytechnique de Grenoble (Grenoble – France) in the area of biorefinery and development of the enzymatic hydrolysis process of lignocellulosic materials.

Dr. Minaxi Sharma is currently working as Senior Researcher at CARAH (Ath), Belgium. She has expertise in the area of food science and technology, nanoencapsulation, food waste valorization and food chemistry.

Dr. Rachid Benhida is currently the director of research at CNRS (DR1), co-director of the ICN institute for more than 8 years, and team leader of Bioactive Molecules' team. He received his PhD from the University of Paris XI, Orsay, France.

Dr. Vijai Kumar Gupta received his PhD in Microbiology in the year 2009 from Dr. RML Avadh University, India. Currently, he is working as Senior Fellow and Group Leader at *Center for Safe and Improved Foods & Biorefining and Advanced Materials Research Center*, SRUC, Edinburgh, UK.

Prof. Ramesh Chander Kuhad is MSc, MPhil, and PhD in Microbiology and has been working as Vice-Chancellor at the Central University of Haryana, Mahendergarh. He is also the Chairman of the Indian Academy of Microbiological Sciences and has served as a Member at the University Grants Commission and the Joint Director of the Institute of Life Long Learning (ILLL), University of Delhi, Delhi.

1 Sustainable Management of Agri-food Waste Valorisation: Current Status

Praveen Kumar Dikkala
Department of Food Science and Technology, College of Agriculture, Punjab Agricultural University, Ludhiana, Punjab, India

Supta Sarkar
Department of Foods & Nutrition, College of Community Science, Professor Jayashankar Telangana State Agricultural University, Hyderabad, Telangana, India

Monika Sharma
Department of Botany, Sri Avadh Raj Singh Smarak Degree College, Gonda, Uttar Pradesh, India

Sridhar Kandi
Department of Food Technology, Karpagam Academy of Higher Education (Deemed to be University), Coimbatore, Tamil Nadu, India

Ekta Belwal
Department of Foods and Nutrition, GB pant University of Agriculture and Technology, Pantnagar, Uttarakhand, India

Srinu Dhanavath
College of Food and Dairy Technology, TNUVAS, Chennai, Tamil Nadu, India

Pradeepa Roberts
Millet Processing and Incubation Centre, PJTS Agricultural University, Rajendra nagar, Hyderabad, Telangana, India

DOI: 10.1201/9781003128977-1

Kairam Narsaiah
AS & EC Division, ICAR-Central Institute of Post-Harvest Engineering and Technology, Ludhiana, Punjab, India

Amarjeet Kaur
Department of Food Science and Technology, College of Agriculture, Punjab Agricultural University, Ludhiana, Punjab, India

Zeba Usmani and Minaxi Sharma
Department of Applied Biology, University of Science and Technology, Ri-Bhoi, Baridua, Meghalaya, India

CONTENTS

1.1 INTRODUCTION

Innovative agricultural practices are increasing day by day to cope up with rising demand of food which accompanies constant growth in population. From the literature, it was clear that around 30–40% of food is wasted every year. From 30–40% it was reached to 50%, which is an undesirable impact to food industry and environment as well. The manufacture of food is inefficient, multifaceted, and emission intensive (Gustavsson et al., 2011). With different agriculturally based food wastes, stress is implied on the environment. In today's scenario, the AFWs are utilised in the production of bioactive components for the potential application in the areas of pharmaceutical, cosmetic, and food industrial use. These are providing additional income to the dependent sectors. Along with this, the valorisation of AFWs and their byproducts ensures food security and sustainability. In the food processing chain, food wastage is reached from raw material procurement site to the retail market. Globally, the food processing industries are generating a huge amount of food waste (Gustavsson et al., 2011). Specificity and seasonal production are the drawbacks for the industrial implementation of agri-food wastes recovery. In the food supply chain, the food waste is exclusively processed to produce animal feed, compost, fertiliser and also the anaerobic digestion to produce biogas (Lin et al., 2014). The current waste management techniques are disturbing the atmosphere with the surrounding landscape and releasing methane with decomposition (EEA, 2016). The inefficiency in waste disposal system posed an unnecessary cost for the consumers in several ways. So, the present system is adopting different ways to utilise the benefits from AFW. The AFW are deemed to be unavoidable and they are converted into different bioactive components. The shelf life of the different food products could be improved with the valorisation processes (Jones, 2019). Waste valorisation processes play an important role in the different recyclable packaging materials production (Lin et al., 2014). The bioactive substances extracted from wastes, with antioxidant, antimicrobial activity could be incorporated into the biodegradable packages. These are produced through the fermentation process with food waste streams. The benefits from this process are: the development of process with low carbon foot print, biodegradable packaging development and food products shelf-life development. The selected food waste residues can be used in extraction of different value-added products such as essential oils, bioactive components, whey protein concentrates, edible oils and other bioactive nutrients. With the above premises, integrated valorisation system of food supply chain is playing an important role in the production of different biomaterials ranging from bioactive components to chemical components (Lin et al., 2014).

1.2 OVERVIEW OF GLOBAL PRODUCTION OF AGRI-FOOD WASTE (AFW)

The agri-food waste is generated in the complete food processing chain. The production of huge quantities of food wastes (in any form) has become a global issue nowadays (Dahiya et al., 2016). The United Nation Environment Programme (UNEP) estimated around one-third of the food produced for the global consumption is wasted, which is almost 1.3 billion tonnes (https://www.unep.org/thinkeatsave/get-informed/worldwide-food-waste). Along with AFW, the non-consumable materials (packaging materials) used in the food processing chain are creating burden to producers, consumers, and environment (Ferreira da Cruz et al., 2012; 2014). Due to population increase, the AFW production is also increased. Globally, the estimated population increase in the next decade is one billion, by 2050 it reaches up to 9.6 billion (FAO, 2014). The different types of AFWs in the year 2019 have been shown in Figure 1.1 (FAO, 2019). Food security in the global way is really a big problem. Without proper mitigation strategies, with the increase in the food production the greenhouse gases are increased with deforestation which directly increased the agri-waste production (FAO, 2015; Lambin & Meyfroidt, 2011; Hoornweg et al., 2013). AFW in relation with environment got attention with different policy makers, analysts, and government organisations. According to United States Environmental Protection Agency (US EPA), the wastage in the food represents single type which is called as landfills (Nishida, 2014). Indeed, the reduction and diversification of wastes on this landfill becomes a priority. The excessive production of food leads to excessive utilisation of land, water, and fossil fuels. This leads to unnecessary depletion of natural resources, which causes environmental change (McLaughlin & Kinzelbach, 2015). Food degradation in the landfills emits methane, carbon dioxide and leads to increase the greenhouse gases in the environment. The different types of the developed and developing countries are taking preventive measures to reduce AFW. Along with the practical solutions, alternative techniques for valorisation of AFW are needed (Morone, 2016).

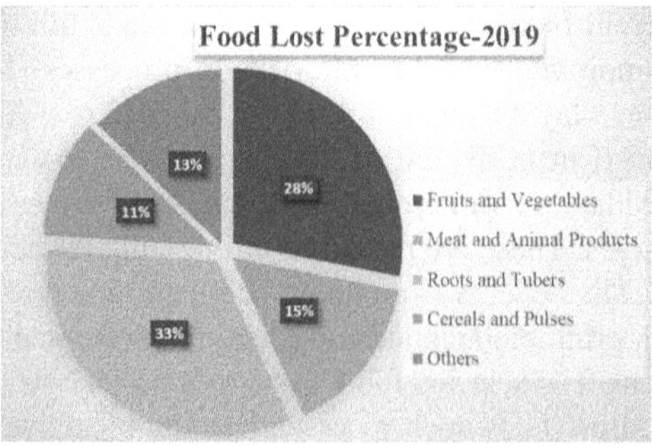

FIGURE 1.1 Food lost percentage during 2019 (data adapted from FAO, 2019).

1.3 DIFFERENT TYPES OF AFW COMPOSITION

The information on the AFW has a great importance in selecting the management technology, that is fundamentally depending on the nature of food components and its composition (Bräutigam et al., 2014). Depending upon the continent and country, the consumption of food varies, so the waste also varies. The composition in the fruits and vegetables, meat, and cereal foods changes, so that the valorised products also change according to their composition. The fruits and vegetable wastes consist of more bioactive components when compared with only vegetable wastes. On the wet weight basis total solids in fruits are ranging from 7.5% to 23%, while in the vegetable they are ranging from 3% to 11%; similarly, the volatile and mineral components are also changing, depending upon the product. In the anaerobic and aerobic biological treatments, the fruit and vegetable wastes gave the balanced composition and specific composition (Hegde et al., 2018). The fruit and vegetable wastes (cores, pits, twigs, spoiled fruits and vegetables, rinds, and skin) from the agricultural processing industries include the parts that are removed during complete processing (includes cleaning, processing, and packaging). Food processing sector produces large amounts of wastewaters that require suitable management practices. In fruit and vegetable processing industry, the wastewaters contain low concentration of the nutrients (both macro and micronutrients). The reduction and utilisation in food waste to obtain the different value-added products increases the efficiency that reduces the costs, improves accessibility towards the food, and further improves security of food (Gustavsson et al., 2011).

1.4 CURRENT VALORISATION AND MANAGEMENT PRACTICES FOR AFW

In the process of AFW valorisation, there are two types of techniques that include conventional techniques and innovative techniques. Figure 1.2 is showing different methods for the utilisation of biomass wastes to obtain different valuable compounds.

1.4.1 Conventional Techniques

The conventional techniques which were used in the process of valorisation management include the animal feeding, land filling/incineration, anaerobic digestion, and gasification.

1.4.2 Animal Feeding

The process of animal feeding should be considered as optional as every animal is unable to take the residues. Moreover, risks are associated with the unbalanced dietary patterns. Sometimes transportation and conservation make it unfeasible (Otles et al. 2015). The rendering of fruit waste safely with different thermal treatments increases cost of production. Under the controlled conditions, the use of waste in the animal feed is advantageous than the other treatments such as anaerobic digestion and landfilling. Some pre-treatments should be required for the preparation of AFW into

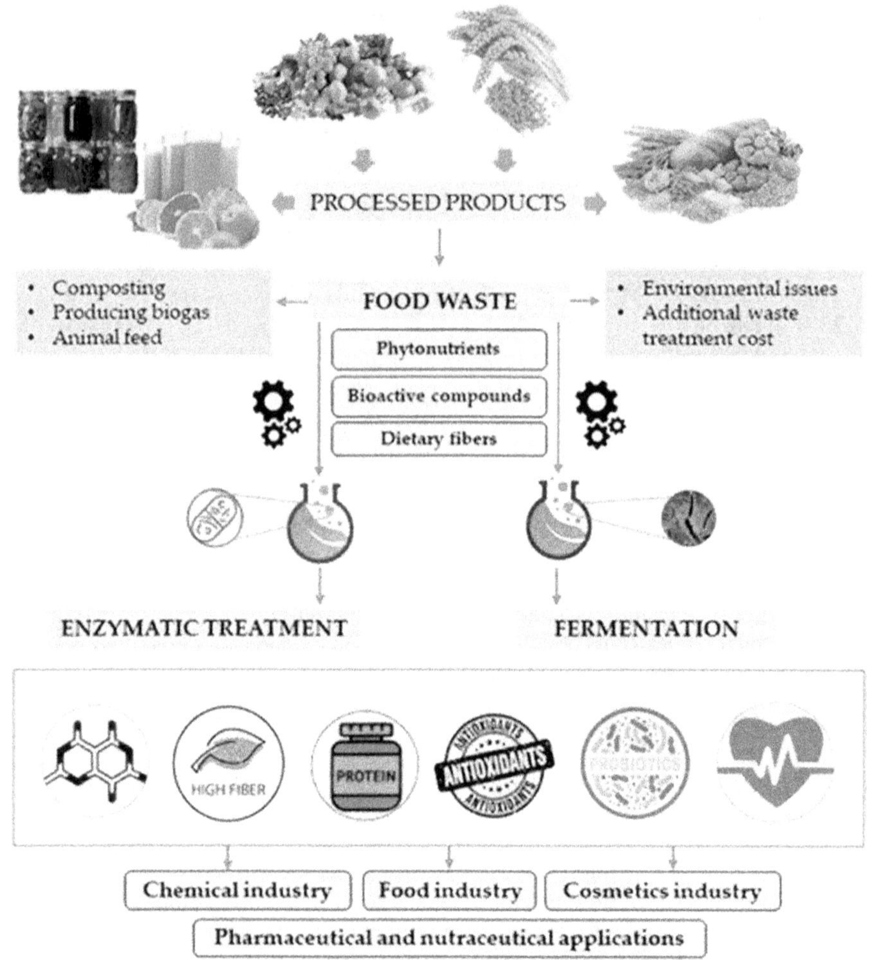

FIGURE 1.2 Methods of utilisation of agri-food wastes for different value-added products (adapted from Tlais et al., 2020).

animal feed. The partial substitution of this feed is having significant impact on the health and environment (Salemdeeb et al., 2017).

1.4.3 Landfilling and Incineration

This technique is the most common and simplest way of disposing solid waste residues that are having environmental impact. Methane is produced with the biological degradation which is the third biggest anthropogenic source that accounts 800 MT of CO_2 equivalent (Breeze, 2018). Micronutrients that are present in organic matter are hydrolysed into the soluble products in the production of methane gas, called as methanogenesis (Pearse et al., 2018). Combustion/incineration practice is regarded as advantageous from the producer point of view, it creates issues to the environment. To sustain the higher operating temperatures, the primary energy is needed (natural gas). The combusted heat is partially used in the process of incineration (to preheat air) or in the process of generation of power. Combustion results in the solid residues which are formed as a part of organic fraction, which is most frequently landfilled and that ashes are used as building materials (Dhir et al., 2018).

1.4.4 Anaerobic Digestion

The product which is formed with the anaerobic degradation of organic matter in landfills is biogas. Anaerobic degradation is very well-established technique in the manure and sludge waste treatment plants (Moral et al., 2018). The implementation of AFW valorisation techniques is very low (Ren et al., 2018; Braguglia et al., 2018). In an anaerobic process of digestion, the organic matter of the fruit wastes takes place same as in the process of landfill. This is carried out in a digester which controls operating parameters. The control of temperatures of anaerobic bacteria such as mesophiles or thermophiles plays a key role, because they are active at a temperature of 30–40 or 50–60°C range, respectively (Kibler et al., 2018). The other parameters that influence include the pH, C/N ratio, metal presence, and organic acid profile (Zhang et al., 2014). From the digesters, the higher yield of methane is obtained than in the landfills. The details about anaerobic digestion are further discussed in the biochemical profiling.

1.4.5 Gasification

The organic matter is converted into the different products in the process of pyrolysis or gasification (De Corato et al., 2018). In both of these processes, the gas consists of hydrogen and also carbon monoxide, further processed in the syngas (mixture of H_2 and CO). This is used as a feedstock for the process of chemical synthesis or further it is used as a fuel for heat and power generation (Kibler et al., 2018). In this, thermal decomposition exists in the non-oxidising atmosphere in the absence of reactant which results in the production of solid, liquid, and gas fraction. This proportion usually depends upon the operating conditions such as processing temperature, residence time, rate of heat, and also vaporisation time (García et al., 2015). The solid residues are having many applications including amendments in the solid and liquid residues as these are used in the production of chemicals and clean syngas (Opatokun et al., 2016).

1.5 EXTRACTION OF ORGANIC ACIDS AND CHEMICALS FROM AGRI-INDUSTRIAL BIOMASS

Different biomass processing units are utilising agri-food industrial wastes for biochemicals such as aspartic acid, lactic acid, succinic acid, citric acid, citrate salts, biopolyester, methanol, sugars, and lignin production (Dessie et al., 2020). This clearly shows the importance of AFW in the production of essential commodities and other bioactive ingredients. In the process of conversion of biomass, pre-treatments play an important role, the first step is conversion of biomass into sugars (hydrolysates) and the second step is conversion of hydrolysates into biofuels (methane, ethane, ethanol, butanol) or in the form of organic acids such as propionic acid, acetic acid, butyric acid, and lactic acid (Liu & Wu, 2016).

1.5.1 Extraction of Biofuels from Agri-industrial Food Wastes

Different industries are utilising AFI (agri-food industry) wastes for biofuels production, including the diesels from fats and cooking oil residues, thermal and electrical power from wood and agricultural biomass, etc. (Dessie et al., 2020; Usmani et al., 2021a). With different agricultural wastes (corn stover, sugarcane bagasse. and carbohydrate biomass), ethanol is majorly produced. Greenhouse gases were reduced with the use of ethanol mixes by 12–19% when compared with the petroleum-based fuels (Irmak, 2017).

1.5.2 Biogas

Biogas is the mix of combination of gases with the gasification of biomass substrates. It is made up of CO, CO_2, H_2, and CH_4. Biogas derived from AFI wastes is a sustainable form of energy, reducing environmental issues with generating lots of business. Nowadays, marine biowastes are the novel innovative materials for the production of biogas (Eswari et al., 2020).

1.5.3 Syngas

Syngas is a mixture of carbon monoxide and hydrogen, which is capable to replace petroleum. It is a valuable renewable resource of energy (De María et al., 2013). With the utilisation of lignocellulosic biomass and process optimisation, syngas is produced (Liu & Wu, 2016).

1.5.4 Bioplastics

The agricultural waste residues produced from forests, marine, and agricultural land are gaining tremendous importance in the production of poly hydroxy alkanoates (PHAs) and poly-3-hydroxybutyrates (PHBs) and are called as bioplastics (Al-Battashi et al., 2019). These are safe substitutes to the modern plastics, mostly produced homopolymer, having good potential as bioplastics (Hamieh et al., 2013).

1.6 EMERGING TECHNIQUES FOR FOOD WASTE VALORISATION

Several types of innovative and green extraction techniques, pre-treatment, and purification strategies are currently utilised by food and biotechnological industries to exploit their potential bioactive components at their utmost chelation. Different strategies are highlighted in Figure 1.3.

Before the innovative valorisation methods, due to some restrictions based on the physical status/condition of the AFWs, there is a need to utilise some pretreatment techniques to increase the storage and handling properties of the food

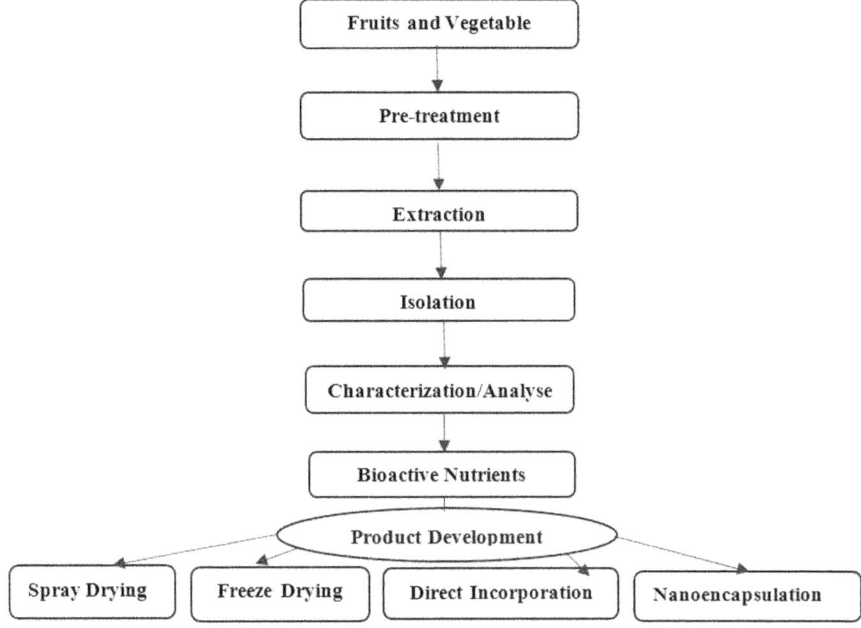

FIGURE 1.3 Different strategies for fruit and vegetable waste valorisation to functional food development.

wastes. There are several pre-treatment methods utilised for the valorisation of AFWs, such as drying, particle size reduction, densification, palletisation, and extrusion, as shown in Figure 1.4.

1.6.1 Pre-treatment Techniques

Pre-treatment techniques are applied to reduce the volume/mass and physical state of the AFWs. For the processing of raw materials which are in the wet state, the drying is necessary. With the increase in the storage time, the wet biomass will start degrading biologically and chemically. If the agri-food processing waste is to be stored for some time, the conversion of larger particles into the smaller particles takes place faster with higher efficiency. In this process, baling, pelletising, and extrusion process play an important role, some of them are discussed below. After the complete recovery of bioactive components, the residual materials should be taken care. The utilisation of the different AFW in the fertiliser production requires drying, palletisation, and densification steps (Arshadi et al., 2016).

For more efficient utilisation of AFW, palletisation pre-treatment plays an important role amongst all. This process consists of one or more roller, with a perforated die. In the process of rotation, the die and rollers are forced through the perforations, which forms a densified pellet. The factors affecting the pelletised products are moisture, particle shape, size, chemical composition, and equipment type. Pressure and temperature gradients play an important role in the production of pellets (Arshadi et al., 2016).

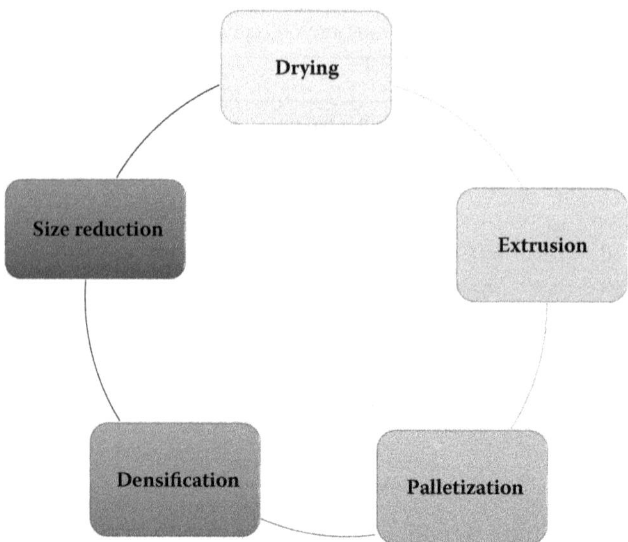

FIGURE 1.4 Pre-treatment techniques for agri-food waste valorisation.

1.6.2 Extraction Technologies for AFW Valorisation

The bioactive components that are extracted from the process of conventional and non-conventional methods consist of both the advantages and disadvantages. With the utilisation of water or other green solvents with the innovative extraction methodologies, the green technologies are playing an important role over the traditional extraction technologies (solvent extraction), also called as non-conventional technologies which are gaining tremendous advancements to valorise the AFWs (Sharma et al., 2021). The higher yield, less processing time, high-quality products, less generation of wastes in green technologies have replaced the traditional conventional technologies in valorisation sector. The effectiveness of these techniques depends upon the processing parameters such as solvent, time, pressure and temperature, and chemical composition of the biomass (Azmir et al., 2013). In this chapter, we only described the novel, innovative, or green extraction technologies that are used to process the AFW which include: nanotechnologies, high hydrostatic pressure, ultrasound extraction, ohmic heating, radiofrequency drying, pulsed electric fields, pressurised fluids, and ultrasound-assisted extraction.

1.6.2.1 Supercritical CO_2 Extraction

This area highlights the main areas in the extraction of different byproducts from AFW. Supercritical fluid extraction is the process of extraction of different bioactive substances in the form of green extraction technologies. The fluids that are used in the process of supercritical fluid extraction are water and CO_2 (Herrero et al., 2006). It has been used vigorously due to many benefits over traditional methods of organic solvent extraction. It is renewable, recyclable, and non-toxic. As it is highly adjustable extraction solvent, there are some slight adjustments that are applied to the temperature and pressure that change the solvent density, which in turn change the density-dependent properties such as solubility, partition coefficient, and dielectric constant. This is the reason that supercritical carbon is selected as a solvent. The addition of small amounts of polar modifiers leads to extraction of more yields, due to increase in the solvent polarity. It is possible to carry out the process of supercritical fractionation with reduction in downstream processing. In this process, no solvent residues were remained after the extraction, so that it can be used in both the food and pharmaceutical industries. With these technologies the high valued waxes, sesquiterpenes, and essential oils are extracted as these are insoluble in water. The different lignocellulosic biomasses from wheat straw, flax straw, corn wastes, and sugarcane bagasse were optimised to extract different hydrophobic materials (Kerton & Marriott, 2013). These are optimised for extraction and characterisation of hydrophobic constituents which consist of waxes. The typical hydrophobic extracted compounds from the supercritical carbon dioxide extracts consist of fatty acids (both the saturated and unsaturated fatty acids), hydrocarbons (long chain), fatty aldehydes, sterols, waxy esters, steroid ketones, and also triterpenoids (Arshadi et al., 2016). These molecules are widely used in the different applications such as nutraceuticals, medicinal products, and flavour and fragrance products. From the supercritical wax extractions, corn stover can be used as anti-foam component in the detergent formulations. These are potential replacers for the current non-renewable anti-foams having negative impacts on eutrophication, carcinogenesis and which will have very less impact on the environment (Attard et al., 2015).

1.6.2.2 Subcritical Water Extraction

Subcritical water is referred as liquid at room temperatures between the boiling point and critical temperature (374°C) and pressure required to maintain it in liquid state. As carbon dioxide is a non-polar solvent, it is ideal for lipophilic components. Water acts as an attractive solvent in particularly subcritical extraction process as it is polar solvent. Here, water grabbed the attention due to its safe, selectivity, economic viability and environmentally friendly nature. As like CO_2, it offers high selectivity, polarity of the solvent, and also targets the compounds (Haghighi & Khajenoori, 2013). With the application of heat, the critical temperature decreases and changes the polarity range (13–80). Subcritical water acts as reactive medium in the attractive application of AFW management. Water in the subcritical and critical states follows the potential hydrolysis, oxidation, and decomposition reactions; these reactions are exploited for the valorisation of AFW. From different AFWs, different phenolic compounds from different types of plant wastes such as phenolic compounds from different apple byproducts, potato, mango peels, and bitter melon extracts are produced. Some researchers focused on the optimisation of extraction techniques for the efficient isolation of different phenolic substances from several plants. Cvetanović et al. (2015) investigated the effect of different extraction techniques such as soxhlet, ultrasound-assisted, microwave-assisted, and subcritical water extractions on the bioactivities of chamomile (*Matricaria chamomilla* L.) flower extracts. They reported that the subcritical water extraction showed highest yield (49.70%) and total phenolic content (151.45 mg ECA/mL) among all the extracts obtained from the different tested techniques.

1.6.2.3 Ultrasound-assisted Extraction

This technology disrupts the tissues in the different cells and allows the deeper penetration of solvent into the raw materials. The degradation in the structure of cells further extracts the important target compounds like antioxidants and other bioactive substances (Roselló-Soto et al., 2015). For example, the utilisation of ultrasound-assisted extraction is successfully applied in coconut shells valorisation to produce phenolic compounds, has been studied by Rodrigues and Pinto (2007).

1.6.2.4 Microwave-assisted Extraction

It is an endothermic and spontaneous process, the electromagnetic radiations are in the range of 300 MHz–300 GHz, in which most commonly used frequency is 2450 MHz. In this technology, electromagnetic radiations are converted into thermal energy which facilitates the food processing (Zhang et al., 2011). In this technique, when the sample comes in contact with the solvent, the hydrogen bonds break. When the sample is heat treated, it results in dipole movement of molecules that migrates the ions which dissolute the components (Datta et al., 2005). In this mechanism, due to polarity, the water molecules heated-up that induce a pressure on the cell wall which change the physical properties and porosity of the cell membrane, thus improve the penetration of the solvent in the cells and improve the yield of targeted biomaterials (Routray & Orsat, 2011). In the process of extraction, the temperature, solvent, and feed ratio influence very strongly in the yield of flavonoids. The hesperidin extraction is about 47.7 mg/g from the citrus fruit skin, the immature peels consist of higher amounts of hesperidin content (3.1 times more) (Inoue et al., 2010). The phenolic content yield in the choke berries is 420.1 equivalents mg of gallic acid/ 100g at 300 watts for 5 minutes (Simić et al., 2016).

1.6.2.5 Electro-technologies

The electro-technologies include the pulsed electric field, high voltage electrical discharges, non-pulsed electric fields (ohmic heating and moderate electric field), radio frequency drying, high hydrostatic pressure, pressurised fluids, and nanotechnology (Zhu et al., 2020). These inactivate the microbes in the liquid foods, while low pulsed electric fields were used in the extraction of different bioactive substances (Gabric et al., 2018). This approach is helpful in the extraction of different nutraceuticals that allow the selective extraction in the intracellular spaces (Koubaa et al., 2016; Parniakov et al., 2016). High voltage pulses are passed through materials which are placed in between the electrodes for a short period of time (microseconds to milliseconds) in the pulsed electric field (Saini et al., 2019; Puértolas et al., 2010, 2016). Some of the recovery compounds which are sensitive to heat, light is needed to be protected and delivered in human body in a controlled way (Gómez et al., 2018). This process is useful in reduction of reactivity in the bioactive components (light and heat), increase in solubility, bioavailability, masking, and controlled release. The physicochemical properties of bioactive substances are always interlinked with valorisation in the nano-micro encapsulation process (Vincekovic et al., 2017).

1.6.2.6 Enzyme-assisted Extraction

It is an alternative technology for the conventional processes to extract various valuable components with water as a solvent. In this process, water is used as a solvent and enzyme is used as an extracting medium. The concentration of the enzymes, particle size, water–solid ratio, and hydrolysis time are the major factors influencing the extraction efficiency of the targeted compounds (Puri et al., 2012). Some studies deliberately extracted carotenoids from pumpkin, and anthocyanins from *Crocus sativus* and grape skin using this process (Ghosh & Biswas 2015; Lotf et al., 2015; Muñoz et al., 2004).

1.6.2.7 Laser Ablation

This process is allowed to improve heat and mass transfer processes to recover the macromolecular substances. This method is used in the extraction of pectin and other aromatic substances from fruit matrices (Panchev et al., 2011).

1.7 BIOACTIVE COMPONENT RECOVERY FROM THE AFWS

Enormous production of food waste is becoming a global concern nowadays (Dahiya et al., 2016). The world needs to feed around 10 billion by 2050; this should reduce the environment burden and balance the economy (Zabaniotou & Kamaterou, 2018).

1.7.1 Potential Recovery from AFW

Amongst all the AFWs, fruit and vegetable processing wastes consist of large number of suspended solids (SS), chemical oxygen demand (COD), and biological oxygen demand (BOD) that influence recovery in many ways. The BOD ranges from 3.2 g/L in the bakery products, 0.53 g/L in meat specialties while COD ranges from 7–0.9 g/L (UNIDO, 2013). The organic wastes are composed of 75% hemicellulose, 9% of cellulose, and 5% lignin (Kosseva, 2011). The extraction procedures of different components involve in the potential production of different bioactive substances. Table 1.1 shows various kinds of bioactive nutrients extracted from the fruit and vegetable wastes. The recovery and reuse should not affect the environment and new production process. It is preferable to promote reuse practices without any manipulations which evaluates the biorefinement with the adoptions in the environment assessment methodologies which encompasses the entire lifecycle of the by product (Mirabella et al., 2014).

1.8 BIOCHEMICAL PROFILING OF AFW

The major biochemical processes that are involved in biowaste degradation are anaerobic degradation and composting. There are many factors that influence the biochemical processes of agricultural and food waste (Arshadi et al., 2016).

TABLE 1.1
The Different Types of the Bioactive Components Extracted from AFW

Technology	Source	Bioactive Substances	Solvent Used	Yield	References
Supercritical Fluid extraction (SFE)	Orange pomace (dry)	Phenolic component	Pure ethanol	21.2 GAE/g of extract	Espinosa-Pardo et al. (2017)
Ultrasound- and microwave-assisted extraction	Pumpkin peel waste	Carotenoids	Corn oil	UAE-peel: 38.03; MAE-peel: 34.94 μg/g of oil extracts	Sharma and Bhat (2021)
Ultrasound-assisted extraction	Seabuckthorn pomace	Dietary fibre	Citric acid and ethanol	16%	Hussain et al. (2021)
Pulsed electric field	Grape byproduct	Anthocyanin	Water and ethanol	14.05 mg Cy-3-glu eq./g dry matter	Corrales et al. (2008)
Supercritical Fluid extraction (SFE)	Orange pomace (dry)	Phenolic component	Ethanol/water (9:1)	20.7 GAE/g of extract	Espinosa-Pardo et al. (2017)
Supercritical fluid extraction (SFE)	Skin fraction (Grape pomace)	Polyphenol content	–	11.9 mgGAE/g	Manna et al. (2015)
Microwave-assisted extraction	Red grape pomace	Polyphenols	Water	52,645 ppm GAE (dry extract)	Drosou et al. (2015)
Pulsed electric field	Blueberry press cake	Polyphenols	50% ethanol and 0.5	+63%	Bobinaitė et al. (2015)
Ultrasound-assisted extraction	Tomato pomace	Carotenoids (all-translycopene, β-carotene)	Hexane and ethanol	7.49 to 14.08 mg/100 g dw	Luengo et al. (2014)
Microwave-assisted extraction	Gac peels	Carotenoids	Ethyl acetate	262.3±3.5 mg/100 g DW	Chuyen et al. (2017)
Microwave-assisted extraction	Peel waste from *C. sinensis*	Polyphenols	Acetone in water (20–80%)	12.09 mgGAE/g DW	Nayak et al. (2015)

Pulsed electric field	Grape seeds	Polyphenols	Ethanol	9 g/100 g GAE	Boussetta et al. (2012)
Supercritical Fluid extraction (SFE)	Tomato waste	Lycopene	Liquid CO2	729.98 mg/kg	Kehili et al. (2017)
Ultrasound-assisted extraction	Grape seeds	Phenols, antioxidant,	Ethanol	5.41 mg GAE/100 ml	Ghafoor et al. (2009)
Enzyme-assisted extraction	Pumpkin	Carotenoids	Pectinex UltraSP	2 mg/100 g	Ghosh and Biswas (2015)
Ultrasound-assisted extraction	Grape solid waste	Naringin (Flavonoid)	Ethanol	24–36 mg/g dw	Garcia-Castello et al. (2015)
Enzyme-assisted extraction	*Crocus sativus*	Anthocyanins	Pectinex	6.7 mg/g	Lotf et al. (2015)
Ultrasound-assisted extraction	Rind waste from pomegranate	Natural colour	Water	20%	Sivakumar et al. (2011)
Enzyme-assisted extraction	*Capsicum annuum*	Carotenoids	Cellulose and pectinase	41.72–279.83 mg/100 g	Nath et al. (2016)

1.8.1 The Possible Biochemical Processes

Low-temperature chemical process is a choice that can be used in the combination or alone in the biochemical or thermal process. The low-temperature hydrolysis allows to obtain soluble polymeric substances like lignin and others (Rosso et al., 2015). In the other bio- and thermo-chemical processes, the low-temperature hydrolysis doesn't disrupt the natural molecular structure. This process saves the original structures with original carbon types without change in the functional groups. The initial investments and production costs are very less for this process. It doesn't require any secondary treatment of waste. Different soluble biological wastes are produced in the range from 14 to >100 KD and have been obtained with the acid or alkaline hydrolysis at 60–100 °C after the process of aerobic or anaerobic degradation (Rosso et al., 2015; Franzoso et al., 2015).

1.8.2 Environmental Aspects

Anaerobic process of digestion and composting are classified as recycling material when compost is used. If it is not used for the purposes, it should be pre-treated before incineration or land filling. Anaerobic degradation can be seen in the energy recovery, which produces biogas for energy production (95% of biological treatment) (Arshadi et al., 2016).

1.8.3 Economic Aspects

In the circular biological economy, target towards economic viability with sustainability is very important. The current practices that are involved in treating the biowastes are burden to the manufacturers, even though with the value-added products. The estimated valuable input for the potential compost production is around 35–40 MT (Montoneri et al., 2011). The marketability for this compost is very poor. Instead of chemical fertilisers, the public and private organisations should focus on biocompost, which is the major solution. This product is also used in the restoration of land by filling, helps in the development of agricultural crops. In the agricultural waste treatment, biowaste treatment plants play a vital role that integrates all the processes, as it contains four sections: anaerobic digestion, aerobic digestion, sewage wastewater treatment, and land fill area. The municipal solid waste, entered the anaerobic process is fermented to yield biogas and organic matter. By using the same way, we can effectively valorise the AFW to produce valuable biofuel and can preserve the fossil fuel resources. This ultimately optimise the economy as well as reduce the impact on the environment (Arshadi et al., 2016).

1.9 AFW VALORISATION IMPROVES SUSTAINABILITY OF FOOD SYSTEMS

A huge amount of food is wasted in the agricultural and food processing system. Degradation in the environment, economical differences, and distress in the society are the main challenges on the development and reconsideration (IPCC, 2012). The sustainable development in food security, energy efficiency, and protection of environment are the main drivers for food waste valorisation. The idea on harvesting the high value-added products from different agricultural and food wastes are aligned with the sustainable development, which aims to achieve food security, environmental protection, and energy efficiency (Ong et al., 2018).

Biodiversity on the entire globe is declining with ongoing loss in the different species, habitats, and populations (UNEP, 2012). As population is increasing day by day, the different variants of plants are lost due to the land clearance for the cultivation of different food crops. So, utilisation of agricultural and food wastes is playing a key role in the recovery of macro- and micronutrients, which has a positive impact on environment. The policies need to strengthen public's perception towards the nature and humanity as they are interdependent and interacting. These adaptive changes acknowledge the systematic and dynamic nature of global changes (Allen & Prosperi, 2016). Sustainable food system is one which provides food safety and nutritional security along with environmental sustainability which provides food for many generations with minimum adverse effect. The process of sustainability in the agricultural system encourages the local production of food and its distribution and availability of nutritious food to consumers and other communities (Story et al., 2009).

1.10 AFW VALORISATION FOR SUSTAINABLE NUTRITIONAL SECURITY DEVELOPMENT

The process of agri-food wastes valorisation increases the interest by developing the different sustainable value-added products (biorefinery and bioactive components). The biorefinery components include the biofuels, bio stimulants, and fine chemicals (Nayak & Bhushan, 2019; Usmani et al., 2021b). In AFW management, sustainability is a guiding strategy for the society. The sustainable development is a key goal for present and future generations with certain effects in the socio-economic and environmental progress. There is an inter relationship between the food system sustainability and security which involves in the reduction of food waste. The waste reduction has been emphasised recently in many ways at global level with many strategic documents (Capone et al., 2014). The protection of natural resources which are strongly linked up with food supply is a major priority nowadays. The ever-growing population creates a serious issue on the food and energy. All of these created an uncertainty in the global food processing sector. The generation of huge amounts of AFW is the major indicator for this instability. Food processing sector can produce enough food with safety with very less impact on the environment. The improvement in the efficiency, prevention in food waste generation, and valorisation play a key role for sustainability development goals (SDGs). In this case, the valorisation techniques improved the demand for different bioactive materials and are solution for restoring the bioactive potential of the AFW (Otles & Kartal, 2018). Appropriate waste management practices such as prevention of unsustainable natural resource utilisation and reutilisation strategies for the recovery of bioactive nutrients, biobased chemicals, and biofuel should be discussed and elaborated by the sharing efforts of farmers, industrialists, consumers, and policy makers.

1.10.1 The Legal, Hygienic, and Sanitary Requirements

The main aspect of valorisation technologies is to make feasible study on the various valorisation techniques with legal, sanitary, and hygienic proposed solutions. Some legislations are specific to promote some AFW and some are made for some foods; European Union legislations considered fruit and vegetable wastes as a biodegradable waste with specific regulations (European Commission, 2010). The major elements in this type of regulation include: (a) registration is compulsory for all food business operators, (b) perfect hygienic practices should be implemented for all food and feed business operators, (c) introduction of policies related to HACCP (hazard analysis of critical control point), and (d) compulsory requirements should be introduced at farm level (Martin et al., 2017).

1.11 AFW Valorisation for Industrial and Economic Sustainability Improvements

The development which is undergoing to meet the needs of present generation without compromising the future generation meets the sustainability. In order to improve sustainability of the food production system, it should focus on components such as environment, society, and economic activities. Recovery from the AFWs is a wide way to extract different valuable components with high market value (Rahmanian et al., 2014). It dramatically reduces the environmental pollution, generates new opportunities, and benefits to the public directly/indirectly. The sustainability in the food production systems can be achieved by recycling its valuable components by reducing the food waste. Food products must be sustainable politically, economically, ecologically, socially, and technologically. In a supply chain, the manufacturer's responsibility of the product is till the consumer's satisfaction. So, everyone should take responsibility for health safety and environmental impacts of the products across the product life cycle. Achieving the sustainable supply chain requires taking into account about political, economic, physical, financial, social, and technical impacts on the food waste from production to the disposal, even after the disposal stage. Efficient process of waste management system plays an important role that benefits all the members in the supply chain. Different food organisations should take the long-term sustainable approaches for the optimisation of the shared value and opportunities to create innovative products that develop the social well-being (European Commission, 2010). In order to increase the degree of profitability of food chain members, effective waste management is essential. The less utilisation of raw materials and energy substantially improves the environmental efficiency of the food systems, which is achieved with the effective utilisation of raw materials and resources for the production. Effective use of materials can be done in two ways in the process of recycling of food waste. For example, grape marc obtained after pressing can undergo the distillation with ethanol or solvent extraction in order to extract the phenolic compounds with retained antioxidant activity (Lin et al., 2014). Through these processes, the solid remains of the grapes can be transformed into high-value new materials instead of throwing them away which would simply add to the landfills. Other examples could be valorising tomato peel from fruit processing industries to obtain β-carotene, lycopene via solvent or supercritical extraction methods; orange peels to obtain citric acid, β-carotene, polygalacturonates, limonene, pectin via different methods like solvent extraction, fermentation, hydro distillation, etc. (Teigiserova et al., 2019). Recovering of monosaccharides and oligopeptides from the cheese industrial wastes is another way of sustainable improvement of wastes (Madureira et al., 2010). It is beneficial to the industry financially, as these materials will not be needed to purchase. The reuse of wastewater is another economic improvement that could be accomplished by recovering valuable components of food waste in the wastewater.

1.12 SOCIAL AND ENVIRONMENTAL SUSTAINABILITY IMPROVEMENTS

Owing to their unsustainable usage of resources, increased global competition, population growth, and other environmental problems, planet is facing depletion of natural resources. Valorisation might provide win-win solutions to reduce wastage, stimulate income growth and job opportunities, and thereby improve sustainable local development. Besides, the waste reduction practices are the most preferred choice since they have greater potential for better environmental and socioeconomic outcomes, even though they are considered as the most challenging (Imbert, 2017). Using the suitable waste valorisation techniques, different high-quality bioactive components were produced, which could be used in the different food processing sectors to improve the shelf life of the food (Oreopoulou & Tzia, 2007). This improves the availability of food by delaying the food deterioration so that the product improves the health and livelihood of the people as a whole. The high valued components recovered from the AFW will help to foster the viability and diversity in the rural and urban economics. New job opportunities could be provided by current and new methods of efficient recovery of food waste materials.

1.13 FUTURE ASPECTS AND CONCLUSIONS

Many developing countries are processing food and storing for further use that produces huge amounts of agricultural waste including household and retail, which poses harmful effect to the environment; so, valorisation techniques are required to protect the environment and livelihood. In the developed countries, most of the losses occur at the retailer and the consumer level due to its inadequate planning or bad purchasing habits. The nonreusable food waste is having negative consequences in the country's GDP and to environment which is contributing to the climate changes. In the different developing nations, the major loss in the primary production is caused due to deficiencies in the production systems. All the actions taken against AFW should be classified into three categories: reduction, reuse or re-entry, recycling. In the hierarchy of valorisation, fruit and vegetable wastes should be prioritised as they can be reused for the extraction of bioactive compounds for the development of functional foods which is an interesting area for many food industries. These bioactive food ingredients could be a valuable platform in the food fortification sector and future functional food development. The food industry wastes, which are going to be utilised for further food production, must fulfil the principles which are subjected to the economical, technical, and legal requirements that should be environmentally and socially sustainable. It should be very important to know about every aspect of the innovative valorisation techniques, which gives a way forward to highlight the opportunities of high valued products. The different bioengineered micro-organisms also play a vital role to scale up the value-added products from the fruit and agricultural wastes. Scalable waste management practices at production, marketing, transportation, retailer, house-hold, and at consumer level should be more practiced, designed, and elaborated. So, much of research is required to reach the suitable progress in the areas of food waste valorisation. Therefore, advancement in food-biotechnological processes, biochemical, bioengineering, and microbiological interventions are needed for the valorisation of food waste biomass towards zero waste utilisation in the framework of circular economy.

REFERENCES

Al-Battashi, H. S., Annamalai, N., Sivakumar, N., Al-Bahry, S., Tripathi, B. N., Nguyen, Q. D., & Gupta, V. K. (2019). Lignocellulosic biomass (LCB): a potential alternative biorefinery feedstock for poly hydroxy alkanoates production. *Rev Environ Sci Biotechnol* **18**: 183–205.

Allen, T., & Prosperi, P. (2016). Modelling sustainable food systems. *Environ Manag* **57**: 956–975.

Arshadi, M., Attard, T. M., Bogel-Lukasik, R. M., Brncic, M., da Costa Lopes, A. M., Finell, M., Geladi, P., Gerschenson, L. N., Gogus, F., Herrero, M., Hunt, A. J., Ibanez, E., Kamm, B., Mateos-Aparicio Cediel, I., Matias, A., Mavroudis, N., Montoneri, E., Morais, A. R. C., Nilsson, C., Papaioannou, E. H., Richel, A., Ruperez, P., Škrbi, B., Bodroza Solarov, M., Švarc-Gaji, J., Waldron, K., & Yuste, F (2016). Pre-treatment and extraction techniques for recovery of added value compounds from wastes throughout the agri-food chain. *Green Chem.* **18**: 6160–6204.

Attard, T. M., Theeuwes, E., Gomez, L. D., Johansson, E., Dimitriou, I., Wright, P. C., Clark, J. H., McQueen-Mason, S. J., & Hunt, A. J. (2015). Supercritical extraction as an effective first-step in a maize stover biorefinery. *RSC Adv* **5**: 43831–43838.

Azmir, J., Zaidul, I. S. M., Rahman, M. M., Sharif, K. M., Mohamed, A., Sahena, F., Jahurul, M. H. A., Ghafoor, K., Norulaini, N. A. N., & Omar, A. K. M. (2013). Techniques for extraction of bioactive compounds from plant materials: a review. *J Food Eng* **117**: 426–436.

Bobinaitė, R., Pataro, G., Lamanauskas, N., Šatkauskas, S., Viškelis, P., & Ferrari, G. (2015). Application of pulsed electric feld in the production of juice and extraction of bioactive compounds from blueberry fruits and their by-products. *J Food Sci Technol* **52**: 5898–5905.

Boussetta, N., Vorobiev, E., Le, L. H., Cordin-Falcimaigne, A., & Lanoisselle, J. L. (2012). Application of electrical treatments in alcoholic solvent for polyphenols extraction from grape seeds. *LWT-Food Sci Technol* **46**: 127–134.

Braguglia, C. M., Gallipoli, A., Gianico, A., & Pagliaccia, P. (2018). Anaerobic bioconversion of food waste into energy: a critical review. *Bioresour Technol* **248**: 37–56.

Bräutigam, K. R., Jörissen, J., & Priefer, C. (2014). The extent of food waste generation across EU-27: different calculation methods and the reliability of their results. *Waste Manag Res* **32**(8): 683–694.

Breeze, P. (2018). Landfill waste disposal, anaerobic digestion, and energy production. In *Energy from waste* (pp. 39–47). Academic Press – Elsevier.

Capone, R., Bilali, H. E., Debs, P., Gianluigi, C., & Noureddin, D. (2014). Food system sustainability and food security: Connecting the dots. *J Food Security* **2**(1): 13–22.

Chuyen, H. V., Nguyen, M. H., Roach, P. D., Golding, J. B., & Parks, S. E. (2017). Microwave assisted extraction and ultrasound-assisted extraction for recovering carotenoids from gac peel and their effects on antioxidant capacity of the extracts. *Food Sci Nutr* **6**: 189–196.

Corrales, M., Toepf, S., Butz, P., Knorr, D., & Tauscher, B. (2008). Extraction of anthocyanins from grape by-products assisted by ultrasonics, high hydrostatic pressure or pulsed electric felds: A comparison. *Innov Food Sci Emerg Technol* **9**: 85–91.

Cvetanović, A., Švarc-Gajić, J., Mašković, P., Savić, S., & Nikolić, L (2015). Antioxidant and biological activity of chamomile extracts obtained by different techniques: perspective of using superheated water for isolation of biologically active compounds. *Indus Crops Prod* **65**: 582–591.

Dahiya, S., Kumar, A. N., Shanthi Sravan, J., Chatterjee, S., Sarkar, O., & Mohan, S. V. (2016). Food waste biorefinery: sustainable strategy for circular bioeconomy. *Bioresour Technol* **215**: 2–12.

Datta, A. K., Sumnu, G., & Raghavan, G. S. V. (2005). Dielectric properties of foods. In Rao, M. A., Rizvi, S. S. H., & Datta, A. K. (Eds.), *Engineering properties of foods* (3rd edn.). Boca Raton: Taylor & Francis.

De Corato, U., De Bari, I., Viola, E., & Pugliese, M. (2018). Assessing the main opportunities of integrated biorefining from agro-bioenergy co/by-products and agro-industrial residues into high-value added products associated to some emerging markets: a review. *Renew Sustain Energy Rev* **88**: 326–346.

de María, R., Díaz, I., Rodríguez, M., & S´ aiz, A. (2013). Industrial methanol from syngas: kinetic study and process simulation. *Int J Chem Reac Eng.* **11**: 469–477.

Dessie, W., Luo, X., Wang, M., Feng, L., Liao, Y., Wang, Z., Yong, Z., & Qin, Z. (2020). Current advances on waste biomass transformation into value-added products. *Appl Microbiol Biotechnol* **104**: 4757–4770.

Dhir, R. K., De Brito, J., Lynn, C. J., & Silva, R. V. (2018). Municipal solid waste composition, incineration, processing and management of bottom ashes. In *Sustainable construction materials* (pp. 31–90). Elsevier.

Drosou, C., Kyriakopoulou, K., Bimpilas, A., Tsimogiannis, D., & Krokida, M. (2015). A comparative study on different extraction techniques to recover red grape pomace polyphenols from vinification by-products. *Ind Crops Prod* **75**: 141–149.

Espinosa-Pardo, F. A., Nakajima, V. M., Macedo, G. A., Macedo, J. A., & Martínez, J. (2017). Extraction of phenolic compounds from dry and fermented orange pomace using supercritical CO_2 and cosolvents. *Food Bioprod Process* **101**: 1–10.

Eswari, A. P., Meena, R. A. A., Kannah, R. Y., Sakthinathan, G., Karthikeyan, O. P., & Banu, J. R. (2020). Bioconversion of marine waste biomass for biofuel and value-added products recovery. *Refining biomass residues for sustainable energy and bioproducts* 481–507.

European Commission. (2010). Preparatory study on food waste across EU 27– Technical report (Accessed on Dec 23).

European Environment Agency (EEA). (2016). Circular economy in Europe-Developing the knowledge base, EEA Report No 2/2016. (Accessed on Dec 23).

Ferreira da Cruz, N., Ferreira, S., Cabral, M., Simoes, P., & Cunha Marques, R. (2014). Packaging waste recycling in Europe: Is the industry paying for it? *Waste Manag* **34**(2): 298–308.

Ferreira da Cruz, N., Simoes, P., & Cunha Marques, R. (2012). Economic cost recovery in ~ the recycling of packaging waste: the case of Portugal. *J Clean Prod* **37**: 8–18.

Food and Agricultural Organisation (FAO). (2014). Food Wastage Footprint. Full-Cost Accounting e Final Report. *Rome* (Accessed on Dec 20, 2020).

Food and Agricultural Organisation (FAO). (2015). The State of Food Insecurity in the World 2015. *Rome* (Accessed on Dec 20, 2020).

Food and Agricultural Organisation (FAO). (2019). Key Facts on Food Loss and Waste. http://www.fao.org/save food/resources/keyfindings/en/ (Accessed on Dec 20, 2020).

Franzoso, F., Antonioli, D., Montoneri, E., Persico, P., Tabasso, S., Laus, M., Mendichi, R., Negre, M., & Vaca-Garcia, C. (2015). Films made from poly(vinyl alcohol-*co*-ethylene) and soluble biopolymers isolated from postharvest tomato plant. *J Applied Polymer Sci* **132**(18): 41935 (1–11).

Gabric, D., Barba, F., Roohinejad, S., Gharibzahedi, S. M. T., Radoj'cin, M., Putnik, P., & Bursac Kova´cevi´c, D. (2018). Pulsed electric fields as an alternative to thermal processing for preservation of nutritive and physicochemical properties of beverages: a review. *J Food Proc Engineer* **41**(1): e12638.

García, L., Abrego, J., Bimbela, F., & Sanchez, J. L. (2015). Hydrogen production from catalytic biomass pyrolysis. In Fang, Z., Smith Richard, J. L., & Qi, X. (Eds.), *Production of hydrogen from renewable resources* (pp. 119–147).

Garcia-Castello, E. M., Rodriguez-Lopez, A. D., Mayor, L., Ballesteros, R., Conidi, C., & Cassano, A. (2015). Optimization of conventional and ultrasound assisted extraction of flavonoids from grapefruits (*Citrus paradisi L.*) solid wastes. *LWT Food Sci Technol* **64**: 1114–1122.

Ghafoor, K., Choi, Y. H., Jeon, J. Y., & Jo, I. H. (2009). Optimization of ultrasound-assisted extraction of phenolic compounds, antioxidants, and anthocyanins from grape (*Vitis vinifera*) seeds. *J Agric Food Chem* **57**: 4988–4994.

Ghosh, D., & Biswas, P. K. (2015). Enzyme-aided extraction of carotenoids from pumpkin tissues. *Indian Chem Eng* **58**: 1–11.

G6mez, B., Barba, F. J., Domínguez, R., Putnik, P., Kovačević, D. B., Pateiro, M., Toldrá, F., & Lorenzo, J. M. (2018). Microencapsulation of antioxidant compounds through innovative technologies and its specific application in meat processing. *Trends Food Sci Technol* **82**: 135–147.

Gustavsson, J., Cederberg, C., Sonesson, U., Van Otterdijk, R., & Meybeck, A. (2011). *Global food losses and food waste: extent, causes and prevention*. Rome (Italy): Food and Agriculture Organisation of the United Nations (FAO) (Accessed on Dec 23).

Haghighi, A., & Khajenoori, M. (2013). Subcritical water extraction. *Mass Transfer – Advances in Sustainable Energy and Environment Oriented Numerical Modeling*. 10.5772/54993

Hamieh, A., Olama, Z., & Holail, H. (2013). Microbial production of poly hydroxy butyrate, a biodegradable plastic using agro-industrial waste products. *Glob Adv Res J Microbiol* **2**: 54–64.

Hegde, S., Lodge, J. S., & Trabold, T. A. (2018). Characteristics of food processing wastes and their use in sustainable alcohol production. *Renew Sustain Energy Rev* **81**: 510–523.

Herrero, M., Cifuentes, A., & Ibañez, E. (2006). Sub- and supercritical fluid extraction of functional ingredients from different natural sources: plants, food-by-products, algae and microalgae: a review. *Food Chem* **98**: 136–148.

Hoornweg, D., Bhada-Tata, P., & Kennedy, C. (2013). Environment: waste production must peak this century. *Nature* **502**: 615–617.

https://www.unep.org/thinkeatsave/get-informed/worldwide-food-waste

Hussain, S. Sharma, M., & Bhat, R. (2021). Valorisation of Sea Buckthorn pomace by optimization of ultrasonic-assisted extraction of soluble dietary fibre using response surface methodology. *Foods* **2021**(10): 1330.

Imbert, E. (2017). Food waste valorisation options: opportunities from the bioeconomy. *Open Agricul.* **2**(1): 195–204.

Inoue, T., Tsubaki, S., Ogawa, K., Onishi, K., & Azuma, J. I. (2010). Isolation of hesperidin from peels of thinned *Citrus unshiu* fruits by microwave-assisted extraction. *Food Chem* **123**: 542–547.

IPCC. (2012). Summary for policymakers. In *Managing the risks of extreme events and disasters to advance climate change adaptation. A special report of working groups I and II of the intergovernmental panel on climate change.* Cambridge: Cambrige University Press.

Irmak, S. (2017). Biomass as raw material for production of high-value products. *Biomass Volume Estimation and Valorisation for Energy.* **9**. doi:10.5772/65507

Jones, K. (2019). Agri-food waste valorisation. *Food Sci Technol* **33**(4): 60–63.

Kehili, M., Kammlott, M., Choura, S., Zammel, A., Zetzl, C., Smirnova, I., Allouche, N., & Sayadi, S. (2017). Supercritical CO_2 extraction and antioxidant activity of lycopene and β-carotene-enriched oleoresin from tomato (*Lycopersicum esculentum* L.) peels by-product of a Tunisian industry. *Food Bioprod Process* **102**: 340–349.

Kerton, F. M., & Marriott, R. (2013). *Alternative solvents for green chemistry.* Cambridge, UK: Royal Society of Chemistry.

Kibler, K. M., Reinhart, D., Hawkins, C., Motlagh, A. M., & Wright, J. (2018). Food waste and the food-energy-water nexus: A review of food waste management alternatives. *Waste Manag* **74**: 52–62.

Kosseva, M. R. (2011). Comprehensive biotechnology, management and processing of food wastes. **6**: 557–593. 10.1016/b978-0-08-088504-9.00393-7

Koubaa, M., Barba, F. J., Grimi, N., Mhemdi, H., Koubaa, W., Boussetta, N., & Vorobiev, E. (2016). Recovery of colorants from red prickly pear peels and pulps enhanced by pulsed electric field and ultrasound. *Inno Food Sci Emer Technol* **37**: 336–344.

Lambin, E. F., & Meyfroidt, P. (2011). Global land use change, economic globalization, and the looming land scarcity. *Proc Natl Acad Sci USA* **108**: 3465–3472.

Lin, C. S. K., Koutinas, A. A., Stamatelatou, K., Mubofu, E. B., Matharu, A. S., Kopsahelis, N., Pfaltzgraff, L. A., Clark, J. H., Papanikolaou, S., & Kwan, T. H. (2014). Current and future trends in food waste valorisation for the production of chemicals, materials and fuels: A global perspective. *Biofules, Bioproducts Biorefining* **5**: 686–715.

Liu, C.-M., & Wu, S.-Y. (2016). From biomass waste to biofuels and biomaterial building blocks. *Renew Energy* **96**: 1056–1062.

Lotf, L., Kalbasi-Ashtari, A., Hamedi, M., & Ghorbani, F. (2015). Effects of enzymatic extraction on anthocyanins yield of saffron tepals (*Crocos sativus*) along with its colour properties and structural stability. *J Food Drug Anal* **23**: 210–218.

Luengo, E., Condón-Abanto, S., Condón, S., Álvarez, I., & Raso, J. (2014). Improving the extraction of carotenoids from tomato waste by application of ultrasound under pressure. *Sep Purif Technol* **136**: 130–136.

Madureira, A. R., Tavares, T., Gomes, A. M. P., Pintado, M. E., & Malcata, F. X. (2010). Invited review: psychological properties of bioactive peptides obtained from whey proteins. *J Diary Sci* **93**(2): 473–455.

Manna, L., Bugnone, C. A., & Banchero, M. (2015). Valorisation of hazelnut, coffee and grape wastes through supercritical fluid extraction of triglycerides and polyphenols. *J Supercrit Fluid* **104**: 204–211.

Martin, D. S., Bald, C., Cebrian, M., Iñarra, B., Orive, M., Ramos, S., & Zufía, J. (2017). Principles for developing a safe and sustainable valorisation of food waste for animal feed: Second generation Feedstuff. *Handbook of Famine, Starvation, and Nutrient Deprivation.* Cham: Springer, 1–20.

McLaughlin, D., & Kinzelbach, W. (2015). Food security and sustainable resource management. *Water Resour Res* **51**: 4966–4985.

Mirabella, N., Castellani, V., & Sala, S. (2014). Current options for the valorisation of food manufacturing waste: A review. *J Clean Prod* **65**: 28–41.

Montoneri, E., Mainero, D., Boffa, V., Perrone D. G., & Montoneri, C. (2011). Biochemenergy: A Project to turn an urban wastes treatment plant into biorefinery for the production of energy, chemicals and consumer's products with friendly environmental impact. *Int J of Global Environmen.* **11**: 170–196.

Moral, A., Reyero, I., Alfaro, C., Bimbela, F., & Gandía, L. M. (2018). Syngas production by means of biogas catalytic partial oxidation and dry reforming using Rh-based catalysts. *Catal Today* **299**: 280–288.

Morone, P. (2016). The times they are a-changing: making the transition towards a sustainable economy. *Biofuels Bioprod Biorefin.* 10.1002/bbb.1647

Muñoz, O., Sepulveda, M., & Schwartz, M. (2004). Effects of enzymatic treatment on anthocyanic pigments from grapes skin from Chilean wine. *Food Chem.* **87**: 487–490.

Nath, P., Kaur, C., Rudra, S. G., & Varghese, E. (2016). Enzyme assisted extraction of carotenoid-rich extract from red capsicum (*Capsicum Annuum*). *Agric Res.* **5**: 193–204.

Nayak, A., & Bhushan, B. (2019). An overview of the recent trends on the waste valorisation techniques for food wastes. *J Environ Manag* **233**: 352–370. doi:10.1016/j.jenvman.2018.12.041

Nayak, B., Dahmoune, F., Moussi, K., Remini, H., Dairi, S., Aoun, O., & Khodir, M. (2015). Comparison of microwave, ultrasound and accelerated-assisted solvent extraction for recovery of polyphenols from citrus sinensis peels. *Food Chem* **187**: 507–516.

Nishida, J. (2014). Reducing food waste and promoting food recovery globally. EPA Connect. *The Official Blog of the EPA Leadership.* https://blog.epa.gov/blog/2014/10/reducing-food-waste-and-promoting-food-recovery-globally/ (Accessed on Dec 20, 2020).

Ong, K. L., Kaur, G., Pensupa, N., Uisan, K., & Lin, C. S. K. (2018). Trends in food waste valorization for the production of chemicals, materials and fuels: case study South and Southeast Asia. *Bioresour Technol* **28**: 100–112.

Opatokun, S. A., Kan, T., Al Shoaibi, A., Srinivasakannan, C., & Strezov, V. (2016). Characterization of food waste and its digestate as feedstock for thermochemical processing. *Energy Fuel* **30**(3): 1589–1597.

Oreopoulou, V., & Tzia, C. (2007). Utilization of plant by-products for the recovery of proteins, dietary fibers, antioxidants, and colorants. In Oreopoulou, V., & Russ, W. (Eds.), *Utilization of by-products and treatment of waste in the food industry* (pp. 209–232). New York: Springer Science+Business.

Otles, S., & Kartal, C. (2018). Sustainable food systems from agriculture to industry. *Food Waste Valorization* **11**: 371–397. doi:10.1016/B978-0-12-811935-8.00011-1

Otles, S., Despoudi, S., Bucatariu, C., & Kartal, C. (2015). Food waste management, valorisation, and sustainability in the food industry. In Kosseva, M. R., & Webb, C. (Eds.), *Food waste recovery* (pp. 3–23). Academic Press.

Panchev, I. N., Kirtchev, N. A., & Dimitrov, D. D. (2011). Possibilities for application of laser ablation in food technologies. *Innov Food Sci Emer Technol* **12**(3): 369–374.

Parniakov, O., Barba, F. J., Grimi, N., Lebovka, N., & Vorobiev, E. (2016). Extraction assisted by pulsed electric energy as a potential tool for green and sustainable recovery of nutritionally valuable compounds from mango peels. *Food Chem* **192**: 842–848.

Pearse, L. F., Hettiaratchi, J. P., & Kumar, S. (2018). Towards developing a representative biochemical methane potential (BMP) assay for landfilled municipal solid waste – A review. *Bioresour Technol* **254**: 312–324.

Puértolas, E., Koubaa, M., & Barba, F. J. (2016). An overview of the impact of electrotechnologies for the recovery of oil and high-value compounds from vegetable oil industry: energy and economic cost implications. *Food Research Inter* **80**: 19–26.

Puértolas, E., López, N., Saldaña, G., Álvarez, I., & Raso, J. E. (2010). Evaluation of phenolic extraction during fermentation of red grapes treated by a continuous pulsed electric field process at pilot-plant scale. *J Food Eng* **98**:120–125.

Puri, M., Sharma, D., & Barrow, C. J. (2012). Enzyme-assisted extraction of bio-actives from plants. *Trends Biotechnol* **30**: 37–44.

Rahmanian, N., Jafari, S. M., & Galanakis, C. M. (2014). Recovery and removal of phenolic compounds from olive mill wastewater. *J Am Oil Chem Soc* **91**: 1–18.

Ren, Y., Yu, M., Wu, C., Wang, Q., Gao, M., Huang, Q., & Liu, Y. (2018). A comprehensive review on food waste anaerobic digestion: Research updates and tendencies. *Bioresour Technol* **247**: 1069–1076.

Rodrigues, S., & Pinto, G. A. S. (2007). Ultrasound extraction of phenolic compounds from coconut (*Cocos nucifera*) shell powder. *J Food Eng* **80**(3): 869–872.

Roselló-Soto, E., Barba, F. J., Parniakov, O., Galanakis, C. M., Lebovka, N., Grimi, N., & Vorobiev, E. (2015). High voltage electrical discharges, pulsed electric field, and ultrasound assisted extraction of protein and phenolic compounds from olive kernel. *Food Biopro Technol* **8**(4): 885–894.

Rosso, D., Fan, J., Montoneri, E., Negre, M., Clark, J., & Mainero, D. (2015). Conventional and microwave assisted hydrolysis of urban biowastes to added value ligninlike products. *Green Chem* **17**(6): 3424–3435.

Routray, W., & Orsat, V. (2011) Blueberries and their anthocyanins: factors affecting biosynthesis and properties. *Comp Rev Food Sci Food Saf* **10**: 303–320.

Saini, A., Panesar, P. S., & Bera, M. B. (2019). Valorisation of fruits and vegetables waste through green extraction of bioactive compounds and their nano emulsions-based delivery system. *Bioreso Bioprocess* **6**(26): 1–12.

Salemdeeb, R., zu Ermgassen, E. K. H. J., Kim, M. H., Balmford, A., & Al-Tabbaa, A. (2017). Environmental and health impacts of using food waste as animal feed: a comparative analysis of food waste management options. *J Clean Prod* **140**: 871–880.

Sharma, M., & Bhat, R. (2021). Extraction of carotenoids from pumpkin peel and pulp: comparison between innovative green extraction technologies (ultrasonic and microwave-assisted extractions using corn oil). *Foods* **10**(4): 787.

Sharma, M., Usmani, Z., Gupta, V. K., & Bhat, R. (2021). Valorization of fruits and vegetable wastes and by-products to produce natural pigments. *Crit Rev Biotechnol* **41**(4): 535–563.

Simić, V. M., Rajković, K. M., Stojičević, S. S., Veličković, D. T., Nikolić, N. Č., Lazić, M. L., & Karabegović, I. T. (2016). Optimization of microwave assisted extraction of total polyphenolic compounds from chokeberries by response surface methodology and artificial neural network. *Sep Purif Technol* **160**: 89–97.

Sivakumar, V., Vijaeeswarri, J., & Anna, J. L. (2011). Effective natural dye extraction from different plant materials using ultrasound. *Ind Crops Prod* **33**: 116–122.

Story, M., Hamm, M. W., & Wallinga, D. (2009). Food systems and public health: linkages to achieve healthier diets and healthier communities. *J Hunger Environ Nutr* **4**: 219–224.

Teigiserova, D. A., Hamelin, L., & Thomsen, M. (2019). Review of high-value food waste and food residues biorefineries with focus on unavoidable wastes from processing. *Resour Conser Recyc* **149**: 413–426.

Tlais, A. Z. A., Fiorino, G. M., Polo, A., Filannino, P., & Di Cagno, R. (2020). High-value compounds in fruit, vegetable and cereal byproducts: an overview of potential sustainable reuse and exploitation. *Molecules* **25**: 2987. 10.3390/molecules25132987

UNEP. (2012). *Measuring progress: environmental goals and gaps*. Nairobi: UNEP.

United Nations Industrial Development Organization (UNIDO). (2013). Pollution from Food Processing Factories and Environmental Protection. Available at: http://www.unido.org/fileadmin/import/32129_25PollutionfromFoodProcessing.7.pdf (accessed Sept 2013).

Usmani, Z., Sharma, M., Awasthi, A. K., Lukk, T., Tuohy, M. G., Gong, L., & Gupta, V. K. (2021a). Lignocellulosic biorefineries: The current state of challenges and strategies for efficient commercialization. *Renew Sustain Energy Rev* **148**: 111258. 10.1016/j.rser.2021.111258

Usmani, Z., Sharma, M., Awasthi, A. K., Sivakumar, N., Lukk, T., Pecoraro, L., Thakur, V. K., Roberts, D., Newbold, J., & Gupta, V. K. (2021b). Bioprocessing of waste biomass for sustainable product development and minimizing environmental impact. *Bioresour Technol* **322**: 124548.

Vincekovic, M., Viskic, M., Juric, S., Giacometti, J., Bursac Kovacevic, D., Putnik, P., Donsi., F., Barba., F. J., & Rezek Jambrak, A. (2017). Innovative technologies for encapsulation of Mediterranean plants extracts. *Trends Food Sci Technol* **69**: 1–12.

Zabaniotou, A., & Kamaterou, P. (2018). Food Waste valorisation advocating circular bioeconomy – a critical review of potentialities and perspectives of spent coffee grounds biorefinery. *J Cleaner Production.* doi 10.1016/j.jclepro.2018.11.230

Zhang, C., Su, H., Baeyens, J., & Tan, T. (2014). Reviewing the anaerobic digestion of food waste for biogas production. *Renew Sustain Energy Rev* **38**: 383–392.

Zhang, W., Yao, Y., Sullivan, N., & Chen, Y (2011). Modelling the primary size effects of citrate-coated silver nanoparticles on their ion release kinetics. *Environ Sci Technol* **45**: 4422–4428.

Zhu, Z., Gavahian, M., Barba, F. J., Roselló-Soto, E., Kovacevic, D. B., Putnik, P., & Denoya, G. I. (2020). Valorization of waste and by-products from food industries through the use of innovative technologies. *Agri-Food Industry Strategies for Healthy Diets and Sustainability, New Challenges in Nutrition and Public Health.* **11**: 249–266.

2 Microbial Processing on Agri-wastes to Volatile Fatty Acids

Fateme Asadi
Department of Environmental Health Engineering, Hamadan University of Medical Sciences, Hamadan, Iran

Fatemeh Nouri
Department of Pharmaceutical Biotechnology, School of Pharmacy, Hamadan University of Medical Sciences, Hamadan, Iran

Mohammad Taheri
Department of Medical Microbiology, Faculty of Medicine, Hamadan University of Medical Sciences, Hamadan, Iran

CONTENTS

DOI: 10.1201/9781003128977-2

2.1 INTRODUCTION

Volatile fatty acids (VFAs) are organic acids containing six or fewer carbon molecules, e.g., acetic acid, propionic acid, butyric acid, isobutyric acid, and isovaleric acid (1). VFAs contain several utilizations in the cosmetics, textiles, bioenergy, food industry, and pharmaceuticals (2). The commercial generation of VFAs counts on the chemical production utilizing nonrenewable raw petroleum as supply matter (3), but petroleum raw material production is a developing distress owing to global warming and the reduction of nonrenewable fuel source (4). VFAs can, in addition, be formed through microbial fermenting and these acids are the last production of different fermentation processes (5); and biological processes are achieving concentration for VFA generation since renewable carbon origin sources can be applied as raw material, building alike methods further environmentally friendly. Consumers must besides attend that food additives and pharmaceutical goods starting from the biological route are healthy, and they are frequently set to suggest an upper value for like products (6). Different microbes have been investigated to make VFAs under anaerobic conditions, e.g., Acetobacter, Clostridium, Kluyveromyces, Propionibacterium, and Moorella using different raw materials as carbon origin (7). Most recently, investigators have discovered various procedures to generate VFAs via fermentation by using a rate of carbon origin as precursors (8). Mainly VFA generation is based upon utilizing net sugars, i.e., glucose, xylose, etc., and these routes lead to an upper efficiency with least byproducts (9). In any case, the advantage of an upper efficiency of such methods is perhaps invalidated by the upper price of the raw materials. To decrease generation price, investigators are now considering utilizing richly obtainable lignocellulose biomass as a carbon origin for VFA generation (10). Microbes are not capable of directly used complex lignocellulose biomass; therefore, the biomass must be primary treated utilizing different physical, chemical, and enzymatic processes, so addition to the price (10). Now, investigators have begun to employ waste sewage to make VFAs, and various efforts have been made to investigating various kinds of waste (11). Waste sewage fermentation is a cost-effective method for VFA generation and can be a varied key for waste purify (12). VFAs formed utilizing waste sewage can be more utilized like raw material for polyhydroxyalkanoate, electricity, and biogas generation (1). The employ of pure commercial sugar is cost-efficient, but treatment coming about the product is low-cost. Because of fewer side products, also, the employ of other raw materials, such as biomass, whey, and waste sewage, is cost-effective, but the treatment route is expensive (13). To choose a bioprocess to create VFAs, there is a necessity to regard all variables associated with the efficiency, performance, precursors price, byproducts, and ease of downstream method. A metabolic pathway method can be utilized to raise the efficiency of the VFAs, inhibit the byproducts, and use varied carbon origin. In any case, because of the inaccessibility of genetic devices to generate different VFAs in anaerobic micro-organisms, it does not appear to be an easy job to operate their metabolic engineering. Therefore, this section describes the approaches to VFA generation, their utilization, and solutions for more progress in VFA generation.

2.2 DEVELOPMENT OF VFAS

2.2.1 Anaerobic Digestion Process

Firstly, utilized principally for food and beverage generation, anaerobic decomposition deals with the oldest biological method technologies applied by humans. They have been used and expanded over many years, even though significant progress has been obtained in the previous years with the overview of different forms of high-rate treatment technologies, especially for manufacturing wastewater. Anaerobic digestion includes the conversion of organic matter by a consortium of bacterial and oxygen-free conditions. During this process, a gas is formed that mainly contains methane and carbon dioxide. The steps of anaerobic digestion are shown in Figure 2.1. The primary stage of anaerobic digestion is the hydrolysis of animal or plant material. This stage breaks down biopolymers and other organic matter and produces smaller molecules, as below molecules (Figure 2.2):

Lipids → fatty acids
Protein → amino acids
Nucleic acids → purines and pyrimidines
Polysaccharides → monosaccharides

The next stage is the transformation, by acetogenic bacteria, of products of the primary stage to organic acids, carbon dioxide, and hydrogen. In addition to acetic acid, acetogenic bacteria produce other organic acids. The most important organic acids that are formed include acetic acid (CH_3COOH), propionic acid (CH_3CH_2COOH), and butyric acid (CH3CH2CH2COOH). Ethanol (CH3CH2OH) is too generated. The last stage is the methanogenesis. Methane and carbon dioxide are created from acetate, ethanol, and other byproducts according to Equations (2.1)–(2.3):

$$CH_3COOH \rightarrow CH_4 + CO_2 \tag{2.1}$$

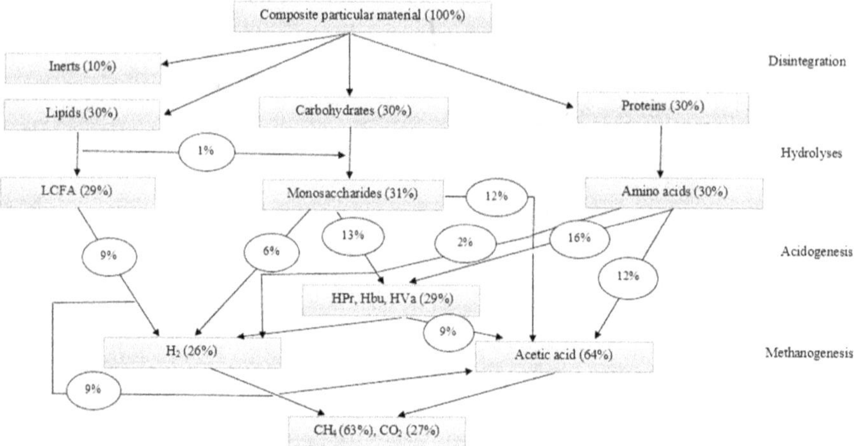

FIGURE 2.1 COD for a spatial degradation comprised of 10% inference and 30% all carbohydrates, proteins, and lipids (in the parameter of COD). All organic proportions of surviving things are depressed to volatile fatty acids (a mix of C2–C6 acid consisting of acetic acid, propionic acid/lactic acid, butyric acid, valeric acid, and caproic acid), which transfer methane and CO_2 gas with anaerobic hydrolysis. Methane will be chemically or biologically oxidized into CO_2 and H_2O. Nutrients (Nitrogen and phosphate) in surviving such as animals, plants, and microbial biomass can be recovered finely into N_2 gas and phosphate salts, via several chemical or biological reactions. Sulfur products digested anaerobically are transferred into H_2S gas.

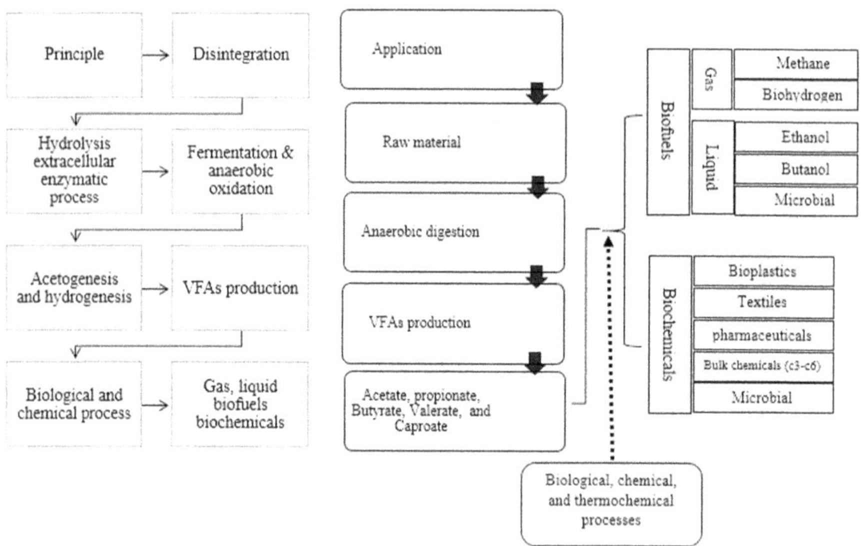

FIGURE 2.2 Concept of VFA platform.

$$2C_2H_5OH + CO_2 \rightarrow CH_4 + 2CH_3COOH \quad (2.2)$$

$$CO_2 + 4H_2 \rightarrow CH_4 + 2H_2O \quad (2.3)$$

The advantage of this process is that accompanies the treatment of waste with energy generation (methane).

2.2.2 Mixed VFA Fermentation

Technology usually requires a sterilization method to grow a particular micro-organism. Nevertheless, the mixed VFA fermentation progression does not need a sterilization procedure and instead uses a varied microbial community. The consortium micro-organisms can prepare energy reserves and is a cost-effective process (14). The composition of created mixed VFAs in anaerobic mixed acid fermentation is various. Depending on the mix of the VFAs, the mass balance may too differ. At high temperature, the usual composition is the ratio of acetate/butyrate/propionate = 6:3:1 or 7:2:1. For a ratio of 7:2:1, the theoretical mass balance is achieved as Equation (2.4):

$$2.3C_6H_{12}O_6 \rightarrow 0.8C_3H_7COOH + 0.4C_2H_5COOH + 3.6CH_3COOH + 2.2CO_2 + 0.2H_2O + 2.4H_2 YVFA/G = 76.3 \quad (2.4)$$

2.2.3 VFA Platform Development

Homoacetogenic bacteria like *Moorella thermoacetica* were separated as an obligate heterotroph bacteria and were obtained to change glucose to acetate without carbon loss due to the CO_2 created through oxidation that was then applied in the synthesis of the acetate; the stoichiometry of this mechanism is stimulated by reaction (2.5):

$$C_6H_{12}O_6 \rightarrow 3CH_3COOH \quad (2.5)$$

According to good convert acetate to carbon from glucose without any loss, three moles of ethanol can be converted to one mole of glucose by a hydrogenation mechanism as depicted in reaction (2.6):

$$3CH_3COOH + 6H_2 \rightarrow 3CH_3CH_2OH + 3H_2O \quad (2.6)$$

due to little production of CO_2 throughout fermentation, mixed VFA fermentation gives a more carbon performance than direct ethanol fermentation utilizing *Zymomonas mobilis* or yeast (15). Moreover, the Anaerobic Digestion procedure, also, transformed proteins and lipid carbohydrates into mixed VFAs. So, if VFAs can be transformed into fuels and chemicals such as ethanol and butanol through economical methods, mixed VFA fermentation could supply an innovative platform with various uses for the generation of biofuels and biochemical (Figure 2.1).

2.3 MICROBIAL PRODUCTION OF EACH VFA

2.3.1 Acetic Acid

Acetic acid (ethanoic acid; CH_3COOH) is the main usually applied organic acid in early times. More than 18 million tons of acetic acid are used annually in the world (16). Generally, the acetic acid is formed by chemical matter including acetaldehyde, methanol, butane, or ethylene (17), and nowadays, fermentation-based acetic acid generation is attaining the notice of the researchers and requires to be commissioned at a full scale. Based on studies, different micro-organisms can produce acetic acid, such as *Acetomicrobium*, *Thermoanaerobacter*, *Acetothermus*, *Clostridium*, and *Acetobacter* (16, 18, 19). According to the literature, most acetic acids were produced by commercial sugars as raw material. Free sugars are digested and then via the glycolysis pathway are metabolized into pyruvate, finely converted into acetyl-CoA by oxidative decarboxylate. Micro-organisms correlated to *Acetobacter* are usually extracted in the industrial generation of acetic acid. These micro-organisms can use different sugars such as glucose, mannose, arabinose, trehalose, galactose, ribose, melibiose, and xylose (20). Commercial sugars are expensive, and to make the production cost-effective, scientists are looking for abundantly obtainable matters as carbon sources. Ehsanipour et al. (2016) created a bioprocess for acetic acid generation from lignocellulose sugars utilizing *Moorella thermoacetica* and generated 17g/L of acetic acid (Table 2.1) (7). According to the Ravinder et al. study, *Clostridium lentocellum* SG6 can produce 30.98 g/L of acetic acid using cellulose as the precursor (21). Another way to reduce the cost of production is the simultaneous production of other products with acetic acid. Mounir et al. (2016) used thermal *acetobacter* to develop a process for the production of gluconic acid and acetic acid (22). Other sources of carbon are cheese that reached the source and is easily available during the year from the dairy industry. Nayak and Paul (2013), from the continuous combined membrane hybrid method, used cheese whey as a carbon origin for the fermentation of *Acetobacter aceti* and after concentration generated 96.9 g/L of acetic acid (18). Wang et al. (2013), by using *saccharomyces cerevisiae* and *Acetobactor pasteurianus*, generated acetic acid and utilized glucose as well as carbon origin, so the initial strain was fermented glucose to ethanol and then generate acetic acid (89.3g/L) from ethanol (Table 2.1) (23). In a similar study, Li et al. (2015), using food waste and microaerobic fermentation as a carbon source, produced 25.88 g/L acetic acid (24). The mixed culture processes have a restriction in that it is difficult to operate and maintain the favorite situations necessary for coculture growth, leading to hard application at an industrial scale. Jourdin et al. (2015) found out the acclimatized and augmented microbial culture with anew synthesized electrode material EPD-3D that can produce acetic acid from CO_2 (25). For using CO, *Clostridium acetium*, a *Cchemolithotrophic* bacterium, was applied to generate acetic acid in the presence of decreasing cysteine-HCl, H_2O, and sodium sulfide. The bacteria were cultured for 64 h and 1.28 g/L acetic acid was converted to CO at 100% yield (19). Different metabolic engineering techniques have been additionally endeavored to develop micro-organisms and upgrade the generation of acetic acids, like the more expression of alcohol dehydrogenase (*adh*) and acetic acid supplier, Fukaya et al. (1992) illustrated the expression of the *aldh* gene of *Acetobacter polyoxogenes* NBI1028 (26). According to Table 2.2, aceti raised the

TABLE 2.1
Summary of VFA Generation Utilizing Different Micro-organisms

VFA	Micro-organism	Substrate	Fermentation Condition	Productivity (g/L/h)	Yield (g/L)	Titer (g/g)	References
Acetic acid	*Clostridium acetium*	CO	Batch fermentation	0.021[c]	1.28	–	(19)
	Moorella thermoaceticm	Sugarcane straw hydrolysate	Flask fermentation	0.238	17.2	–	(7)
	Saccharomyces lactis and *Clostridium formicoacetium*	Whey lactose	Co-culture at 35°C	0.375[c]	30	–	(37)
	Acetobacter aceti	Date extract	Continuous CSTR	2.8–11	0.46–0.5	0.46–0.5	(38)
	Acetobacter aceti	Cheese way	Continuous membrane-integrated hybrid process	4.6	0.96	0.96	(18)
	Clostridium thermoaceticm	Glucose	Batch CSTR	0.5	39	0.39	(39)
	Clostridium lentocellum SG6	Paddy straw	Vials level	0.1	30.98	0.44	(21)
	Kluyveromyces fragilis	Whey	Shake flask	0.14[c]	25.85	0.5	(40)
	Saccharomyces cerevisiae and Acetobacter pasteurianus	Glucose	Fed batch with coculture	–	89.3	–	(23)
Propionic acid	*Propionibacterium acidipropionici*	Lignocellulose hydrolysate	Batch fermentation	–	18	–	(41)
	Propionibacterium acidipropionici	Glycerol	Fed batch	0.2	44.62	–	(42)
	Propionibacterium acidipropionici	Jerusalem artichoke hydrolysate	Free cell fibrous fed bioreactor	0.19	40.6	0.379	(43)
	Propionibacterium shermanii	Glycerol- glycerol	Batch	–	9	–	(44)
	Propionibacterium freudenreichii	Glycerol	Multipoint fibrous- bed bioreactor (fed batch)	0.14	67.05	–	(45)
	Propionibacterium acidipropionici	Jerusalem artichoke hydrolysate	Immobilizes cell fibrous fed bioreactor	1.55	68.5	0.434	(9)

(Continued)

TABLE 2.1 (Continued)
Summary of VFA Generation Utilizing Different Micro-organisms

VFA	Micro-organism	Substrate	Fermentation Condition	Productivity (g/L/h)	Yield (g/L)	Titer (g/g)	References
	Propionibacterium acidipropionici	Lactate	Batch fermentation	0.11	15	–	(8)
	Propionibacterium acidipropionici	Sugarcane molasses	Batch fermentation	0.061	8.23	–	(46)
	Propionibacterium acidipropionici	Glycerol	Batch fermentation	0.05	6.77	–	(47)
	Propionibacterium acidipropionici	Chees way	Continuous fermentation	0.98	19.7	–	(43)
	Propionibacterium acidipropionici	Glycerol- glucose-lactate	Fibrous bed bioreactor	–	100	–	(8)
Butyric acid	*Clostridium butyricum* ZJUCB	Glycerol	Fed batch	–	16.74	–	(48)
	Clostridium tyrobutyricm	Xylose	Immobilizes cell fibrous fed bioreactor	3.19	18.4	0.38–0.59	(49)
	Clostridium butyricum	Sucrose	Fed batch	–	20	0.3	(48)
	Clostridium thermoaceticm	Glycerol	Continuous culture	2.4	12.25	–	(50)
VFA	**Micro-organism**	**Carbone source**	**Fermentation condition**	**Productivity (g/L/h)**	**performance (g/L)**	**Titer (g/g)**	**References**
	Clostridium tyrobutyricm	Glucose	Immobilizes cell fibrous fed bioreactor	–	57.9	–	(51)
	Clostridium tyrobutyricm	Cane molasses	Batch fermentation	4.13	86.9	0.47	(52)
	Clostridium tyrobutyricm	Sugarcane bagasse	Batch fermentation	0.51	26.2	0.48	(53)
Isobutyric acid	*Propionibacterium freudenreichii*	Complex media	Fermentation	–	20.9	–	(54)

TABLE 2.2
Summary of the Metabolic Process Approach Applied to Formation and Develop VFAs Generation

VFA Production	Microbes	Expression	Knockout	Effect	References
Acetic acid	*A. aceti*	aatA		A putative ABC transporter increase resistance for acetic acid	(27)
		aldh		Increase oxidation of ethanol to acetic acid	(26)
		aarC		Role in acetic acid assimilation and resistance	(77)
		aarA, aarB, aarC		Increase resistance to acetic acid	(78)
Propionic acid	*P. freudenreichii* subsp. shermanii	CoAT		Increase transfer rate of co-enzyme A (CoA) from propionyl-CoA to succinic acid	(79)
	P. freudenreichii	ppc		Increase conversion of phosphoenolpyruvate to oxaloacetate	(80)
		gldA, mdh, fumC		Increase conversion rate	(76)
	P. acidipropionici		Δack	Decrease acetate production and increase propionic acid production	(81)
		arcA, arcC, gadB, gdh, and ybaS		Protect propionibacteria against propionic acid stress by maintaining intracellular pH homeostasi	(82)
		ppc	ldh and poxB	Decrease side product production, i.e., acetate and lactate	(83)
	P. jensenii	gldA		Increase conversion of glycerol to dihydroxyacetone	(84)

(Continued)

TABLE 2.2 (Continued)
Summary of the Metabolic Process Approach Applied to Formation and Develop VFAs Generation

VFA Production	Microbes	Expression	Knockout	Effect	References
Butyric acid	*C. tyrobutyricum* Decrease		pta, ack	acetate production	(85)
		ta-ctfB, adhE1			(86)
	E. coli	phaA, hbd, crt, ter, atoAD	frdA, ldhA,adhE, pta	New pathway constructed to produce butyric acid	(87)
		acc, fadR, fabA, fabZ, tes,	fade, fadD, fadR	Enhanced butyric acid production through fatty acid pathway	(88)
	C. cetobutylicum	bukII	acetobutylicum pta, buk, tfB, adhE1	Selective production of butyric acid with reduction of acetate	(89)
Isobutyric acid	*Pseudomonas* sp. strain VLB120	kivD	ilvE/leuA, panB, pycAB	Valine synthesis pathway diverted towards isobutyric acid	(90)
	E. coli	ilvD-alsS, kivD-X	yqhD	Biosynthetic pathway for isobutyric acid	(91)
Isovaleric acid	*E. coli*	leuA, leuB, leuC, leuD, ilvD, alsS		New biosynthetic pathway for isovaleric acid production	(92)

acetic acid productivity from 68.4 to 96.6 g/L. Another researcher from Japan founded an ATP-binding cassette (ABC) transformer encoded by the *aatA* gene responsible for acetate persistence. Acetic acid production increased from 103.7 to 111.7 g/L by overexpression of *aatA* in *A. aceti* in batch fermentation (27). Numerous fermentation factors, including microaerobic fermentation, pH regulated fed batch fermentation, and a fibrous bed bioreactor, have been used to progress the generation of acetic acid, and moreover researches are essential to further improve the procedure (24, 28, 29). Acetic acid in concentrated form is cited as glacial acetic acid and vinegar at 5–20% concentration. Industrially, vinegar is formed during slow and quick fermentation (30, 31). Various kinds of vinegar have been demonstrated agreeing to their source, e.g., palm vinegar, date vinegar, apple vinegar, cane vinegar, fruit vinegar, rice vinegar, malt vinegar, wine vinegar, distilled vinegar, etc. (32–35). Vinegar is formed during a sequence of strains depending on the concentration of acetic acid through fermentation. *Acetobacter* is the predominant strain, at a lower concentration of acetic acid, and after its concentration increments over 5% other micro-organisms as *Gluconacetobacter intermedius* and *Komagataeibacter europaeus* take over the procedure (30, 36).

2.3.1.1 Critical Factors Affecting the Acidogenic Process

Among the series of anaerobic biological reactions, hydrolysis is evaluated to be the ratio drawback stage. For the anaerobic digestion process with isolated steps, there are three great targets in the hydrolysis–acidogenesis stage: (1) developing hydrolysis yield to obtain more soluble chemical oxygen demand (sCOD), (2) accelerating acidogenesis to generate more VFAs, and (3) decreasing inhibitors. Commonly, variables such as OLRs, gas mass transfer limitation, hydrogen partial pressure, pH, temperature, and other operating conditions all affect the performance of hydrolysis–acidogenesis.

2.3.1.1.1 OLR

Conceptually, the VFA generation could be improved by an enhancement in OLR due to the rising availability of soluble substrate. The study of Wijekoon et al. (2011) revealed that rising OLR changes the prevalent VFA fermentation pathway from acetic acid to butyric acid generation and so the total VFA concentration enhanced with rising OLR from 5–12 kg COD m^{-3} d^{-1} (55). Moreover, a study carried out by Lim et al. (2008) emphasizes the influence of various OLRs on acidogenesis of food waste and they identified that total VFA concentration raised with rising OLR (56). Although, the more OLR rate in reactors could lead to unsteady efficiency due to the viscous solid fermentative substrates subsequent from relatively more OLR in the reactor. Another result like above has also been shown by Jiang et al. (2013), who investigated the effect of OLR on VFA composition, and in addition, revealed that low OLR favored propionate and butyrate generation whereas the suggestion of acetate and valerate enhanced with rising OLR.

2.3.1.1.2 Gas Mass Transfer Limitation

Even though initial period performance enhancement with properly particular advantages of process variables and progressive process, hydrogen generation development is still problems for the FW anaerobic digestion procedure, particularly, when treating the solid-state substrate with high OLR. García-Bernet et al. (2011) carried out a study about water distribution in biowaste and digestion of dry anaerobic digestion and they reported that the rheological behavior of the biomass was various after the amount of water was reduced (57). So, the transmission of side products at the macroscopic level from the liquor medium to the solid substrate was decreased (58). Improved mass shift yield helps to achieve object gas, such as hydrogen generation at the acidogenic stage. The last-mentioned could be described by the mechanism of mass transfer. Based on the explain by Giovannini et al. (2016), there are three mass transfer procedures for hydrogen production reactors: (1) mass transfer between gas and liquor which is usually explained by the mass transfer coefficient K_{La} (/d); (2) interior gas transmission that takes place within the cell of microbes; and (3) exterior gas dissemination that happens outside the cell among hydrogen application microbes (59). Pauss et al. (1990) performed laboratory and theoretical evaluation on CO_2 and H_2 gas mass transfer limitations and concluded that the hydrogen concentration of the liquid stage was 80 times more than the value of the thermodynamic equilibrium model (60). In addition, Abbasi et al. (2012) proposed that the CO_2 and H_2 gas mass transfer between the gas and liquor stage was the main inhibitor (61). Similarly, interior and exterior mass transfer inhibitions were considered as two independent activities, causes the produced hydrogen gas can be transmitted via exterior gas diffusion and then be used with hydrogen consumers (62). If the hydrogen generators and consumers are close to each other, the inhibition of interior and exterior mass transfer is low (63). Although, in the more solid stage anaerobic digestion process without good mixing conditions, the effect of these three mass transfer constraints cannot be avoided.

2.3.1.1.3 Hydrogen Partial Pressure

Several literatures have illustrated that the yield of the anaerobic biological would be inhibited by hydrogen partial pressure. It is recognized that the hydrogen generation method is a reversible reaction depending on the oxidization and decreases the potential of the reaction source. Hydrogen generation diminishes when the oxidation reactions are not good, and the reduction method of ferredoxin happens under rising hydrogen partial pressure (64). Hydrogen efficiency increments by 30–71% when the working pressure is reduced (65). Level a minor vacuum of 0.03 atm can slowly develop hydrogen efficiency (66). The hydrogen concentration impacts both hydrogen generation and fermentation products through the hydrolysis–acidogenesis method. Jones and Greenfield (1982) show that adaptable metabolic pathways during controlling headspace pressure comes about in gas dissolved in the reactor. Several literatures have found that a headspace medium containing headspace pressure and partial pressure is a valuable factor for managing the metabolic pathways and controlling particular metabolic generates on the acidogenic stage (67). In addition, the hydrogen generation is diminished With a reduction in hydrogen part pressure, therefore variations the underlying metabolic processes. In this condition, the acetogenesis procedure is good thermodynamically, coming about in an improvement of hydrogen-producing processes and raised a generation of soluble metabolic products (68, 69). As discussed previously, when hydrogen part pressure increments, metabolic pathways will be further favorable for the generation of decreased products like ethanol, lactate, and acetone (70). Furthermore, homoacetogenesis and the acetate oxidation reaction can be regulated by hydrogen part pressure, and so control the efficiency of acetate and hydrogen (Equation 2.7) (71). Moreover, the syntrophic oxidation method of propionate and butyrate requires +76.2 and +48.4 kJ, respectively; it can take place positively under continuous low hydrogen partial pressure conditions. The change of ethanol to acetate requires +9.6 kJ and can be induced at lower hydrogen partial pressure. So, the transfer of metabolic pathways for alcohols and VFAs' generation could be impacted by the changes of headspace medium, particularly the hydrogen partial pressure. Even though hydrogen has a good exposed impact on dynamic variations of metabolic generates, the exact influence on microorganisms in controlling the metabolic mechanism under various hydrogen partial pressure remains unknown:

$$C_6H_{12}O_6 \rightarrow CH_3CH_2OH + 2CO_2 \tag{2.7}$$

2.3.2 Propionic Acid

Propionic acid is directly applied as a metabolite to generate numerous commercially principal chemical materials. The main feedstock constituent of probiotic acid is petrochemicals through chemical production with a yearly generation of 770 million pounds (72). Now, the biological fermentation procedure of propionic acid biosynthesis is achieving a considerable volume of notice while it is eco-friendly. Table 2.1 shows that the details of propionic-acid-generating bacteria belong to *Propionibacterium* spp. and include various strains: *P. freudenreichii*, *P. shermanii*, *P. thoenii*, and *P. acidipropionici*. Propionic acid generation has been detailed from numerous carbon origin, as well as glucose, lactose, and xylose (8). Compared to marketable sugars, crude glycerol generated by the biodiesel industry as an intermediate is a low-cost feedstock to generate propionic acid (42). The mutant strain of *P. acidipropionici* (ACK-Tet) can use glycerol as a carbon origin, and this strain can generate propionic acid with an upper yield in the range from 0.54 to 0.71 g/g via glycerol like the carbon source like compared to glucose producing 0.35 g/g propionic acid just (42). Glycerol fermentation enhanced the treatment of propionic acid with the low generation of byproducts, and other low-cost feedstocks include whey and hemicellulose hydrolyzed cornmeal has been moreover evaluated for application in propionic acid generation. A study carried out by Liang et al. (2012) emphasizes that the immobilized *P. acidipropionici* cells are steady and able to generate 26.2 g/L propionic acids at over to eight cycles repeated fermentation (43). In addition, cane molasses, the side product of the sugar commercially, including ~50% (w/w) entire sugar (sucrose, glucose, and fructose) can be a low-cost precursor. Quesada-Chanto et al. (1994) reported that *P. acidipropionici* can apply sugarcane molasses as a carbon origin to generate 30 g/L of propionic acid (73). Propinobacterium similarly produces other organic acids through with the major product of propionic acids, so lessening the performance and productivity. Based on researches, propionate is to prevent fermentation even at a lesser concentration, for example 10 g/L (74), and genetic engineering may be a potential methodology to among with this problem. Luna-Flores et al. (2016) used a genome shuffling methodology to develop *P. acidipropionici* tolerance to propionate, and the coming about strain formed 25% more propionic acid than in comparison to wild strain (75). Zhang et al. applied a mutant strain of *P. acidipropionici* (ACK-Tet), immobilized, and then modified it in a fibrous-fed bioreactor (FBB) to tolerate an upper concentration of propionate. The progressed strain formed 100 g/L of propionic acid (42). Liu et al. (2015) was increased propionic acid production from 26.39 g/L to 39.43 g/L with used P. jenseniis strain engineered plus glycerol dehydrogenase (gdh) expression and malate dehydrogenase (mdh) (76). To decrease the generation of lactate and acetate side products, *P. jenseniis* was engineered by removing the lactate dehydrogenase (*ldh*) and pyruvate oxidase (*poxB*), and over-expressing the phosphoenolpyruvate carboxylase (*ppc*) gene to directly change phosphoenolpyruvate to oxaloacetate, avoiding pyruvate intermediate (Table 2.2).

2.3.3 Butyric Acid

Butyric acid is a futuristic building block for commercially valuable chemical materials. Butyric acid is formed via a chemical route beginning from crude oil as the feedstock (93). Butyric acid from a biological source is in a considerable request for food and pharmaceutical usage (94). Butyric acid fermentation is an intricate procedure that is difficult to control; hence, it is inhibited by the final product (94). Different microbial strains belonging to various genera have been isolated from unlike media to produce butyric acid, e.g., Clostridium, Butyrivibrio, Butyribacterium, Fusobacterium, Eubacterium, Sarcina, and Megasphera. Clostridium is the principal studied group of bacteria that are enabled to using various carbon origins, i.e., hexoses, pentoses, and saccharides. Baroi et al. (2015) reported that an adjusted strain of *Clostridium tyrobutyricum* can be used to cultivate on 80% wheat straw hydrolysate, application glucose, and xylose at the same time to generate butyrate by an upper productivity of 0.37–0.46 g/g (95). Fermentation with immobilized cells of *C. tyrobutyricum* in an FBB leads to 26.2 and 20.9 g/L butyric acid, so cane molasses and sugar bagasse were applied as the carbon origin, respectively (Table 2.1) (52, 53). Ai et al. (2014) applied an indistinct mixed culture to ferment NaOH-pretreated rice straw lacking any cellulolytic enzymes, producing 6 g/L butyric acids (96). Dovidar et al. (2013) used a microbial mixed culture of Bacillus strain using sucrose with *C. tyrobutyricum* ATCC 25755 to produce the enzyme levansucrase. Based on the results, sucrose was hydrolyzed to glucose and fructose, and *C. tyrobutyricum* fermented these sugars to 34.2 g/L butyric acid (97). Zhu and Yang (2004) expressed the function of the pH in founding the ultimate product of fermentation in *C. tyrobutyricum*. Xylose fermentation at 6.3 pH enhanced butyric acid generation (57.9 g/L), while fermentation at pH 5.7 come about in lactate and acetate as the important product with low production of butyric acid (49). This metabolic transfer in acid fermentation is related to enzymatic activity. At pH 6.3, *C. tyrobutyricum* cells have upper activities of enzymes phosphotransbutyrylase (PTB) and NAD-independent lactate dehydrogenase (*iLDH*) responsible for administrating the creation of butyrate and shift of lactate to pyruvate, respectively. At a lesser pH, phosphotransacetylase (PTA) enzyme, which controls the acetate creation and enzyme lactate dehydrogenase (LDH), catalyzes pyruvate to lactate transformation and is further active. Zhu and Yang (2004) propose the conversion in the metabolic ratio that NADH balance is the main key in controlling the metabolic pathway through fermentation (49). Jiang et al. (2011) illustrated that the fed batch-adapted *C. tyrobutyricum* strain leads to increase tolerance to butyric acid and becomes decreased inhibition affecting butyrate formation enzymes, i.e., phosphotransbutyrylase (PTB) and ATPase (51). The nutrient products, like trace elements, were demonstrated to affect the butyrate generation from glycerol in *C. butyricum*. Reimann et al. (1996) showed that higher phosphate content increased butyrate production and decreased acetate, while iron had the opposite effect (98). One of the main byproducts of butyrate fermentation is acetate, which was integrated by mutagenesis to disrupt the genes responsible for acetate production in *C. tyrobutyricum* ATCC 25755. Zhu et al. (2005) reported that a mutant strain was enabled to create 15% higher butyric acid with 14% decrease in the acetate generation (99). *C. acetobutylicum* produces both butyrate and acetate. To enhance the selective generation of butyric acid, the butyrate kinase-II (*BK-II*) gene was applied in its place of butyrate kinase-I (*BK-I*). Jang et al. (2014) show that different metabolic pathways able to increase the NADH driving force were furthermore engineered, and the subsequent strain *C. acetobutylicum* HCBEKW (pta-, buk, ctfBand adhE1-) generated 32.5 g/L of butyric acid (Table 2.2) (89). A genetic engineering methodology was in addition useful to other nonproducer strains, e.g., *E. coli*, to produce butyrate by removing major NADH-dependent reactions, reproducing the pathway for butyryl-CoA, and overexpressing *atoAD* need to transfer butyryl-CoA to butyrate and removing acetate-synthesis pathways. Table 2.2 shows a constructed strain to generate 10 g/L butyrate starting from glucose and acetate as carbon origin (87).

2.3.4 Isobutyric Acid and Isovaleric Acid

Isobutyrate is a beneficial platform chemical, and it is formed through chemical reactions because there is no biological procedure yet published for microbes (91). Isobutyric acid is synthesized by acid-catalyzed carbonation Koch of propylene (100). Propylene is formed by cracking huge hydrocarbon molecules, such as petroleum and natural gases, which perhaps is the reason to harm to the environment. Synthesis analysis can apply to investigate the isobutyric acid pathway via isobutyraldehyde, which is a metabolite of the Ehrlich pathway and can be formed via the metabolic pathway of glucose (91). Glucose is metabolized through glycolysis to pyruvate, which catalyzes a series of reactions performed by acetolactate synthase (*alsS*), keto acid reductoisomerase (*ilcC*), dihydroxy acid dehydratase (*ilvD*), and keto acid decarboxylase. The complete isobutyrate biosynthesis pathway in *E. coli* was formed using two plasmids containing the operon ilvD-alsS and kivD-padA (Table 2.2). The engineering strain of *E. coli* formed isobutyrate 4.8 g/L. Any more deletion mutant of this engineered *E. coli* strain was arranged to discontinue the generation of byproducts by erasing yqhD, which competes with the aldehyde substrate to create isobutanol. A constructed strain was applied to perform shake-flask fermentation utilizing glucose like the carbon substrate, coming about in 11.7 g/L of isobutyric acid generation. Also, *Pseudomonas* sp. strain VLB120 has a valine degrading pathway, which can lead to the generation of isobutyric acid channel through 334 isobutyl-CoA (90, 91). on the other hand, the engineered strain of *Pseudomonas* sp. strain VLB120 overexpressing 2-ketoacid decarboxylase (*kivd*) from *Lactococcus lactis* by removal of different genes formed various side products, able just 2.3 g/L isobutyric acids with less generation of isobutanol (90). Most microbes such as *E. coli*, Bacillus, Corynebacterium, and Streptomyces under anaerobic and microaerobic conditions produce mixed acids directly from pyruvate to maintain the redox equilibrium to support isobutyric acid generation (101). Another advantage of utilizing Pseudomonas is its simplified downstream processing since it can utilize glucose with the generation of side products, e.g., acetate or ethanol (102). A flavoring compound also found in Swiss cheese, generated mostly by *Propionibacterium freudenreichii*, is isovlerate (54). A mutant strain of *Clostridium saccharoperbutylacetonicum* Nl-4 ATCC 13564, without thiolase activity, is developed by Ahn et al. (103). Ahn and Hayashida (1990) show that thiolase-deficient mutant AS2-1 accumulates an excess of electrons, which causes the reductive creation of isovaleric acid from pyruvate (103). Xiong et al. engineered the metabolic chain amino acid pathway through *E. coli*. In this pathway, ketovaline is produced via the condensation of two pyruvate molecules by a run of enzyme-catalyzed reactions, i.e., AlsS, IlvC, and IlvD (Table 2.2). Ketovaline is elongated to ketoleucine with a series of reactions performed through 2-isopropylmalate synthase (*leuA*), isopropylmalate isomerase complex (*leuCD*), and isopropylmalate dehydrogenase (*leuB*), and it is further shifted to isovaleric acids by different dehydrogenases. Xiong et al. (2012) used an engineered strain to function a bioreactor fermentation, coming about in 32 g/L of isovaleric acid (92).

2.3.5 Methane Production

Methane formation happens in the methanogenesis period with the aid of methanogens, which belong to the prevail Archaea, phylum Euryarchaeota (104). Table 2.3 shows the commonly methanogenic reactions that occur during anaerobic digestion. Compared to other microbes during the anaerobic digestion method,

TABLE 2.3
Methanogenic Reaction by Gibbs Free Energy

Substrate	Reaction	Gibbs Free Energy ΔG (kJ/mol)	No. Equation
Acetic acid	$CH_3COO^- + H^+ \rightarrow CO_2 + CH_4$	−27.5	1
Methanol	$4CH_3OH + 3CH_4 + CO_2 + 4H_2O$	−544.8	2
Carbon monoxide	$4CO + 2H_2O \rightarrow CH_4 + 3H_2O$	−185.9	3
Formic acid	$4CHOO^- + 2H^+ \rightarrow CH_4 + 2CO_2 + H_2O$	−302.6	4
Methanol-hydrogen	$CH_3OH + H_2 \rightarrow CH_4 + H_2O$	−149.8	5
Hydrogen	$4H_2 + CO_2 \rightarrow CH_4 + 2H_2O$	−139.1	6

methanogens are hard anaerobic and are susceptible to environmental conditions such as pH and temperature. Methanogens cannot directly have used the organic source in the anaerobic digestion process containing carbohydrates, proteins, and lipids. Even though the types of methanogens are very varied, they can obtain energy from partial types of substrates, such as acetate, CO_2/H_2 mixture, and methyl-group including products (105). Acetate is the main kind of substrate for methanogens and the method is recognized as acetoclastic methanogenesis, consequentially, in which the carboxyl oxidation of acetate in acetate decreases (105). More than half of the methane generation in the anaerobic digestion method comes about from the decarboxylation of acetic acid by acetotrophic methanogenesis. The mixture of CO_2/H_2 is the second essential carbon substrate for methanogens. This procedure is recognized as hydrogenotrophic methanogenesis, in which CO_2 is dramatically decreased to methane when hydrogen is utilized as the electron donor. Moreover, the hydrogenotrophic methanogens can also apply format as the carbon source. The methyl group including products, containing methanol, monomethyl amine, dimethylamine, trimethylamine, tetramethylammonium, methyl sulfide, and methanethiol is considered to be the third type of methanogenesis substrates (106), but the last two substrates counting just a lesser quantity of methane efficiency.

2.4 OTHER RECENT STRATEGIES FOR MIXED VFA

2.4.1 Production from Cost-effective Carbon Sources

To obtain the cost-effective generation of VFAs, different kinds of the waste source from municipal solid waste, food, agriculture, dairy, and paper industry were evaluated (1, 107). Municipal wastewater is generated at the widest amount and has a chemical oxygen demand (COD) range of 14,800–23,000 mg/L, creating it the best substrate for VFA generation. Luo et al. (2011) showed that the integrated adding of sodium dodecyl sulfate (SDS) and varied enzyme promotes sludge hydrolysis and VFA generation (108). The suspended solid COD range of these kinds of wastewater is hundred times lower than the total COD value, thus gradual the hydrolysis of organic solid waste (109). This is a limiting stage in municipal wastewater fermentation, thus there is a requirement to investigate wastewater pretreatment. Food leftovers are one more commonly utilized feedstock, but their recovery from other domestic waste such as glass and plastic accomplishes the process more challenging (1, 110). To accomplish this process, public training and commitment to the separation of organic matter and other household waste are essential. In any case, this is not simple. Yin et al. (2014) applied hydrothermally pretreated food waste for VFA generation and obtained 0.294–0.411 g of VFAs, including butyrate and acetate as the major organic acids followed by propionate and valerate (111). The use of liquor effluent generated in agriculture, food, and paper industries has also been studied for VFA production (112). Food and paper industry wastes contain a lot of organic matter and are suitable for VFA fermentation, while the petroleum industry wastes are rich in many toxic compounds that are harmful to microbial activity (1, 113). Bengtsson et al. (2008) make acidogenic fermentation of cheese whey and paper mill industrial wastewater, generating acetate, propionate, and butyrate as the major VFAs (114). Pretreatment is necessary to degrade toxic components from industrial wastewater, thus it is not an attractive method for VFA generation. Many investigators have suggested cofermentation for mixed wastewater for VFA generation. Cofermentation of containing starch industrial wastewater with initial sludge at a 1:1 ratio increased the VFA (45 mg VFA/g VSS/day) when compared to initial sludge fermentation (31 mg VFA/g VSS/day) (115). Yang et al. (2016) investigated a cofermentation procedure for b-cyclodextrins and alkaline purified food wastewater and sludge, coming about in VFAs with 8631.7 mg/L of productivity (116). Increasing VFA generation was due to the more hydrolysis of the mixture sludge. The use of different waste as feedstock for VFA generation can be a fine method just to generate nonfood grade mixed VFAs. The treatment and segregation of clean organic acid from mixed VFAs is challenging because several side products are too generated throughout waste fermentation. So, a VFA combination can be applied as precursor to generate other industrially valuable compounds like PHAs and biofuel (117).

2.4.2 Fundamental Factors Affecting the Methanogenic Phase

The methanogenesis yield is correlated to methanogenic activities that are additionally susceptible to environmental changes compared to other microbes included in the methanogenic stage. When the methanogenic stage is worked under mesophilic conditions, the powerful correlative variables impacting VFA generation require to be considered. These contain hydraulic retention time (HRT), VFAs' profile, and syntrophic oxidation, all of which finely lead to methane generation.

2.4.2.1 HRT

The variations in HRT often disorder the activity of methanogenic microbes which after impact methane generation. Conceptually, a low HRT infers more OLR. The removal performance of sCOD upgraded by enhancing HRT led to more methane generation (118). Furthermore, VFAs concentration various due to the variations of HRT, OLR, and temperature (119). The result of Dareioti and Kornaros's (2015) study shows that decreasing the HRT leads to the instability of methanogenesis and a reduction in biogas and methane yield largely due to enhancing VFA concentrations (120).

2.4.2.2 VFA Profile

The VFAs, which are formed by microbes throughout acidogenesis, can be finally shifted into methane and CO_2 by syntrophic acetones and methanogenic archaea. In any case, accumulation of VFAs typically happens at high OLR, which leads to a reduction in pH less than the methanogenic rate, so leading to a probable methanogenic failure in the anaerobic digestion method (121, 122). Methanogens are especially reactive to various pH, ordinarily culturing methanogenic bacteria between pH 6 and 8.5 while a preferably limited range of 6.5–7.2 is more favorable. Among the five kinds of VFAs formed as well as acetate, propionate, butyrate, valerate, and caproic acid, acetic acid is mainly a VFA for biogas generation. Propionic acid is more toxic than other VFAs and its accumulation often

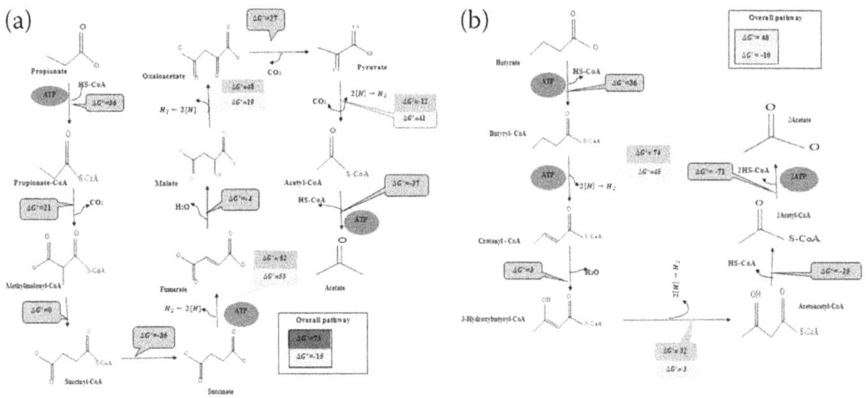

FIGURE 2.3 Biochemical pathways of syntrophic propionate and butyrate oxidation: (a) propionate oxidation and (b) butyrate oxidation. The ΔGo′ and ΔG′ (at 1 Pa hydrogen) of each metabolic stage is demonstrated (kJ/mol) (126).

recommends an imbalance of metabolic pathways in anaerobic digestion (123). The rate of propionate to acetate concentration is significant as a favorable indicator of the methanogenic imbalance and a rate greater than 1.4 leads to methanogenesis failure (124).

2.4.2.3 Syntrophic Oxidation

As mentioned earlier, just three types of carbon sources can be directly utilized by methanogens. The side products from the initial period such as lactate, propionate, butyrate, valerate, and ethanol must be transformed into acetate, hydrogen, and CO_2 by the acetones. Usually, lactate and valerate are not steady and would be shifted to propionate via the fermentative method. Ultimately, the propionate and butyrate removal stages lead to the most challenging processes whereas there in order oxidation pathways are energetically fewer favorable, that is, the oxidation of succinate to fumarate and butyryl-CoA to crotonyl-CoA, respectively (Figure 2.3). Numerous bacteria are identified to the degradation of propionate in syntrophic coloration with methanogens which contain *Syntrophobacter fumaroxidans*, *Syntrophobacter wolinii*, *Syntrophobacter pfennigii*, *Syntrophobacter sulfatireducens*, *Pelotomaculum thermopropionicum*, *Pelotomaculum schinkii*, *Pelotomaculum propionicicum*, *Smithella propionica*, and *Desulfotomaculum thermobenzoicum* subsp. *Syntrophomonas erecta*, *Syntrophomonas curvata*, *Syntrophomonas zehnderi*, and *Thermosyntropha lipolytica* are butyrate oxidizers, which belong to the family Syntrophomonadaceae (125).

2.5 FERMENTATION APPROACHES FOR VFA GENERATION

The manufacturing generation of different commercially valuable VFAs depends upon the chemical production technology applied, which includes wide application of energy, manpower, and coproduction of different byproducts that are a matter of distress (3). The employ of the microbial fermentation process can be varied for

VFA generation beginning from low-cost and commercially available raw materials. Different fermentation technologies have been investigated to improve economic progression with the greatest product performance. Fed-batch, batch, and continuous fermentation processes are generally (8, 50, 53). Batch system fermentation is easy, but due to repeated cleaning, determining the appropriate media, inoculation, cell lag phase, and generate inhibition of the microbial culture at a later phase, it is time consuming (96). Using a batch system and repeated fermentation may decrease this nonreproductive time as a portion of the prior broth growth is utilized as inoculation (127). repeated-batch operation process, mixed culture fermentation reduced in every one stage owed to transformed into the microbial community profile (96). A continuous fermentation process leads to higher productivity compared to fed-batch and batch. In any case, the generate concentration is little, and the anaerobic procedure is more cost-effective than the aerobic process because the anaerobic process does not require aeration and can produce VFA with higher efficiency (128). Microaerobic fermentation is dominated over aerobic and anaerobic technology because it increments the degradation of raw materials by including aerobic and anaerobic microbes (129). The advantages of thermophilic fermentation include a higher reaction rate and resistance to contamination (130). A cell recycling system can be utilized, but by every cycle, cells mislay activity like achieved in the production of generally acids (42). The use of immobilized cells leads to produce VFAs with reducing separation costs (131). Fibrous fed bioreactor with immobilized cells enables them to keep productivity for a long time because of the higher mass shift capabilities (93). The pH of the fermentation media is, in addition, an important factor in the metabolic transfer of the VFA, and at pH 6, butyric acid is the ideal generate, while at a lesser pH, for example 5, acetic acid is formed (52). Each fermentation process has advantages and drawbacks, so it is difficult to choose the exact fermentation pathway. Various kinds of process have been investigated for VFA generation, i.e., packed bed reactor (PBR), fluidized bed reactor (FBR), continuous stirred tank reactor (CSTR), and upflow anaerobic sludge blanket (UASB) (1, 132). The PBR reactor is a plug flow reactor (PFR) in which the biofilm grows on a porous packaging material (zeolite or granular activated carbon) so the biofilm does not float in liquid (Figure 2.4a) (133). In the FBR reactor, the biomass is attached to fine particle matter and transported by the current of the liquid. Fine particle matter supply a great space for microbe attachment and progress oxygen and nutrient accessibility (Figure 2.4b) (134). At the most concentration of the total suspended solids, PBR is clogged, and in that case, FBR may perhaps be helpful. CSTR is a complete mixed reactor that a biomass enters and generates are continuously degradation (Figure 2.4d). It permits for complex of the biomass, and various properties are controlled, especially, temperature and density, during the process. These reactors can operate in three modes: continuous, batch or semi-continuous (135). UASB relies on the creation of a blanket of granular suspended sewage in the tanks (Figure 2.4c). The inflow into the UASB reactor from the bottom to up, moves through the substrate in the sludge bed, and then contacts the substrate with the sludge blanket (Figure 2.4c). The fundamental drawback of the UASB is a lengthy setup time in reactor (136).

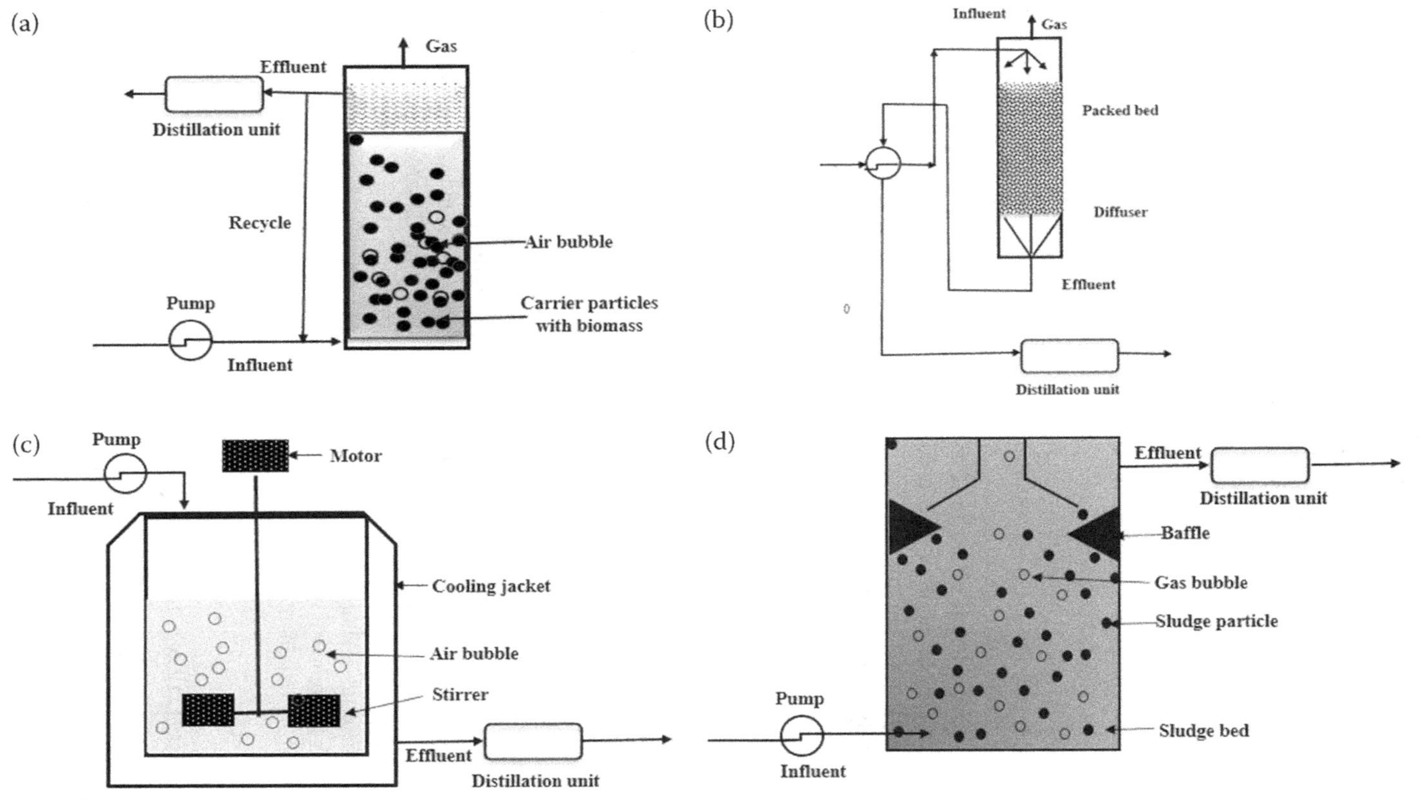

FIGURE 2.4 Different kinds of reactors applied in VFA fermentation. (a) Fluidized bed reactor. (b) Packed bed reactor. (c) Continuous stirred tank reactor (CSTR). (d) Upflow anaerobic sludge blanket reactor.

2.6 DOWNSTREAM PROCESSING

The producing of VFA from a fermentation broth is a most important bottleneck, and a more effort has been used to develop a possible, cost-efficient process for VFA regeneration. Numerous approaches, such as precipitation, chromatography, membrane separation, liquid–liquid extraction, and distillation have been investigated for CFA regeneration.

2.6.1 PRECIPITATION

The precipitation process is applied to extract organic acids from fermentation broth at a bulk scale. The fermentation broth is separated to eliminate the cell substrate, and Ca $(OH)_2$ or $CaCO_3$ is added to the supernatant with mixing. Calcium salts with organic acids, and the coming about salts are filtered off and after refine with sulfuric acid to set free the organic acids (137). The important drawback of this method is the production of waste ($CaSO_4$), which is costly to dispose of (138). Cyclodextrin can in addition be used for precipitation because it can form unsolvable crystalline complexes with organic acids. β-Cyclodextrin is enabled to precipitate butyric acid with 100% yield (139).

2.6.2 LIQUID–LIQUID EXTRACTION

The liquid–liquid extraction process is according to the relevant solubility of organic matter in two unmixable solvents. Mkhize et al. (2014) used dichloromethane (DCM) and methyltert-butyl-ether (MTBE) solvent to extract different VFAs such as, acetic, propionic, butyric, isobutyric, valeric, and isovaleric acids (140). The principal disadvantage of this process is the solvent toxicity and energy depletion throughout distillation, which forbids its application for bulk chemical production. Liquid–liquid extraction is affected by various factors, counting the kind of VFA, pH, and concentration of the extractant (141). Several solvents, such as alcohols, aliphatic hydrocarbons, ethers, ketones, and organophosphates, can be utilized for VFA extraction, and various solvents have their drawbacks because extraction with hydrocarbons can result in low performance, organophosphates are toxic and valuable, and ether is flammable (142). To make the extraction method comfortable, the generate is reacted with other materials. Alkaya et al. (2009) reported a reactive liquid–liquid extraction process for VFA renewal applying trioctylphosphine oxide (TOPO) in kerosene, and it is enabled to obtain the most VFA regenerations (61–98%) at 20% TOPO in kerosene (143).

2.6.3 DISTILLATION

The distillation process of the dissociation depends on the variation in the volatility of different materials in attendance in the mixture. The distillation method is of great effect at a lesser concentration while the performance is reduced at a more concentration (144). The important drawback of the distillation process is its weak regenerate, as Siedlecka et al. (2008) recovered just (53–58%) VFA (145).

2.6.4 Chromatography

This system is contingent on the ion exchange and adsorption kind of resin. Resins applied in chromatography have steady conditions and are unsolvable in acid, alkali, and organic solvents. Ion exchange resins and macroporous adsorption resins are the greatly applied resins because of their good option for organic acids (146). The fundamental drawback of ion-exchange chromatography is the attendance of other anions in the analytic combination that can compete for active sites in the resin and may perhaps decrease its adsorption performance. Another type of method, like membrane-based solvent extraction (MBSE) and electrodialysis, are important for study. In MBSE, the process biomass shift between two immiscible solvents happens during the immobilized L/L interface (147). Zhang et al. (1993) applied electrodialysis to regenerate propionic acid and acetic acid from the fermented broth of the biomass (148). There is in addition a requirement to evaluate new procedures further to apply in bulk scale downstream process.

2.7 THE MOST RELEVANT PARAMETERS INFLUENCING VFAS' PRODUCTION

2.7.1 Influence of Carbon-source Nature in FWs

The food waste component is basically in VFAs biosynthesis, enabling to impact both their quantity and their chemical as acetic acid, butyric acid, and propionic acid distribution. Moreover, the food waste types decide in addition the selection of the operational parameters. Commonly, lipids in food wastes are less appropriate for fermentation than carbohydrates and proteins. Although lipids contribute to high levels of COD in the substrate, they have slower biodegradation kinetics (149). In addition, hydrolysis of lipids leads to generate glycerol and long-chain fatty acids (LCFAs). Furthermore, glycerol can be utilized as a fermentation substrate. LCFAs are enabled to stick to the cellular wall, influencing the transfer of nutrients, that shows that limitation the metabolism of anaerobic bacteria (150). As a result, carbohydrates are comfortably transformed by microbial enzymes into glucose, which is instantly obtainable for glycolysis and fermentation into VFA. Yin et al. (2016a) evaluated the pure glucose fermentation in a batch reactor, at a mesophilic rate, under pH 6, utilizing granular activated sludge as seed. Based on the results, fermentation of pure carbohydrates, like glucose, leads to the generation of acetic acid, butyric acid, and propionic acid; also, authors achieved a maximal VFA generation of 38.2 g COD/L, with acetic acid, butyric acid, propionic acid, and valeric acid obtained for 17, 30, 50, and 3%, respectively (151). They presumed that, despite a higher theory high transition yield of glucose into acetic acid, it was not accumulated into the reactor as a result of its depletion for the generation of H_2 or microbial culture. Fermentation of starch achieving as a raw material in a packed-bed reactor operating at the mesophilic rate and unregulated pH caused to a total VFA generation of about 18 g/L, with acetic acid, butyric acid, and propionic acid demonstrating around the 45, 28, and 14% of the total, respectively. However, total VFAs and their profiling transmitted doubling the loading organic rate; VFA generation comes to a lower rate, of about 17 g/L, with acetic acid demonstrating around 70% of the total (152). The fermentation of synthetic substrates, combined with row food rich in carbohydrates, obtained a concentration of butyric acid close to that of acetic acid, with a butyric acid/acetic acid rate of about 0.8 (149). Consequently, acetic acid, butyric acid, and propionic acid substrates as major carbohydrates fermentation generates. The various in this substrate of single rate in VFAs depends on the complexity and heterogeneity of the substrates and of the various working variables, which can activate or inactivate a special metabolic pathway (153). Alternatively, proteins are usually described by a lesser biodegradability, caused by their tertiary and quaternary arrangement, which make them low sensitive to protease action (154). The consequence is that the yield of carbohydrate hydrolysis from food wastes is more than 80%, whereas from protein is at the rate 40–70. Because of this, food waste protein hydrolysis is known as a ratio-inhibition stage throughout acidogenic fermentation (150). Decomposition of proteins occurs effectively via three biochemical pathways: (1) Stickland reaction, (2) oxidative deamination of amino acids, and (3) reductive deamination of amino acids. Firstly, the pathway permits a minimum of 90% of protein decomposition; the next reaction, which leads to a hydrogen-like special end generate, is not thermodynamically ideal and requires the lowest values of hydrogen partial pressure. Ultimately, the third pathway is good energetically and consumes hydrogen (149). However, fermentation of proteins leads to the generation of the acetic acid, propionic acid, and butyric acid achieved from sugar metabolism, but the relative ratio among them is dissimilar. A study carried out by Yin et al. (2016a) emphasizes that the acetic acid was the fundamental VFA achieved from fermentation of peptone, was obtained for the 70% of total VFAs formed, while butyric acid and propionic acid showed about the 10 and 15% (151). Moreover, valeric acid generation is mostly correlated with proteins fermentation, consequently of redox Stickland pathway between combine of amino acids and of decrease deamination, of concentration only amino acids (152) was the 5% of total VFAs formed (151). As for carbohydrates, VFA generation appears in addition to be influenced by source of proteins. Newly, it was illustrated that fermentation of animal or vegetable proteins leads to various VFAs profiling.

2.7.2 Influence of Working Parameters on VFAs Production

Working variables, such as temperature, pH, HRT, and OLR, illustrate principal impacts on VFAs generation from FWs fermentation, both in terms of efficiency and relative distribution among various products. However, these variables have a synergetic impact on the microbial consortium included in fermentation procedures as acting on cellular metabolism, mainly investigating their impact one at a time.

2.7.3 pH

Among the working variables, pH has the most strong influence on VFAs generation from food wastes fermentation. Jiang et al. (2013) investigated the impact of various pH rates, comprised of acidic variety (5–7), on VFAs generation from synthetic kitchen waste (155). They observed that a pH range comprised between 6 and 7 brought to a rise of about 20% of hydrolysis rate, obtaining a rate for soluble COD of 82 g/L. At lesser unregulated pH, the obtained sCOD was 60 g/L. This raise of solubilization permitted to double VFAs generation in the batch reactor with pH regulated to a rate of 6 and 7 and a raise of fermentation products by 10 times compared to the unregulated pH, as a result of a higher hydrolytic enzyme activity and prevention of inhibition caused to acidification of environment reactor. Also, a pH near to neutrality brought to a various distribution among VFAs, by butyric acid, acetic acid, and propionic acid obtaining for about 50, 25, and 15% of total fermentation products. Zhang et al. (2005) investigated about pH effect on fermentation efficiency in a reactor operational in continuous type regulating pH to a rate of 5, 7, 9, and 11 (156). They observed that pH near neutrality leads to a better VFA performance. Actually, in a pH rate of 7, fermentative metabolism is favored, and a VFA performance of 0.27 g VFA/g TS was obtained, while it was 0.15 g VFA/g TS in the control reactor, where the pH was not regulated. The positive influence of low acid-neutral conditions on microbial metabolism and so on fermentation generation was illustrated in another research in which, moreover VFAs generation, carbohydrates, and proteins application ranges were evaluated. Below a pH adjusted at a rate of 6, VFAs generation and VFAs performance raised 17 and 7.5 times respectively, concerning pH 4 (157). Dahiya et al. (2015) illustrated the correlation between pH and individual VFAs achieved from acidogenic fermentation (110). They observed when food waste was fermented from the canteen in bench-scale batch reactors, under the below conditions: 28°C, OLR 15 kg COD/m3d, 10% w/w inoculum. pH was set up at the starting of each test to 5, 6, 7, 8, 9, 10, and 11. The most total VFAs production was observed into reactor operating with a first pH of 10, achieved pH 9 (5.17 g VFA/l), pH 6 (4.5 g VFA/L), pH 5 (4.2 g VFA/L), pH 7 (4.1 g VFA/L), pH 8 (3.8 g VFA/L), and pH 11 (3.5 g VFA/L). The slight VFAs generation was obtained in response to the least and most pH value examined, and can be reasonable due to acidogenic bacteria cannot survive under very acidic (pH 3) or alkaline (pH 12) conditions. It is necessary to take into description that the utilization of the formed VFAs happened for methane generation performed by acetoclastic (pH, 6–8) or by hydrogenoclastic (pH, 9–10) *archaebacteria*. Thus, it is not achievable to determine that alkaline pH is perfect for VFAs generation as their concentrations were greater before their change. According to VFAs distribution model, an alkaline pH (10) appeared to good acetic acid generation, which come to the most rate of 4.2 g/L subsequent to 36 h, and after that it reduced up to about 3.6 g/L then 48 h, perhaps as a result of methane generation. Butyric acid was the major fermentation product under a pH of 5 (1.8 g/L), and its concentration did not illustrate a significant reduction between 36 and 48 h. Propionic acid obtained a concentration of about 1.4 g/L under all pH rates examined, and it was not consumed afterward the most concentration was reached.

2.7.4 Temperature

Temperature is an important variable throughout acidogenic fermentation because of its direct implication both in microbial culture and metabolism. Each microbial taxon has an optimal variety of temperature for its reproduction, thus a variation of operational temperature can vary the microbial arrangement of the microbial community included in acidogenic fermentation. He et al. (2012) reported that a raising of working temperature from mesophilic (35°C) to thermophilic rate (55°C) showed to a reduction of total VFAs generation from the greatest concentration of 17 g/L to 11 g/L, under acidic unregulated pH (158). The raising of temperature into hyperthermophilic rate (70°C) lead to decrease in total VFAs generation (around 13 g/L). Komemoto et al. (2009) investigated the influence of temperature on acidogenic fermentation at a bigger rate, alternating from psychrophilic (15, 20°C) to hyperthermophilic (65°C) (159). They reported that at 55 and 65°C the sCOD into reactor significantly raised at the starting of test time up to a rate of about 40 g/L, and after it dropped quickly to 30 g/L. In its place, in mesophilic vary sCOD reached a comparative rate (30 g/L), but it remained steady until the end of the test trial. This dissimilarity is caused to microbial hydrolysis activity; at greater temperature, solubilization is the consequence of the chemical-physics effect, while in mesophilic vary there is an active action of microbial enzymes. Concerning VFAs generation acetic acid was formed at the starting of the test, coming to a rate of 1 and 2 g/L at 35 and 45°C, respectively, and then its concentration reduced as a result of biogas generation. By contrast, butyric acid was achieved at the middle and late test time, and it appeared to a great concentration (6.2 and 5.7 g/L at 35 and 45°C respectively) despite biogas generation. Under psychrophilic conditions (T = 20°C), VFAs generation was greatly lesser and this variety of temperature can be considered inappropriate for any applications. Lee et al. (2008) illustrated how the selection of the optimum working temperature is correlated to pH with remarkable VFAs efficiency. They purified substrates in two various fermentation stages (160). First was at 55°C with unregulated pH. Under these conditions, total VFAs generation obtained a rate of around 12 g/L, with iso-butyric acid as a quite special fermentation product. In the second experimental step, fermentation was performed at 65°C, both with unregulated pH and set up at a value of 7. The modification of pH permitted to achieve a VFAs generation of about 18 g/L, twofold larger for the experimental stage carried out under unregulated pH, but a little larger than the generation obtained throughout the first experimental stage at a lower temperature. In any case, the temperature rise led to a nice relation delivery of fermentation products, by acetic acid, propionic acid, and butyric acid obtaining for the 39, 28, and 17%, respectively. Additional rise of temperature to 70°C, with a pH-controlled at 7, brought to an upper total VFAs generation of about 35 g/L. Under these conditions, acetic acid was yet the fundamental fermentation generate, obtaining 55% of total VFAs, but propionic acid was formed in smaller quantitative demonstrating the 14% of total generation, while butyric acid was slightly further abundant, accounting for 31% of total VFAs achieved. A more increase of temperature until 80°C led to a progressive reduction of VFAs generation, up to a rate similar to that accounting at 65°C. Consequently, it is clear that a higher working temperature, around thermo- and hyperthermophilic range (40–80 °C), leads to an increase in hydrolysis rate, giving hydroxylates theoretically available for fermentative microbial metabolism.

In any case, an increment of VFAs generation is probable just agreeing to optimum bacterial growing temperature, as a lot of acidogens cannot survive at high temperature (161).

2.7.5 HRT

HRT can be described as the mean length of time that substrate and biomass remain in a reactor. Thus, it is the most important variable in a full-scale perspective, as it sets up the flow rate daily treated into the reactor. It ought to be long adequate to obtain solubilization of complex organic material, so favoring consequent acidogenic fermentation of hydrolysates. At the same point, a very great HRT decreases the amount of substrate controllable per day and partiality methanogens at suitable pH ranges (6.5–7.5). As several studies on acidogenic FWs fermentation were carried out in batch reactors, some data are obtainable that investigates the influence of HRT on general VFAs generation. Lim et al. (2008) explored the effect of rising HRT on acidogenesis, beginning from food wastes, in a semi-continuous fed fermenter, under mesophilic conditions, by a pH-controlled to 5.5, and an OLR of 5 g TS/Ld. They experienced three various HRTs: 4, 8, and 12 days. Total VFAs concentration augmented with HRT, from 5.5 to 13 g/L, and lastly to 22 g/L (56). Also, by the maximum HRT, a change in comparative VFAs distribution was achieved. Commonly, acetic acid illustrated the fundamental fermentation generate under shorter HRTs, whereas under an HRT of 12 days propionic acid was the main VFA. Bearing in mind the efficiency accounting at the three HRTs examined, it augmented as a consequence of HRT increase, but no considerable variation between those at HRTs of 8 and 12 days was achieved (from almost 0.34 to 0.39 respectively). So, an HRT of 8 days comes about sufficient to allow a suitable stage of substrate removing, at the slightest in the OLR examined. Another attractive study was conducted by Han and Shin (2002), who investigated the influence of various HRT rates on fermentation yield (1.00, 0.50, 0.33, and 0.25 d) (162). They performed fermentation of FWs from a cafeteria in a 2 L leach bed reactor (LBR), under mesophilic conditions, and adjusted pH. The effect of two inocula were also investigated: ruminal bacteria from the stomach of a cow and an anaerobic bacterial community taken from another reactor fermenting food wastes. The most VFAs concentration of 202 and 181 mmol/L was accounted for by an HRT of 1 with rumen and anaerobic bacteria, respectively. On the opposing, the lesser VFAs concentration was accomplished by an HRT of 0.25 d, accounting for VFA concentrations of 53 and 47 mmol/L, for rumen and anaerobic, and mesophilic bacteria respectively. Concurring to the authors, this consequence was caused to the microbial washout, which occurred at low HRT (0.25 d) coming about in a resulting total VFAs concentration reduction. Moreover, the optimum HRT was 0.33 days for each inoculum, in terms of fermentation performance, expressed as the range between hypothetical VFA of the substrate and real VFA formed: 71.2 and 59.8% at 0.33 d for rumen microbes and mesophilic acidogens, respectively. This demonstrated that rumen bacteria had an enrichment influence on FWs fermentation capability.

2.7.6 OLR

(OLR) shows the number of food wastes fed into the reactor per day and unit of operational volume. Lim et al. (2008) investigated the influence of OLR on acidogenesis, in a semi-continuous reactor (56). The anthers founded out that total VFAs generation increment with augment of OLR, accomplishing the highest concentrations of about 14.0, 24.0, and 30 g/L, with OLRs of 5.0, 9.0, and 13.0 g TS/Ld, respectively. Even though the more concentration at 13.0 g/Ld, the VFAs efficiency was lesser with regard to lesser OLR values. It was caused to more viscosity of average that adversely affects fermentation (154). The study of the VFAs distribution showed that acetic acid was the highest abundant generate, and its concentration raised after the OLR augmented, while the other generates, propionic acid, butyric acid, and HVa, diminished. Furthermore, by rising OLR, the rising of HCa, succinate, and lactate was achieved. Among the three various OLRs examined, an OLR of 9.0 g TS/Ld is further proper for a suitable VFAs generation, approximately to the working variables startup. An alike outcome was achieved from synthetic FWs, fermented in semi-continuous with the subsequent working set up: pH 6, T 35°C, and HRT 5 days. By an increase of OLR from 5.0 to 11.0 g TS/Ld, total VFAs concentration increments from around 13.0 to 21.0 g/ L. In either case, a homogeneous VFAs distribution was obtained, but at lesser OLR butyric acid was the major product, while acetic acid was further abundant with the upper OLR. Nevertheless, the VFA performance (g VFA/g TS) was 13% improved after an OLR of 5.0 g TS/Ld was adopted. It should be noted that for a further increase in OLR from 11.0 to 16.0 g TS/Ld, the reactor must reach steady state (155). Gou et al. (2014) explored a probable correlation between OLR and operational temperature from a cofermentation of waste activated sludge and food wastes (mixed in a 2:1 ratio in terms of TS) conducting in a semi-continuous process (163). A rising OLR was examined, from 1.0 up to 8.0 g TVS/Ld, in three the same CSTR, working at 35, 45, and 55°C. The authors carried out that a steady total VFA is accomplished with upper temperature (4 g/L), while under mesophilic conditions OLR must be reserved under 5 g TVS/Ld to acquire a steady total VFA generation of about 3.5 g/L. So, the selection of an optimum OLR is important for coming to a steady VFA generation. To make sure a good efficiency in terms of VFAs, OLR must be abundantly sufficient to supply a sufficient amount of carbon origin to fermentative metabolism, based on operational temperature. Actually, under mesophilic conditions an OLR upper than 5 g TS/Ld may cause the apparent viscosity increment of the average, decreasing the biomass and heat shift and considerably the substrates transfer into VFAs. This restriction can be risen above using thermo or hyperthermophilic conditions.

2.7.7 Effect of Low Oxygen Concentration on Anaerobic Fermentation

It was illustrated that low concentration of oxygen makes to a great VFA efficiency, causing to the good effects on facultative acidogens, and on the generation of great value of extracellular enzymes (164). Yin et al. (2016b) investigated the effect of exactly anaerobic conditions and middling aerated conditions on VFAs generation from the fermentation of thermal pretreated synthetic food wastes (165). The author's start-up oxidation decrease yield (ORP) in two various rates: −100/−200 mV and −200/−300 mV, which is consistent with inhibited O_2 attendance and exactingly O_2 lack respectively to carry out the optimum fermentation conditions. Two types of inocula were applied: anaerobic granular sludge and aerobic activated sludge. Food wastes were fermented in a batch reactor, at mesophilic condition, pH regulated to 6, for 21 days. the most VFA concentration of 29.4 g/L was achieved on day 17, in the reactor inoculated by anaerobic sludge, in ORP of −100/−200 mV. In its place, VFA accumulation augmented at a low range working by an ORP of −200/−300 mV, and the maximum VFA generation (18.36 g/L) was obtained on day 17, stressing the negative influence of anaerobic conditions acidogenic fermentation. Among the VFAs formed, in all the reactors examined, 80% of VFAs included acetic acid and butyric acid, and butyric acid achieved for up to 60%. Furthermore, the other two VFAs (iso-butyric acid, valeric acid) were detected under lower aeration, but they concentration were fewer than 1.0 g/L in anaerobic conditions. These come about appeared ORP levels, in addition, would influence both total VFAs generation and relative VFAs complex.

2.8 APPLICATIONS OF VFAS

VFAs used for many purposes, such as medical science, food, bioenergy, cosmetics, biomaterials, and the textile industry (Table 2.4). Nowadays, such as energy request is growing, so investigators are searching the proper options for energy sources. VFAs can be applied as substrates to generate electricity using microbial fuel cells (MFC), anywhere microbes oxidize the organic acids (166). Some particular kinds of micro-organisms can store PHAs in response to nutrition reduction or under several environmental problems properties (4). VFAs can be utilized like raw materials and as carbon origins for PHAs, accumulating microbes to generate different copolymers with various conditions. One of the factors affecting the composition and properties of PHA is the length of the VFA chain. Throughout the microbial fermentation in PHA accumulating micro-organisms, acetic acid promotes 3-hydroxybutyrate (3-HB) generation, propionate, and valerate act in 3-hydroxy valerate (3HV) producing, while butyrate plays a role as a raw material for hydroxyhexonate (167). VFAs have also been utilized as precursors to make biogas via anaerobic fermentation carried out by methanogenic microbes. Levin et al. (2004) founded out that purple nonsulfur bacteria are enabled to generate hydrogen from VFAs in photofermentation (168), and the alkyl esters of VFAs, e.g., ethyl acetate, methyl butyrate, and ethyl isobutyrate are applied in the food industry to upgrade flavor. They play a role like raw material in the generation of different kinds of polymers, including vinyl acetate monomer, cellulose-based plastic matter, and polymethyl metacetic acidrylate.

2.9 FUTURE PERSPECTIVE AND CONCLUSION

The industrial generation of VFAs yet depends on chemical material, and investigators are continuously producing attempts to develop and progress the innovation process to produce VFA generation more cost-effective and applied through

TABLE 2.4
Applications of Different VFAs and Their Metabolites

VFAs	Annual Production (million ton)	Company Involved in VFA Related Business	Application	References
Acetic acid	13	Hoechst Celanese (Irving, Texas, USA)	Food preservation: 5–20% acetic acid (vinegar) is used as a condiment to pickle vegetables	(169)
		Monsanto Co. (Texas), BP Chemicals (Hull, UK), Samsung-BP (Ulsan, South Korea)	Cosmetic industry: ethyl acetate is used to enhance aroma in cosmetics Textile industry: polyvinyl acetate used in acrylic fibers	(16)
		BP Petronas (Kertih, Malaysia), Shandong Hualu Hengsheng Chemical Co. Ltd (China)	Another use: ethyl acetate, butyl acetate, propyl acetate is used as solvents for ink. Calcium magnesium acetate (CMA) is used as a deicing salt	(170)
Propionic acid	0.24	Chemische Werke Hu¨ls (Germany)	Food preservation: inhibits the growth of Aspergillus flavus, aerobic Bacillus, Salmonella, and yeast	(171)
		Distillers Company (Britain), USSR (Russia)	Herbicide synthesis: sodium 2,2-dichloropropionate	(93)
		Celanese Chemical Company (USA)	Perfume intermediate: propionic ether and benzyl propionate	(93)
		And Eastman Chemical (USA)	Pharmaceuticals intermediates: synthesis of propionic anhydride and chloropropionic acid	(93)
Butyric acid	0.05	Polymer Plastics Company, LC (Carson City, USA),	Food additive: it prevents colonization of Salmonella in poultry. Methyl butyrate for pleasant aroma and taste	(172)
		Rotuba (Linden, USA),	Riot controlling agent: Due to its unpleasant smell used as a stink bomb Textile: butyrate and its derivative mixture with other compounds like cellulose and acetate used to produce plastic material and fibers	(94)
		Finoric LLC (Houston & Midland, Texas USA), Jiangsu Ruijia Chemistry Co., Ltd (China)	Bioenergy: used as a precursor to produce biofuel e.g., ethyl butyrate and butyl butyrate Pharmaceutical: As an intermediate to produce various drugs used in treatment of cancers and hemoglobinopathies, including leukemia and sickle cell anemia (SCA	(173)

Isobutyric acid	0.05	Eastman Chemical Company (Kingsport, USA),	Food additive: methyl isobutyrate and ethyl isobutyrate are used for the fruity smell in juice	(91)
		Jayshree Aromatics Pvt Ltd (Gujrat, India),	Biomaterial: used as a precursor to synthesize methacrylate which is further used for polymethacrylate production	(91)
		Fooding group Ltd (Shanghai, China)	Emulsifier: used to prepare sucrose acetate isobutyrate which is used as an emulsifier for printing ink Cosmetic: phenoxy ethyl isobutyrate is used as a perfuming agent Coating material: isobutyl isobutyrate is used as a coating material for plastic substrates	(174)
Isovaleric acid	NT	Elan Chemical Company Inc. (Newark, USA),	Food additive: menthyl isovalerate, methyl isovalerate, and ethyl isovalerate used for fragrance	(175)
		CTC Organics (Ellenwood, USA), Inoue Perfumery MFG. Co. Ltd (Okusawa, Tokya)	Pharma: methyl isovalerate and ethyl isovalerate have spasmolytic activity	

fermentation. To create VFA generation an actuality during fermentation, there is a requirement to emphasize on an undoubted area as conversed here. Generally, fermentation happens utilizing a mixed culture, which may outcome in low efficiency and manufacture of mixed acids with other byproducts (176). Recognition and separation of novel microbial growing for upper efficiency is necessary, and developmental process and metagenomics procedure may be useful in this subject (177, 178). VFA generation is mainly according to treated sugar fermentation, which makes the generation procedure expensive (9, 51). Lignocellulose is the mainly great carbon substrate, but just a few micro-organisms are capable to use these (179). So, different physical and chemical processes are utilized to pretreat biomass and release free sugars. These purification processes are the consequence of the generation of different derivatives of sugars, which affect microbial culture and fermentation (180). Microbial engineering to tolerate like inhibitors may be a probable method (181). Furthermore, the downstream method is in addition a challenge, and superior focus is essential in this subject to recover VFAs with an upper yield and treatment (140). In conclusion, developments procedure are necessary for chemical industrialized, and a practical alternative is ineffective process to secure and effective engineered fermentation. Microbial fermentation perhaps is a suitable option for VFA generation because such strategies are friendlier and can be performed in ambient properties.

REFERENCES

1. Lee WS, Chua ASM, Yeoh HK, Ngoh GC. A review of the production and applications of waste-derived volatile fatty acids. *Chemical Engineering Journal.* 2014;235:83–99.
2. Buehring GC, Shen HM, Jensen HM, Jin DL, Hudes M, Block G. Exposure to bovine leukemia virus is associated with breast cancer: a case-control study. *PloS One.* 2015;10(9):e0134304.
3. Volker AR, Gogerty DS, Bartholomay C, Hennen-Bierwagen T, Zhu H, Bobik TA. Fermentative production of short-chain fatty acids in *Escherichia coli. Microbiology.* 2014;160(7):1513–1522.
4. Bhatia SK, Lee B-R, Sathiyanarayanan G, Song HS, Kim J, Jeon J-M, et al. Biomass-derived molecules modulate the behavior of *Streptomyces coelicolor* for antibiotic production. *3 Biotech.* 2016;6(2):223.
5. Van Lingen HJ, Plugge CM, Fadel JG, Kebreab E, Bannink A, Dijkstra J. Thermodynamic driving force of hydrogen on rumen microbial metabolism: a theoretical investigation. *PLoS One.* 2016;11(10):e0161362.
6. Baumann I, Westermann P. Microbial production of short chain fatty acids from lignocellulosic biomass: current processes and market. *BioMed Research International.* 2016;2016.
7. Ehsanipour M, Suko AV, Bura R. Fermentation of lignocellulosic sugars to acetic acid by *Moorella thermoacetica. Journal of Industrial Microbiology & Biotechnology.* 2016;43(6):807–816.
8. Coral J, Karp SG, de Souza Vandenberghe LP, Parada JL, Pandey A, Soccol CR. Batch fermentation model of propionic acid production by *Propionibacterium acidipropionici* in different carbon sources. *Applied Biochemistry and Biotechnology.* 2008;151(2–3):333–341.

9. Feng X-H, Chen F, Xu H, Wu B, Yao J, Ying H-J, et al. Propionic acid fermentation by *Propionibacterium freudenreichii* CCTCC M207015 in a multi-point fibrous-bed bioreactor. *Bioprocess and Biosystems Engineering*. 2010;33(9):1077–1085.
10. Gottumukkala LD, Sukumaran RK, Mohan SV, Valappil SK, Sarkar O, Pandey A. Rice straw hydrolysate to fuel and volatile fatty acid conversion by *Clostridium sporogenes* BE01: bio-electrochemical analysis of the electron transport mediators involved. *Green Chemistry*. 2015;17(5):3047–3058.
11. Ma H, Chen X, Liu H, Liu H, Fu B. Improved volatile fatty acids anaerobic production from waste activated sludge by pH regulation: alkaline or neutral pH? *Waste Management*. 2016;48:397–403.
12. Hasan SDM, Giongo C, Fiorese ML, Gomes SD, Ferrari TC, Savoldi TE. Volatile fatty acids production from anaerobic treatment of cassava waste water: effect of temperature and alkalinity. *Environmental Technology*. 2015;36(20):2637–2646.
13. Bhatia S, Mehta P, Bhatia R, Bhalla T. Optimization of arylacetonitrilase production from *Alcaligenes* sp. MTCC 10675 and its application in mandelic acid synthesis. *Applied Microbiology and Biotechnology*. 2014;98(1):83–94.
14. Ibrahim MF, Abd-Aziz S, Yusoff MEM, Phang LY, Hassan MA. Simultaneous enzymatic saccharification and ABE fermentation using pretreated oil palm empty fruit bunch as substrate to produce butanol and hydrogen as biofuel. *Renewable Energy*. 2015;77:447–455.
15. Kim N-J, Park GW, Kang J, Kim Y-C, Chang HN. Volatile fatty acid production from lignocellulosic biomass by lime pretreatment and its applications to industrial biotechnology. *Biotechnology and Bioprocess Engineering*. 2013;18(6):1163–1168.
16. Pal P, Nayak J. Acetic acid production and purification: critical review towards process intensification. *Separation & Purification Reviews*. 2017;46(1):44–61.
17. Yoneda N, Kusano S, Yasui M, Pujado P, Wilcher S. Recent advances in processes and catalysts for the production of acetic acid. *Applied Catalysis A: General*. 2001;221(1–2):253–265.
18. Nayak J, Pal P. Transforming waste cheese-whey into acetic acid through a continuous membrane-integrated hybrid process. *Industrial & Engineering Chemistry Research*. 2013;52(8):2977–2984.
19. Sim JH, Kamaruddin AH. Optimization of acetic acid production from synthesis gas by chemolithotrophic bacterium–*Clostridium aceticum* using statistical approach. *Bioresource Technology*. 2008;99(8):2724–2735.
20. Kadere T, Miyamotoo T, Oniango R, Kutima P, Njoroge S. Isolation and identification of the genera *Acetobacter* and *Gluconobacter* in coconut toddy (mnazi). *African Journal of Biotechnology*. 2008;7(16).
21. Ravinder T, Ramesh B, Seenayya G, Reddy G. Fermentative production of acetic acid from various pure and natural cellulosic materials by *Clostridium lentocellum* SG6. *World Journal of Microbiology and Biotechnology*. 2000;16(6): 507–512.
22. Mounir M, Shafiei R, Zarmehrkhorshid R, Hamouda A, Alaoui MI, Thonart P. Simultaneous production of acetic and gluconic acids by a thermotolerant *Acetobacter* strain during acetous fermentation in a bioreactor. *Journal of Bioscience and Bioengineering*. 2016;121(2):166–171.
23. Wang Z, Yan M, Chen X, Li D, Qin L, Li Z, et al. Mixed culture of *Saccharomyces cerevisiae* and *Acetobacter pasteurianus* for acetic acid production. *Biochemical Engineering Journal*. 2013;79:41–45.
24. Li Y, He D, Niu D, Zhao Y. Acetic acid production from food wastes using yeast and acetic acid bacteria micro-aerobic fermentation. *Bioprocess and Biosystems Engineering*. 2015;38(5):863–869.

25. Jourdin L, Grieger T, Monetti J, Flexer V, Freguia S, Lu Y, et al. High acetic acid production rate obtained by microbial electrosynthesis from carbon dioxide. *Environmental Science & Technology*. 2015;49(22):13566–13574.
26. Fukaya M, Park Y, Toda K. Improvement of acetic acid fermentation by molecular breeding and process development. *Journal of Applied Bacteriology*. 1992;73(6): 447–454.
27. Nakano S, Fukaya M, Horinouchi S. Putative ABC transporter responsible for acetic acid resistance in *Acetobacter aceti*. *Applied and Environmental Microbiology*. 2006;72(1):497–505.
28. Schwartz RD, Keller FA. Acetic acid production by Clostridium thermoaceticum in pH-controlled batch fermentations at acidic pH. *Applied and Environmental Microbiology*. 1982;43(6):1385–1392.
29. Talabardon M, Schwitzguébel JP, Péringer P, Yang ST. Acetic acid production from lactose by an anaerobic thermophilic coculture immobilized in a fibrous-bed bioreactor. *Biotechnology Progress*. 2000;16(6):1008–1017.
30. Mas A, Torija MJ, García-Parrilla MDC, Troncoso AM. Acetic acid bacteria and the production and quality of wine vinegar. *The Scientific World Journal*. 2014;2014.
31. Spinosa WA, Santos Júnior VD, Galvan D, Fiorio JL, Gomez RJHC. Vinegar rice (*Oryza sativa* L.) produced by a submerged fermentation process from alcoholic fermented rice. *Food Science and Technology*. 2015;35(1):196–201.
32. Atik D, Atik C, Karatepe C. The effect of external apple vinegar application on varicosity symptoms, pain, and social appearance anxiety: a randomized controlled trial. *Evidence-Based Complementary and Alternative Medicine*. 2016;2016.
33. Budak NH, Ozçelik F, Güzel-Seydim ZB. Antioxidant activity and phenolic content of apple cider. *Turkish Journal of Agriculture-Food Science and Technology*. 2015; 3(6):356–360.
34. Cunha MAAD, Lima KPD, Santos VAQ, Heinz OL, Schmidt CAP. Blackberry vinegar produced by successive acetification cycles: production, characterization and bioactivity parameters. *Brazilian Archives of Biology and Technology*. 2016;59.
35. Ghosh S, Chakraborty R, Chatterjee G, Raychaudhuri U. Study on fermentation conditions of palm juice vinegar by response surface methodology and development of a kinetic model. *Brazilian Journal of Chemical Engineering*. 2012;29(3):461–472.
36. Vegas C, Mateo E, González Á, Jara C, Guillamón JM, Poblet M, et al. Population dynamics of acetic acid bacteria during traditional wine vinegar production. *International Journal of Food Microbiology*. 2010;138(1–2):130–136.
37. Tang I-C, Yang S-T, Okos MR. Acetic acid production from whey lactose by the co-culture of *Streptococcus lactis* and *Clostridium formicoaceticum*. *Applied Microbiology and Biotechnology*. 1988;28(2):138–143.
38. Mehaia MA, Cheryan M. Fermentation of date extracts to ethanol and vinegar in batch and continuous membrane reactors. *Enzyme and Microbial Technology*. 1991;13(3):257–261.
39. Witjitra K, Shah M, Cheryan M. Effect of nutrient sources on growth and acetate production by *Clostridium thermoaceticum*. *Enzyme and Microbial Technology*. 1996;19(5):322–327.
40. Mostafa N. Production of acetic acid and glycerol from salted and dried whey in a membrane cell recycle bioreactor. *Energy Conversion and Management*. 2001;42(9): 1133–1142.
41. Gupta A, Srivastava AK. Continuous propionic acid production from cheese whey using in situ spin filter. *Biotechnology and Bioprocess Engineering*. 2001;6(1):1–5.
42. Zhang C, Yang H, Yang F, Ma Y. Current progress on butyric acid production by fermentation. *Current Microbiology*. 2009;59(6):656–663.

43. Liang Z-X, Li L, Li S, Cai Y-H, Yang S-T, Wang J-F. Enhanced propionic acid production from Jerusalem artichoke hydrolysate by immobilized *Propionibacterium acidipropionici* in a fibrous-bed bioreactor. *Bioprocess and Biosystems Engineering*. 2012;35(6):915–921.
44. Zigová J, Šturdík E, Vandák D, Schlosser Š. Butyric acid production by *Clostridium butyricum* with integrated extraction and pertraction. *Process Biochemistry*. 1999;34(8):835–843.
45. Himmi E, Bories A, Boussaid A, Hassani L. Propionic acid fermentation of glycerol and glucose by *Propionibacterium acidipropionici* and *Propionibacterium freudenreichii* ssp. shermanii. *Applied Microbiology and Biotechnology*. 2000;53(4):435–440.
46. Ramsay J, Hassan M-CA, Ramsay B. Biological conversion of hemicellulose to propionic acid. *Enzyme and Microbial Technology*. 1998;22(4):292–295.
47. Zhu Y, Li J, Tan M, Liu L, Jiang L, Sun J, et al. Optimization and scale-up of propionic acid production by propionic acid-tolerant *Propionibacterium acidipropionici* with glycerol as the carbon source. *Bioresource Technology*. 2010;101(22):8902–8906.
48. He G-Q, Kong Q, Chen Q-H, Ruan H. Batch and fed-batch production of butyric acid by *Clostridium butyricum* ZJUCB. *Journal of Zhejiang University Science B*. 2005;6(11):1076.
49. Zhu Y, Yang S-T. Effect of pH on metabolic pathway shift in fermentation of xylose by *Clostridium tyrobutyricum*. *Journal of Biotechnology*. 2004;110(2):143–157.
50. Canganella F, Kuk S-U, Morgan H, Wiegel J. *Clostridium thermobutyricum*: growth studies and stimulation of butyrate formation by acetate supplementation. *Microbiological Research*. 2002;157(2):149–156.
51. Jiang L, Wang J, Liang S, Cai J, Xu Z, Cen P, et al. Enhanced butyric acid tolerance and bioproduction by *Clostridium tyrobutyricum* immobilized in a fibrous bed bioreactor. *Biotechnology and Bioengineering*. 2011;108(1):31–40.
52. Jiang L, Wang J, Liang S, Wang X, Cen P, Xu Z. Butyric acid fermentation in a fibrous bed bioreactor with immobilized *Clostridium tyrobutyricum* from cane molasses. *Bioresource Technology*. 2009;100(13):3403–3409.
53. Wei D, Liu X, Yang S-T. Butyric acid production from sugarcane bagasse hydrolysate by *Clostridium tyrobutyricum* immobilized in a fibrous-bed bioreactor. *Bioresource Technology*. 2013;129:553–560.
54. Thierry A, Richoux R, Kerjean J-R. Isovaleric acid is mainly produced by Propionibacterium freudenreichii in Swiss cheese. *International Dairy Journal*. 2004;14(9):801–807.
55. Wijekoon KC, Visvanathan C, Abeynayaka A. Effect of organic loading rate on VFA production, organic matter removal and microbial activity of a two-stage thermophilic anaerobic membrane bioreactor. *Bioresource Technology*. 2011;102(9):5353–5360.
56. Lim S-J, Kim BJ, Jeong C-M, Ahn YH, Chang HN. Anaerobic organic acid production of food waste in once-a-day feeding and drawing-off bioreactor. *Bioresource Technology*. 2008;99(16):7866–7874.
57. García-Bernet D, Buffière P, Latrille E, Steyer J-P, Escudié R. Water distribution in biowastes and digestates of dry anaerobic digestion technology. *Chemical Engineering Journal*. 2011;172(2–3):924–928.
58. Bollon J, Benbelkacem H, Gourdon R, Buffière P. Measurement of diffusion coefficients in dry anaerobic digestion media. *Chemical Engineering Science*. 2013;89:115–119.
59. Giovannini G, Donoso-Bravo A, Jeison D, Chamy R, Ruíz-Filippi G, Wouwer AV. A review of the role of hydrogen in past and current modelling approaches to anaerobic digestion processes. *International Journal of Hydrogen Energy*. 2016;41(39):17713–17722.

60. Pauss A, Andre G, Perrier M, Guiot SR. Liquid-to-gas mass transfer in anaerobic processes: inevitable transfer limitations of methane and hydrogen in the biomethanation process. *Applied and Environmental Microbiology*. 1990;56(6):1636–1644.
61. Abbasi T, Tauseef S, Abbasi S. Anaerobic digestion for global warming control and energy generation—an overview. *Renewable and Sustainable Energy Reviews*. 2012;16(5):3228–3242.
62. Thiele JH, Zeikus JG. Control of interspecies electron flow during anaerobic digestion: significance of formate transfer versus hydrogen transfer during syntrophic methanogenesis in flocs. *Applied and Environmental Microbiology*. 1988; 54(1):20–29.
63. McCarty PL, Smith DP. Anaerobic wastewater treatment. *Environmental Science & Technology*. 1986;20(12):1200–1206.
64. Chong M-L, Sabaratnam V, Shirai Y, Hassan MA. Biohydrogen production from biomass and industrial wastes by dark fermentation. *International Journal of Hydrogen Energy*. 2009;34(8):3277–3287.
65. Junghare M, Subudhi S, Lal B. Improvement of hydrogen production under decreased partial pressure by newly isolated alkaline tolerant anaerobe, *Clostridium butyricum* TM-9A: Optimization of process parameters. *International Journal of Hydrogen Energy*. 2012;37(4):3160–3168.
66. Kisielewska M, Dębowski M, Zieliński M. Improvement of biohydrogen production using a reduced pressure fermentation. *Bioprocess and Biosystems Engineering*. 2015;38(10):1925–1933.
67. Merkle W, Baer K, Haag NL, Zielonka S, Ortloff F, Graf F, et al. High-pressure anaerobic digestion up to 100 bar: influence of initial pressure on production kinetics and specific methane yields. *Environmental Technology*. 2017;38(3):337–344.
68. Chen Y, Rößler B, Zielonka S, Lemmer A, Wonneberger A-M, Jungbluth T. The pressure effects on two-phase anaerobic digestion. *Applied Energy*. 2014;116:409–415.
69. Clark IC, Zhang RH, Upadhyaya SK. The effect of low pressure and mixing on biological hydrogen production via anaerobic fermentation. *International Journal of Hydrogen Energy*. 2012;37(15):11504–11513.
70. Bielen AA, Verhaart MR, VanFossen AL, Blumer-Schuette SE, Stams AJ, van der Oost J, et al. A thermophile under pressure: transcriptional analysis of the response of *Caldicellulosiruptor saccharolyticus* to different H2 partial pressures. *International Journal of Hydrogen Energy*. 2013;38(4):1837–1849.
71. Wang J, Liu H, Xu K, Wang A, Chen J. Synergistic effect of syntrophic acetogenesis and homoacetogenesis for volatile fatty acids production from sludge by anaerobic digestion. *Environmental Science*. 2011;32(6):1673–1678.
72. Zhang A, Yang S-T. Propionic acid production from glycerol by metabolically engineered *Propionibacterium acidipropionici*. *Process Biochemistry*. 2009;44(12): 1346–1351.
73. Quesada-Chanto A, Wagner F. Microbial production of propionic acid and vitamin B 12 using molasses or sugar. *Applied Microbiology and Biotechnology*. 1994;41(4): 378–383.
74. Goswami V, Srivastava A. Propionic acid production in an in situ cell retention bioreactor. *Applied Microbiology and Biotechnology*. 2001;56(5–6):676–680.
75. Luna-Flores CH, Palfreyman RW, Krömer JO, Nielsen LK, Marcellin E. Improved production of propionic acid using genome shuffling. *Biotechnology Journal*. 2017; 12(2):1600120.
76. Liu L, Zhuge X, Shin H-D, Chen RR, Li J, Du G, et al. Improved production of propionic acid in *Propionibacterium jensenii* via combinational overexpression of glycerol dehydrogenase and malate dehydrogenase from *Klebsiella pneumoniae*. *Applied and Environmental Microbiology*. 2015;81(7):2256–2264.

77. Fukaya M, Takemura H, Tayama K, Okumura H, Kawamura Y, Horinouchi S, et al. The aarC gene responsible for acetic acid assimilation confers acetic acid resistance on *Acetobacter aceti*. *Journal of Fermentation and Bioengineering*. 1993;76(4):270–275.
78. Fukaya M, Takemura H, Okumura H, Kawamura Y, Horinouchi S, Beppu T. Cloning of genes responsible for acetic acid resistance in *Acetobacter aceti*. *Journal of Bacteriology*. 1990;172(4):2096–2104.
79. Wang Z, Ammar EM, Zhang A, Wang L, Lin M, Yang S-T. Engineering *Propionibacterium freudenreichii* subsp. shermanii for enhanced propionic acid fermentation: effects of overexpressing propionyl-CoA: succinate CoA transferase. *Metabolic Engineering*. 2015;27:46–56.
80. Ammar EM, Jin Y, Wang Z, Yang S-T. Metabolic engineering of Propionibacterium freudenreichii: effect of expressing phosphoenolpyruvate carboxylase on propionic acid production. *Applied Microbiology and Biotechnology*. 2014;98(18): 7761–7772.
81. Zhang A, Yang ST. Engineering *Propionibacterium acidipropionici* for enhanced propionic acid tolerance and fermentation. *Biotechnology and Bioengineering*. 2009;104(4):766–773.
82. Guan N, Li J, Shin HD, Du G, Chen J, Liu L. Metabolic engineering of acid resistance elements to improve acid resistance and propionic acid production of *Propionibacterium jensenii*. *Biotechnology and Bioengineering*. 2016;113(6):1294–1304.
83. Liu L, Guan N, Zhu G, Li J, Shin H-D, Du G, et al. Pathway engineering of *Propionibacterium jensenii* for improved production of propionic acid. *Scientific Reports*. 2016;6:19963.
84. Zhuge X, Li J, Shin H-D, Liu L, Du G, Chen J. Improved propionic acid production with metabolically engineered *Propionibacterium jensenii* by an oxidoreduction potential-shift control strategy. *Bioresource Technology*. 2015;175:606–612.
85. Liu X, Zhu Y, Yang ST. Construction and characterization of ack deleted mutant of *Clostridium tyrobutyricum* for enhanced butyric acid and hydrogen production. *Biotechnology Progress*. 2006;22(5):1265–1275.
86. Jang Y-S, Woo HM, Im JA, Kim IH, Lee SY. Metabolic engineering of *Clostridium acetobutylicum* for enhanced production of butyric acid. *Applied Microbiology and Biotechnology*. 2013;97(21):9355–9363.
87. Saini M, Wang ZW, Chiang C-J, Chao Y-P. Metabolic engineering of *Escherichia coli* for production of butyric acid. *Journal of Agricultural and Food Chemistry*. 2014;62(19):4342–4348.
88. Jawed K, Mattam AJ, Fatma Z, Wajid S, Abdin MZ, Yazdani SS. Engineered production of short chain fatty acid in *Escherichia coli* using fatty acid synthesis pathway. *PLoS One*. 2016;11(7):e0160035.
89. Jang Y-S, Im JA, Choi SY, Im Lee J, Lee SY. Metabolic engineering of *Clostridium acetobutylicum* for butyric acid production with high butyric acid selectivity. *Metabolic Engineering*. 2014;23:165–174.
90. Lang K, Zierow J, Buehler K, Schmid A. Metabolic engineering of *Pseudomonas* sp. strain VLB120 as platform biocatalyst for the production of isobutyric acid and other secondary metabolites. *Microbial Cell Factories*. 2014;13(1):2.
91. Zhang K, Woodruff AP, Xiong M, Zhou J, Dhande YK. A synthetic metabolic pathway for production of the platform chemical isobutyric acid. *ChemSusChem*. 2011;4(8):1068–1070.
92. Xiong M, Deng J, Woodruff AP, Zhu M, Zhou J, Park SW, et al. A bio-catalytic approach to aliphatic ketones. *Scientific Reports*. 2012;2:311.
93. Kumar A, Jianzheng J, Li Y, Baral YN, Ai B. A review on bio-butyric acid production and its optimization. *International Journal of Agriculture and Biology*. 2014;16(5).

94. Zigova J, Šturdík E. Advances in biotechnological production of butyric acid. *Journal of Industrial Microbiology and Biotechnology*. 2000;24(3):153–160.
95. Baroi G, Baumann I, Westermann P, Gavala HN. Butyric acid fermentation from pretreated and hydrolysed wheat straw by an adapted *Clostridium tyrobutyricum* strain. *Microbial Biotechnology*. 2015;8(5):874–882.
96. Ai B, Li J, Chi X, Meng J, Liu C, Shi E. Butyric acid fermentation of sodium hydroxide pretreated rice straw with undefined mixed culture. *Journal of Microbiology and Biotechnology*. 2014;24(5):629–638.
97. Dwidar M, Kim S, Jeon BS, Um Y, Mitchell RJ, Sang B-I. Co-culturing a novel *Bacillus* strain with *Clostridium tyrobutyricum* ATCC 25755 to produce butyric acid from sucrose. *Biotechnology for Biofuels*. 2013;6(1):35.
98. Reimann A, Biebl H, Deckwer W-D. Influence of iron, phosphate and methyl viologen on glycerol fermentation of *Clostridium butyricum*. *Applied Microbiology and Biotechnology*. 1996;45(1–2):47–50.
99. Zhu Y, Liu X, Yang ST. Construction and characterization of pta gene-deleted mutant of *Clostridium tyrobutyricum* for enhanced butyric acid fermentation. *Biotechnology and Bioengineering*. 2005;90(2):154–166.
100. Ruppert W, Siegert H-J. Method for making isobutyric acid. Google Patents; 1991.
101. Vemuri GN, Altman E, Sangurdekar D, Khodursky AB, Eiteman MA. Overflow metabolism in *Escherichia coli* during steady-state growth: transcriptional regulation and effect of the redox ratio. *Applied and Environmental Microbiology*. 2006;72(5):3653–3661.
102. Blank LM, Ionidis G, Ebert BE, Bühler B, Schmid A. Metabolic response of *Pseudomonas putida* during redox biocatalysis in the presence of a second octanol phase. *The FEBS Journal*. 2008;275(20):5173–5190.
103. Ahn BK, Hayashida S. Metabolic mechanism of Ethanol–Isovaleric Acid fermentation by a Clostridium saccharoperbutylacetonicum UV-Mutant. *Agricultural and Biological Chemistry*. 1990;54(2):353–357.
104. Balch W, Fox GE, Magrum LJ, Woese CR, Wolfe R. Methanogens: reevaluation of a unique biological group. *Microbiological Reviews*. 1979;43(2):260.
105. Hedderich R, Whitman WB. Physiology and biochemistry of the methane-producing Archaea. *The Prokaryotes*. 2006;2:1050–1079.
106. Liu Y, Whitman WB. Metabolic, phylogenetic, and ecological diversity of the methanogenic archaea. *Annals of the New York Academy of Sciences*. 2008;1125(1): 171–189.
107. Wu Q-L, Guo W-Q, Zheng H-S, Luo H-C, Feng X-C, Yin R-L, et al. Enhancement of volatile fatty acid production by co-fermentation of food waste and excess sludge without pH control: the mechanism and microbial community analyses. *Bioresource Technology*. 2016;216:653–660.
108. Luo K, Yang Q, Yu J, Li X-M, Yang G-J, Xie B-X, et al. Combined effect of sodium dodecyl sulfate and enzyme on waste activated sludge hydrolysis and acidification. *Bioresource Technology*. 2011;102(14):7103–7110.
109. Ji Z, Chen G, Chen Y. Effects of waste activated sludge and surfactant addition on primary sludge hydrolysis and short-chain fatty acids accumulation. *Bioresource Technology*. 2010;101(10):3457–3462.
110. Dahiya S, Sarkar O, Swamy Y, Mohan SV. Acidogenic fermentation of food waste for volatile fatty acid production with co-generation of biohydrogen. *Bioresource Technology*. 2015;182:103–113.
111. Yin J, Wang K, Yang Y, Shen D, Wang M, Mo H. Improving production of volatile fatty acids from food waste fermentation by hydrothermal pretreatment. *Bioresource Technology*. 2014;171:323–329.

112. Jiang Y, Marang L, Tamis J, van Loosdrecht MC, Dijkman H, Kleerebezem R. Waste to resource: converting paper mill wastewater to bioplastic. *Water Research*. 2012;46(17):5517–5530.
113. Ghosh P, Samanta A, Ray S. COD reduction of petrochemical industry wastewater using Fenton's oxidation. *The Canadian Journal of Chemical Engineering*. 2010; 88(6):1021–1026.
114. Bengtsson S, Hallquist J, Werker A, Welander T. Acidogenic fermentation of industrial wastewaters: effects of chemostat retention time and pH on volatile fatty acids production. *Biochemical Engineering Journal*. 2008;40(3):492–499.
115. Maharaj I, Elefsiniotis P. The role of HRT and low temperature on the acid-phase anaerobic digestion of municipal and industrial wastewaters. *Bioresource Technology*. 2001;76(3):191–197.
116. Yang X, Liu X, Chen S, Liu G, Wu S, Wan C. Volatile fatty acids production from codigestion of food waste and sewage sludge based on β-cyclodextrins and alkaline treatments. *Archaea*. 2016;2016.
117. Chang HN, Kim N-J, Kang J, Jeong CM. Biomass-derived volatile fatty acid platform for fuels and chemicals. *Biotechnology and Bioprocess Engineering*. 2010;15(1):1–10.
118. Kim S-H, Han S-K, Shin H-S. Effect of substrate concentration on hydrogen production and 16S rDNA-based analysis of the microbial community in a continuous fermenter. *Process Biochemistry*. 2006;41(1):199–207.
119. González-Fernández C, García-Encina PA. Impact of substrate to inoculum ratio in anaerobic digestion of swine slurry. *Biomass and Bioenergy*. 2009;33(8):1065–1069.
120. Dareioti MA, Kornaros M. Anaerobic mesophilic co-digestion of ensiled sorghum, cheese whey and liquid cow manure in a two-stage CSTR system: effect of hydraulic retention time. *Bioresource Technology*. 2015;175:553–562.
121. Palacio-Barco E, Robert-Peillard F, Boudenne J-L, Coulomb B. On-line analysis of volatile fatty acids in anaerobic treatment processes. *Analytica Chimica Acta*. 2010;668(1):74–79.
122. Zhang C, Xiao G, Peng L, Su H, Tan T. The anaerobic co-digestion of food waste and cattle manure. *Bioresource Technology*. 2013;129:170–176.
123. Franke-Whittle IH, Walter A, Ebner C, Insam H. Investigation into the effect of high concentrations of volatile fatty acids in anaerobic digestion on methanogenic communities. *Waste Management*. 2014;34(11):2080–2089.
124. Buyukkamaci N, Filibeli A. Volatile fatty acid formation in an anaerobic hybrid reactor. *Process Biochemistry*. 2004;39(11):1491–1494.
125. Müller N, Worm P, Schink B, Stams AJ, Plugge CM. Syntrophic butyrate and propionate oxidation processes: from genomes to reaction mechanisms. *Environmental Microbiology Reports*. 2010;2(4):489–499.
126. Stams AJ, Plugge CM. Electron transfer in syntrophic communities of anaerobic bacteria and archaea. *Nature Reviews Microbiology*. 2009;7(8):568–577.
127. Huang W-C, Chen S-J, Chen T-L. Production of hyaluronic acid by repeated batch fermentation. *Biochemical Engineering Journal*. 2008;40(3):460–464.
128. Lata K, Rajeshwari K, Pant D, Kishore V. Volatile fatty acid production during anaerobic mesophilic digestion of tea and vegetable market wastes. *World Journal of Microbiology and Biotechnology*. 2002;18(6):589–592.
129. Dong C-J, Lu B-N, Chen Z-Q. Characteristic of anaerobic granular sludge and digestion sludge under microaerobic conditions. *Journal-Nanjing University of Science and Technology*. 2005;29(2):216.
130. Zeldes BM, Keller MW, Loder AJ, Straub CT, Adams MW, Kelly RM. Extremely thermophilic microorganisms as metabolic engineering platforms for production of fuels and industrial chemicals. *Frontiers in Microbiology*. 2015;6:1209.

131. Mitchell RJ, Kim J-S, Jeon B-S, Sang B-I. Continuous hydrogen and butyric acid fermentation by immobilized *Clostridium tyrobutyricum* ATCC 25755: effects of the glucose concentration and hydraulic retention time. *Bioresource Technology*. 2009;100(21):5352–5355.
132. Dishisha T, Alvarez MT, Hatti-Kaul R. Batch-and continuous propionic acid production from glycerol using free and immobilized cells of *Propionibacterium acidipropionici*. *Bioresource Technology*. 2012;118:553–562.
133. Beccari M, Bertin L, Dionisi D, Fava F, Lampis S, Majone M, et al. Exploiting olive oil mill effluents as a renewable resource for production of biodegradable polymers through a combined anaerobic–aerobic process. *Journal of Chemical Technology & Biotechnology: International Research in Process, Environmental & Clean Technology*. 2009;84(6):901–908.
134. Haribabu K, Sivasubramanian V. Treatment of wastewater in fluidized bed bioreactor using low density biosupport. *Energy Procedia*. 2014;50:214–221.
135. Ai B, Chi X, Meng J, Sheng Z, Zheng L, Zheng X, et al. Consolidated bioprocessing for butyric acid production from rice straw with undefined mixed culture. *Frontiers in Microbiology*. 2016;7:1648.
136. Emadian SM, Rahimnejad M, Hosseini M, Khoshandam B. Investigation on up-flow anaerobic sludge fixed film (UASFF) reactor for treating low-strength bilge water of Caspian Sea ships. *Journal of Environmental Health Science and Engineering*. 2015; 13(1):23.
137. Li Q-Z, Jiang X-L, Feng X-J, Wang J-M, Sun C, Zhang H-B, et al. Recovery processes of organic acids from fermentation broths in the biomass-based industry. *Journal of Microbiology and Biotechnology*. 2016;26(1):1–8.
138. Lee SC, Kim HC. Batch and continuous separation of acetic acid from succinic acid in a feed solution with high concentrations of carboxylic acids by emulsion liquid membranes. *Journal of Membrane Science*. 2011;367(1–2):190–196.
139. Shity H, Bar R. New approach for selective separation of dilute products from simulated clostridial fermentation broths using cyclodextrins. *Biotechnology and Bioengineering*. 1992;39(4):462–466.
140. Mkhize NT, Msagati TA, Mamba BB, Momba M. Determination of volatile fatty acids in wastewater by solvent extraction and gas chromatography. *Physics and Chemistry of the Earth, Parts A/B/C*. 2014;67:86–92.
141. Singhania RR, Patel AK, Christophe G, Fontanille P, Larroche C. Biological upgrading of volatile fatty acids, key intermediates for the valorization of biowaste through dark anaerobic fermentation. *Bioresource Technology*. 2013;145:166–174.
142. Bekatorou A, Dima A, Tsafrakidou P, Boura K, Lappa K, Kandylis P, et al. Downstream extraction process development for recovery of organic acids from a fermentation broth. *Bioresource Technology*. 2016;220:34–37.
143. Alkaya E, Kaptan S, Ozkan L, Uludag-Demirer S, Demirer GN. Recovery of acids from anaerobic acidification broth by liquid–liquid extraction. *Chemosphere*. 2009;77(8):1137–1142.
144. Huang H-J, Ramaswamy S, Tschirner UW, Ramarao B. A review of separation technologies in current and future biorefineries. *Separation and Purification Technology*. 2008;62(1):1–21.
145. Siedlecka EM, Kumirska J, Ossowski T, Glamowski P, Gołębiowski M, Gajdus J, et al. Determination of volatile fatty acids in environmental aqueous samples. *Polish Journal of Environmental Studies*. 2008;17(3):351–356.
146. Bishai M, De S, Adhikari B, Banerjee R. A platform technology of recovery of lactic acid from a fermentation broth of novel substrate *Zizyphus oenophlia*. *3 Biotech*. 2015;5(4):455–463.

147. Schlosser Š, Kertész R, Marták J. Recovery and separation of organic acids by membrane-based solvent extraction and pertraction: an overview with a case study on recovery of MPCA. *Separation and Purification Technology*. 2005;41(3):237–266.
148. Zhang S-T, Matsuoka H, Toda K. Production and recovery of propionic and acetic acids in electrodialysis culture of *Propionibacterium shermanii*. *Journal of Fermentation and Bioengineering*. 1993;75(4):276–282.
149. Alibardi L, Cossu R. Effects of carbohydrate, protein and lipid content of organic waste on hydrogen production and fermentation products. *Waste Management*. 2016;47:69–77.
150. Shen D, Yin J, Yu X, Wang M, Long Y, Shentu J, et al. Acidogenic fermentation characteristics of different types of protein-rich substrates in food waste to produce volatile fatty acids. *Bioresource Technology*. 2017;227:125–132.
151. Yin J, Yu X, Wang K, Shen D. Acidogenic fermentation of the main substrates of food waste to produce volatile fatty acids. *International Journal of Hydrogen Energy*. 2016;41(46):21713–21720.
152. Parawira W, Murto M, Read JS, Mattiasson B. Volatile fatty acid production during anaerobic mesophilic digestion of solid potato waste. *Journal of Chemical Technology & Biotechnology: International Research in Process, Environmental & Clean Technology*. 2004;79(7):673–677.
153. Feng L, Chen Y, Zheng X. Enhancement of waste activated sludge protein conversion and volatile fatty acids accumulation during waste activated sludge anaerobic fermentation by carbohydrate substrate addition: the effect of pH. *Environmental Science & Technology*. 2009;43(12):4373–4380.
154. Battista F, Almendros MG, Rousset R, Boivineau S, Bouillon P-A. Enzymatic hydrolysis at high dry matter content: The influence of the substrates' physical properties and of loading strategies on mixing and energetic consumption. *Bioresource Technology*. 2018;250:191–196.
155. Jiang J, Zhang Y, Li K, Wang Q, Gong C, Li M. Volatile fatty acids production from food waste: effects of pH, temperature, and organic loading rate. *Bioresource Technology*. 2013;143:525–530.
156. Zhang B, Zhang L, Zhang S, Shi H, Cai W. The influence of pH on hydrolysis and acidogenesis of kitchen wastes in two-phase anaerobic digestion. *Environmental Technology*. 2005;26(3):329–340.
157. Wang K, Yin J, Shen D, Li N. Anaerobic digestion of food waste for volatile fatty acids (VFAs) production with different types of inoculum: effect of pH. *Bioresource Technology*. 2014;161:395–401.
158. He M, Sun Y, Zou D, Yuan H, Zhu B, Li X, et al. Influence of temperature on hydrolysis acidification of food waste. *Procedia Environmental Sciences*. 2012;16:85–94.
159. Komemoto K, Lim Y, Nagao N, Onoue Y, Niwa C, Toda T. Effect of temperature on VFA's and biogas production in anaerobic solubilization of food waste. *Waste Management*. 2009;29(12):2950–2955.
160. Lee M, Hidaka T, Tsuno H. Effect of temperature on performance and microbial diversity in hyperthermophilic digester system fed with kitchen garbage. *Bioresource Technology*. 2008;99(15):6852–6860.
161. Shin H-S, Youn J-H, Kim S-H. Hydrogen production from food waste in anaerobic mesophilic and thermophilic acidogenesis. *International Journal of Hydrogen Energy*. 2004;29(13):1355–1363.
162. Han S-K, Shin H-S. Enhanced acidogenic fermentation of food waste in a continuous-flow reactor. *Waste Management & Research*. 2002;20(2):110–118.
163. Gou C, Yang Z, Huang J, Wang H, Xu H, Wang L. Effects of temperature and organic loading rate on the performance and microbial community of anaerobic co-digestion of waste activated sludge and food waste. *Chemosphere*. 2014;105:146–151.

164. Liu C-G, Xue C, Lin Y-H, Bai F-W. Redox potential control and applications in microaerobic and anaerobic fermentations. *Biotechnology Advances*. 2013;31(2):257–265.
165. Yin J, Yu X, Zhang Y, Shen D, Wang M, Long Y, et al. Enhancement of acidogenic fermentation for volatile fatty acid production from food waste: effect of redox potential and inoculum. *Bioresource Technology*. 2016b;216:996–1003.
166. Du Z, Li H, Gu T. A state of the art review on microbial fuel cells: a promising technology for wastewater treatment and bioenergy. *Biotechnology Advances*. 2007; 25(5):464–482.
167. Bhatia S, Yi DH, Kim HJ, Jeon JM, Kim YH, Sathiyanarayanan G, et al. Overexpression of succinyl-CoA synthase for poly (3-hydroxybutyrate-co-3-hydroxyvalerate) production in engineered *Escherichia coli* BL 21 (DE 3). *Journal of Applied Microbiology*. 2015;119(3):724–735.
168. Levin DB, Pitt L, Love M. Biohydrogen production: prospects and limitations to practical application. *International Journal of Hydrogen Energy*. 2004;29(2):173–185.
169. Budiman AW, Nam JS, Park JH, Mukti RI, Chang TS, Bae JW, et al. Review of acetic acid synthesis from various feedstocks through different catalytic processes. *Catalysis Surveys from Asia*. 2016;20(3):173–193.
170. Todd Jr HE, Walters DL. *Deicing compositions comprising calcium magnesium acetate double salt and processes for their production*. Google Patents; 1990.
171. Balamurugan K, Dasu VV, Panda T. Propionic acid production by whole cells of *Propionibacterium freudenreichii*. *Bioprocess Engineering*. 1999;20(2):109–116.
172. Van Immerseel F, Boyen F, Gantois I, Timbermont L, Bohez L, Pasmans F, et al. Supplementation of coated butyric acid in the feed reduces colonization and shedding of *Salmonella* in poultry. *Poultry Science*. 2005;84(12):1851–1856.
173. Hamer HM, Jonkers D, Venema K, Vanhoutvin S, Troost F, Brummer RJ. The role of butyrate on colonic function. *Alimentary Pharmacology & Therapeutics*. 2008; 27(2):104–119.
174. Verardi CA, Meyers LD, Humphrey WA. *Coating composition for plastic substrates and coated plastic articles*. Google Patents; 1999.
175. Suerbaev KA, Zhaksylykova GZ, Appazov N. Biological active esters of the isovaleric acid. *Eurasian Chemico-Technological Journal*. 2014;16(4):299–302.
176. Weimer PJ, Nerdahl M, Brandl DJ. Production of medium-chain volatile fatty acids by mixed ruminal microorganisms is enhanced by ethanol in co-culture with *Clostridium kluyveri*. *Bioresource Technology*. 2015;175:97–101.
177. Escobar-Zepeda A, Vera-Ponce de Leon A, Sanchez-Flores A. The road to metagenomics: from microbiology to DNA sequencing technologies and bioinformatics. *Frontiers in Genetics*. 2015;6:348.
178. Winkler JD, Kao KC. Recent advances in the evolutionary engineering of industrial biocatalysts. *Genomics*. 2014;104(6):406–411.
179. Bhatia SK, Lee B-R, Sathiyanarayanan G, Song H-S, Kim J, Jeon J-M, et al. Medium engineering for enhanced production of undecylprodigiosin antibiotic in *Streptomyces coelicolor* using oil palm biomass hydrolysate as a carbon source. *Bioresource Technology*. 2016;217:141–149.
180. Bhatia SK, Kim J, Song H-S, Kim HJ, Jeon J-M, Sathiyanarayanan G, et al. Microbial biodiesel production from oil palm biomass hydrolysate using marine *Rhodococcus* sp. YHY01. *Bioresource Technology*. 2017;233:99–109.
181. Jönsson LJ, Martín C. Pretreatment of lignocellulose: formation of inhibitory by-products and strategies for minimizing their effects. *Bioresource Technology*. 2016;199:103–112.

3 Microbial Valorization of Pomegranate Waste: Food and Health Applications

Armando Robledo-Olivo
Fermentation and Biomolecules Lab Bioprocess Agrofood Research Group, Food Science and Technology Department, Universidad Autónoma Agraria Antonio Narro, Calzada Antonio Narro Buenavista, Saltillo, Coahuila, México

CONTENTS

3.1 INTRODUCTION

Nowadays, different anthropogenic activities derive in wastes materials that affect the entire environment. Agricultural and agro-industrial activities are not exempt to generate pollution and wastes, but the main advantage is that most of the agricultural wastes are formed of vegetable biomass materials, which are susceptible to microbial degradation. From the cultivation, when the food is obtained, waste is generated; in the industry, raw materials are subjected to processes for the vegetables and fruits adaptation to consumption or their transformation to give it added value for the consumers. The accumulation of waste materials implies a serious

DOI: 10.1201/9781003128977-3

environmental problem, since most of them are spread to the environment generating the proliferation of pathogenic micro-organisms; waste material is burned in the open, which in both cases increases greenhouse gases.

In this context, bioprocess and biorefinery had emerged as a potential alternative for the use of biomass of nonedible feedstock waste as raw materials to produce a range of products, such as bioactive compounds, industrial biochemicals, biomaterials, and biofuel. Human utilization of biomass as feedstock is not new; but now there is a renewed interest in organic wastes to achieve an effective exploitation, sparked by the aim to reduce eco-footprint and accomplish a more secure supply of renewable resources (Mohan et al. 2016).

Pomegranate is a fruit that has been increasing their consumptions due to its many health beneficial properties, such as antioxidant, anti-inflammatory, and antihypertensive (Kandylis and Kokkinomagoulos 2020). The global pomegranate market, included hole fruit and arils for separate, was valued at USD 8.2 billion in 2018 and is expected to reach USD 23.14 billion by year 2026 according to AgriExchange (2020).

The main products that present either pomegranate or pomegranate derivatives and are currently marketed include capsules, soft-gels, tablets, extract, seed oil, juice, powder, and tea (Martini 2020), which show constant market growth with an upward projection. All these products that use pomegranate compounds generate growing waste residues of vegetable origin, with high value compounds that can be recovered by biotechnological processes to reach a circular bioeconomy. Talekar et al. (2019) mention that from the production of 1 ton of pomegranate juice, between 5 and 5.5 tons of waste pomegranate peel (WPP) is generated.

Micro-organisms are well known as an inexhaustible source of a wide range of useful and bulk biochemical compounds, with diverse applications, such as biostimulants, food additives, enzymes, agro-inputs, biopharmaceuticals, and bioenergetics, through the industry of fermentation or biocatalysis processes. In this chapter, we will evaluate the pomegranate waste and identify the possible added value compounds that can be obtained from it, through microbial transformation.

3.2 TRADITIONAL USES

Pomegranate is a nearly round fruit from a shrub (*Punica granatum* L) that is mainly cultivated in Asia, in the Mediterranean region, and in some parts of America where the climate is adequate for its cultivar (high sunlight-exposed, mild winters, and dry summers). The pomegranate is categorized as a fleshy berry, composed by the arils as the edible portion (40%), protected by the pomegranate peel who comprises around 50% of the whole fruit, and completing with seeds the remaining 10% (Kandylis and Kokkinomagoulos 2020).

The regions with a hot and dry climate, with alkaline soils with good drainage, favor the production of pomegranate; however, the correct irrigation increases the production of the fruit (Marzolo 2015). The pomegranate tree is a shrub that grows up to 8 meters in height and produces between 100 and 200 fruits. Various parts of the pomegranate tree have been used to treat various diseases because the tree in general is a good source of antioxidants and alkaloids (Pathak, Mandavgane, and Kulkarni 2017).

3.2.1 Food

Due to the recent trend of healthy and functional foods, the pomegranate has become a popular fruit of consumption, due to its characteristic flavor, high nutritional levels, and its medicinal properties (Pathak, Mandavgane, and Kulkarni 2017).

Pomegranate (*Punica granatum* L.) is normally consumed directly by removing the peel or making a juice by squeezing its seeds. In addition to juicing, arils have complemented Middle Eastern culinary dishes such as fesejan, khosaf al-rumman, and pomegranate khoresh (Gayle Engels 2013). In Mexico, it is sometimes added to guacamole and is an essential part of "Chiles en nogada".

Due to the increasing demand for functional foods, fermented pomegranate juices have been developed to satisfy various demands for bioactive compounds. A study focused on the technological transformation of pomegranate juice is the one reported by Yavari et al. (2018), where they carried out a fermentation of pomegranate juice using the "kombucha" tea fungus to obtain a drink with a high content of glucuronic acid (17.07 g/L) as a bioactive compound. Moreover Mantzourani et al. (2019) developed a novel fermented pomegranate beverage by the application of *Lactobacillus plantarum* ATCC 14917, with no sensorial significant differences between fermented and nonfermented pomegranate juice. Another attractive food that involves pomegranate is the one reported by Al-Hindi and Abd El Ghani (2020), where they developed a novel fermented milk beverage with a combination of polyphenols extracted from pomegranate peel and probiotic lactic acid bacteria.

3.2.2 Health

Throughout history, various cultures report the use of pomegranate or parts of the tree, for the elaboration of elements with medicinal actions. Some of the diseases that are fought with pomegranate remedies include roundworm and other parasites. Other ailments that are traditionally treated with pomegranate extracts in the form of tea are stomach pain, mouth discomfort, diarrhea, and to treat respiratory and urinary tract infections. Another of the pomegranate compounds that has become popular for health use, are essential oils. Pomegranate seed oil contains around 65–80% conjugated fatty acids, been the punicic acid the most important (Abbasi, Rezaei, and Rashidi 2008), which is a conjugated linolenic acid isomer (Khajebishak et al. 2019). Pomegranate seed oil is rich in steroidal estrogens such as tocopherol, testosterone, stigmasterol, b-estrolsitosterol, and 17-a-estradiol, and in nonsteroidal estrogens such as coumestrol and campestral (Pirzadeh et al. 2020).

Recent studies highlight the benefits of the use of pomegranate against various diseases such as H1N1 and H3N2 influenza viruses (Sundararajan et al. 2010) and Sars-cov-2 (Coronavirus COVID-19 disease) (Ahmad 2020), anti-inflammatory and antiatherogenic (Salama, Ismael, and Bedewy 2020), antimicrobial effects (Javanmard 2020).

3.3 POMEGRANATE WASTE

Pomegranate is one of the most popular fruits consumed worldwide, with a world production around 1.5 million tons, from which peels constitute 50%, arils 40%, and seeds 10% (Kandylis and Kokkinomagoulos 2020). The pomegranate peel is one of the major byproducts of the food processing industry (Pathak, Mandavgane, and Kulkarni 2017).

The pomegranate is a good source of sugars and is characterized by its high content of phenolic compounds, of which polyphenols, tannins, and anthocyanins stand out. From the polyphenols, flavonoids, ellagitannins, and the color molecules anthocyanin are present in high amounts (Bar-Ya'akov et al. 2019). Pomegranate peel contains diverse compounds such as phenolics and flavonoids (kaempferol, quercetin, luteolin, and proanthocyanidin), hydrolysable tannins (ellagitannin, punicalin, punicalagin (PG), and pedunculagin), phenolic acids divided in hydroxycinnamic acids (caffeic acid, chlorogenic acid, and p-coumaric acid), hydroxybenzoic acids (ellagic acid (EA) and gallic acid), and cyclitol-carboxylic acids (quinic acid), anthocyanins (cyanidin, pelargonidin, and delphinidin), and alkaloids (pelletierine) (Pathak, Mandavgane, and Kulkarni 2017, Papaioannou et al. 2020). All these compounds can be extracted by different methods (ethanol, isopropanol, and water). The pomegranate peel main phenolic acids are protocatechuic acid, syringic acid, EA, gallic acid, cinnamic acid, quinic acid, and isoferulic acid (El-Hamamsy and El-khamissi 2020). The main polyphenol compounds that can be found in pomegranate peel are α-punicalagin (1060–1480 g/100 g), β-punicalagin (740–1070 g/100 g), EA glycosides (5.34–7.68 g/100 g), ellagitannin (1.96 g/100 g), granitin B (0.59 g/100 g), lagerstannin C (0.39 g/100 g), PG (1.05–4070 g/100 g), punicalin (0.07–2000 g/100 g), punigluconin (0.38 g/100 g), and pedunculagin I (0.35 g/100 g), mainly obtained by methanol extraction (Pirzadeh et al. 2020). According to Farag et al. (2014) and Sreekumar et al. (2014), in a Pathak, Mandavgane, and Kulkarni (2017) publication, the primary polyphenol compounds in pomegranate peel are protocatechuic acid (14.512 %), gallic acid (14.147 %), caffeine (6.420 %), vanillic acid (3.851 %), coumarin (3.534 %), caffeic acid (2.748 %), chlorogenic acid (2.355 %), ferulic acid (1.857 %), quercetin (0.949 %), and oleuropein (0.590 %). In a recent study were isolated for the first time from the pomegranate peel isohydroxymatairesinol, punicatannin C, phloretin, quercetin glycoside, punigratane, and coutaric acid (Nazeam et al. 2020).

Some amino acids can be found in the pomegranate peel, such the case of glutamine (0.52 g/100 g), glycine (0.41 g/100 g), and aspartate (0.3 g/100 g) (Bar-Ya'akov et al. 2019). The pomegranate peel is a good, abundant, and inexpensive source of polyphenols with remarkable biological activities.

3.4 BIOPROCESS TECHNOLOGIES

Today there are different extraction techniques for various compounds with biological activity of interest. They can be classified into three large groups: physical, biological, and chemical, the latter being the most efficient, but the least desired, due to their corrosive essence and their problems in disposing of waste that are not very friendly to the environment. In this section, we will review the most used biological techniques for the bioconversion of pomegranate residues.

3.4.1 Fermentation Processes

The direct fermentation process of pomegranate waste is an effective and low-cost method for the degradation of the constituents of the pomegranate wastes. Due to its vegetal origin, pomegranate wastes have a varying composition that is often a drawback, such that standardization, pretreatment, and addition of further nutrients are needed. During the fermentation process, enzymes, single-cell proteins, and low molecular weight bioactive compounds can be generated. For the use of pomegranate residues, they can be given in two types of medium: liquid medium and solid medium.

For the liquid medium fermentation process (also known as submerged fermentation), dehydrated and ground pomegranate peels are used in various small particle sizes, solubilized in a free-flow water media, in order to improve the contact surface between the micro-organism and the substrate (Parashar, Srivastava, and Garlapati 2019). In the liquid medium fermentation process, the scale up steps are simpler and easier to manipulate and restrain from factors associated with whole production process. Liquid medium fermentation allows greater control of parameters, such as pH, heat, nutrient conditions, etc. The main disadvantage of liquid fermentation for aerobic micro-organisms is the poor transfer of oxygen in the culture medium (Reihani and Khosravi-Darani 2019), which can decrease the production of some metabolites.

The use of solid-state fermentation technology, where crude pomegranate wastes are directly provided as a substrate and support for microbial growth and production, is an outstanding option for obtaining compounds of agri-food interest. The solid-state (or substrate) fermentation (SSF) is characterized by a low moisture content (lower limit ≈ 12%), occurs in a nonseptic and natural state, is carried out in the absence or near absence of free flowing water in the system, and it mainly utilizes fungal species (Nigam and Pandey 2009). SSF is a low-cost technology, suitable mainly for aerobic micro-organisms that require low water activity. Mycelial fungi are ideal species for bioprocesses in SSF, since they can penetrate the material layer and enter the matrix, resistance to biotic and abiotic stresses, as well as growing in conditions of low water activity.

Also, a two-stage fermentation strategy combining submerged with SSF successfully transforms byproducts from agriculture and food industries for ethanol and fungal biomass production (Gmoser et al. 2019).

3.4.2 Enzymatic Processes

Enzyme-assisted extraction is a technology that focuses on the direct application of specific enzymes for the degradation of plant material to obtain the biomolecules present in plants. Enzyme-based extraction depends on the characteristic properties of the enzyme, such as its specificity and its ability to maintain its activity under moderate process conditions (processes occurring at low temperature values and for short periods of time) (Nadar, Rao, and Rathod 2018, Gligor et al. 2019). The basic principle of enzyme-assisted extraction is the breakdown of the cell wall by hydrolysis, using enzymatic catalysis under optimum experimental conditions to release the bioactive compounds.

3.4.3 Immobilized Microbial Biocatalysts (Cells or Enzymes)

Enzyme immobilization refers to the combination of the enzyme, whether trapped, adsorbed, covalently linked, or cross-linked, with a solid carrier that allows the stability of catalysis to be maximized (Basso and Serban 2019, Zdarta et al. 2018). One of the main advantages of viable cell or enzymatic immobilization is reuse for several process cycles while maintaining biological activity. This allows economic savings if the reaction kinetics and specificity can be maintained in the maximum number of reuse cycles possible.

3.5 BIOPRODUCTS WITH ADDED VALUE IN FOOD AND PHARMACOLOGY

3.5.1 Pomegranate Waste in Health Applications

Pomegranate waste can be used indirectly in the health sector by using bioprocesses. As has mentioned above, pomegranate waste has very important polyphenolic compounds as main polymeric constituents, which when hydrolyzate, they release important antioxidant and antiviral compounds (Gumienna, Szwengiel, and Górna 2016). Varadharajan et al. (2017) use pomegranate rind as substrate for tannase production by *Aspergillus oryzae* in submerged fermentation. The use of tannase in the pharmaceutical sector has been in the synthesis of antibacterial drugs. Another important phenolic antioxidant is the PG, which when hydrolyzes, EA and urolithins (ULs) can be obtained, which are responsible for the antioxidant effect in the body (Talekar et al. 2019, Andrade et al. 2019). EA is a potent antioxidant with anti-inflammatory and antiproliferative capacity that can be obtained by the SSF of pomegranate waste with *Aspergillus niger* GH1 (Robledo et al. 2018, Sepúlveda et al. 2018). However, EA has poor stability in physiological environment and relatively low solubility in water. Regarding its pharmacokinetics, EA shows low blood concentration, short peak time, rapid absorption and elimination after oral administration, and low bioavailability and short effective time (Zheng et al. 2019). Recent studies evaluate the supersaturatable self-microemulsifying drug delivery system (S-SMEDDS) to increase the EA loading and evaluate its antioxidant activity, finding that a small amount (0.5%) of polyvinylpyrrolidone (PVP) K-30 as a precipitation inhibitor increases the solubility of EA and improves its antioxidant capacity (Zheng et al. 2019). Tavares et al. (2020) developed a chitosan/zein films for controlled release of EA and its application as a wound bandaging platform, with inhibitory activity against *Staphylococcus aureus* and *Pseudomonas aeruginosa in vitro*. The antimicrobial action of EA is related to its ability to chelate with proteins from the bacterial wall or through the inhibition of gyrase activity that cleaves the DNA strand during bacterial replication (Zuccari et al. 2020).

ULs are metabolites derived from ellagitannins that possess high antioxidant activity and stimulate autophagy and mitophagy in mammalian cells (Ryu et al. 2016, Panth, Manandhar, and Paudel 2017). The capacity of UL to inhibit the virus protease as well as peptidase has allowed recent studies to prevent the entry of the SARS-CoV-2 (COVID-19) inside the host cell (Ahmad 2020). Ellagitannins present in the pomegranate waste can be converted into EA and further to UL by two human gut bacterial species *Gordonibacter urolithinfaciens* and *Ellagibacter isourolithinifaciens* (García-Villalba et al. 2020); nevertheless, only one in three people have the right microbiota to perform this metabolism with maximum efficiency (Cásedas et al. 2020). Gaya et al. (2018) screened 48 strains of bifidobacterial seeking for UL production, finding *Bifidobacterium pseudocatenulatum* INIA P815 as the only strain able to produce urolithins A and B from EA, opening the possible applications in the development of functional foods and nutraceuticals. Fungi such *Alternaria Alternata* also can biosynthesize UL (dibenzo[*b*,*d*]pyran-6-ones or dibenzo-α-pyrones) through the polyketide pathway (Djedjibegovic et al. 2020) using polyketide synthase (PKS) as core enzyme (Mao et al. 2014). ULs (urolithin A and urolithin B as main isomers) are absorbed to the portal hepatic circulation, where they may serve the role of hormone analogs (Gumienna, Szwengiel, and Górna 2016). UL can potentially be used to modulate oxidative stress and ameliorate tissue damage through different mechanisms (Djedjibegovic et al. 2020).

3.5.2 Pomegranate Waste as Food Additives

Food additives have been used since the first civilizations, with the aim of preserving, or enhancing the color, flavor, and texture of food. The reuse of agro-industrial byproducts can represent a renewable source of food additives with functional compounds and properties for the benefit of human health.

Pomegranate is scientifically proven one of the best sources of health beneficial phenolics (Gumienna, Szwengiel, and Górna 2016), including ellagitannins, flavonoids, anthocyanins, and other polyphenols. However, the best additives beneficial for health enhancement are the monomers released from the hydrolysis of the compounds. Pomegranate waste can be used as substrate and support for diverse antioxidant compounds' production. The highest added value bioactive of pomegranate phenolic compound is the EA, which is present at extremely low levels in pomegranate, occurring mostly in the complex form of ellagitannins, including PG isomers, and granatin (Moccia et al. 2019). EA is a stable molecule, with four rings representing the lipophilic domain, and four phenolic and two lactones groups (which can act as hydrogen bond donor and acceptor, respectively) representing the hydrophilic part (Robledo et al. 2018, Gumienna, Szwengiel, and Górna 2016). Robledo et al. (2008) used the pomegranate peel as a support-substrate in solid fermentation to produce EA with the *Aspergillus niger* GH1 strain. Buenrostro-Figueroa et al. (2020) used ellagitannins from pomegranate peel to induce the synthesis of an enzyme capable of hydrolyzing ellagitannins and releasing the EA to the liquid media. EA can be used as a food additive to improve the shelf life of food through smart packaging. Moccia et al. (2019) evaluated the production of EA by *Saccharomyces cerevisiae* in SSF using pomegranate husk, where the findings were an efficient hydrolysis of EA-releasing punicalagin α and β, as well as granatin B. Vilela et al. (2017) evaluated the production of films with a chitosan/ellagic acid combination, obtaining flexible, transparent, and homogeneous films with effective antioxidant and antimicrobial activity against *Staphylococcus aureus* and *Pseudomonas aeruginosa*.

Most reports of the use of UL as an additive in food mention the presence of UL derived from the content of ellagitannins and EA present in the food. However, direct consumption of urolithin A has been reported to be nongenotoxic, and the US FDA (Food and Drug Administration) has been established UL as a food ingredient at levels up to 1000 mg/serving (Djedjibegovic et al. 2020).

Citric acid is also considered as one of the important organic acids that is used in the food and beverage industry to preserve and add flavor to fruit juices, candy ice cream, etc. (Kamble and Gawai 2019). Roukas and Kotzekidou (2020) investigated the citric acid production from dried and nondried pomegranate peel wastes by the fungus *Aspergillus niger* B60 in SSF under nonaseptic conditions. They obtained highest citric acid (278.5 g/kg dry peel) content using dried (at 45°C for 48 h) pulverized pomegranate peels.

Another important food additives are the colorant compounds, which also are byproducts that can be obtained by micro-organisms valorizing pomegranate wastes. Arikan et al. (2020) evaluated pigment extraction under SSF conditions using pomegranate pulp as substrate and *Aspergillus carbonarius*, where they obtained 9.21 ± 0.59 absorbance unit per gram (AU/g) of dry fermented mass.

3.6 CONCLUSION AND FINAL REMARKS

These days, the concepts of sustainability and circular economy have been gaining strength in the social and scientific field. Part of the circular bioeconomy to reduce waste, includes reusing and recovering materials increasing the value and useful life of products and resources. Under these concepts, the recovery of pomegranate waste is a great option in obtaining compounds of great added value and with great application in health and food.

The compounds present in pomegranate residues can be classified as cosmopolitan for the treatment of serious diseases such as cancer, AIDS or even COVID-19. The various biotechnological processes that have been developed today allow the obtaining of molecules with a great possibility of application as functional food and thereby improve the health of the population.

However, there are still pending studies to be carried out, to improve the data on the production of ellagitannins, PG, EA, and UL, as well as the enzymes responsible for their production, the scale-up processes, and the toxicity that they can represent as food additives.

REFERENCES

Abbasi, Hajar, Karamatollah Rezaei, and Ladan Rashidi. 2008. "Extraction of essential oils from the seeds of pomegranate using organic solvents and supercritical CO_2." *Journal of the American Oil Chemists' Society* 85 (1):83–89.

AgriExchange, APEDA. 2020. "Market Intelligence Report for Pomegranates." Agricultural & Processed Food Products Export Development Authority (APEDA), Ministry of Commerce & Industry, Govt. of India, accessed November 15. https://agriexchange.apeda.gov.in/Weekly_eReport/Pomegranate_Report.pdf.

Ahmad, Varish. 2020. "A molecular docking study against COVID-19 protease with a pomegranate phyto-constituents 'Urolithin' and other repurposing drugs: From a supplement to ailment." *Journal of Pharmaceutical Research International* 32 (11):51–62.

Al-Hindi, Rashad R., and Salem Abd El Ghani. 2020. "Production of functional fermented milk beverages supplemented with pomegranate peel extract and probiotic lactic acid bacteria." *Journal of Food Quality* 2020:4710273. 10.1155/2020/4710273.

Andrade, Mariana A., Vasco Lima, Ana Sanches Silva, Fernanda Vilarinho, Maria Conceição Castilho, Khaoula Khwaldia, and Fernando Ramos. 2019. "Pomegranate and grape by-products and their active compounds: Are they a valuable source for food applications?" *Trends in Food Science & Technology* 86:68–84. 10.1016/j.tifs.2019.02.010.

Arikan, Ezgi Bezirhan, Oltan Canli, Yanis Caro, Laurent Dufossé, and Nadir Dizge. 2020. "Production of bio-based pigments from food processing industry by-products (apple, pomegranate, black carrot, red beet pulps) using *Aspergillus carbonarius*." *Journal of Fungi* 6 (4):240.

Bar-Ya'akov, Irit, Li Tian, Rachel Amir, and Doron Holland. 2019. "Primary metabolites, anthocyanins, and hydrolyzable tannins in the pomegranate fruit." *Frontiers in Plant Science* 10:620.

Basso, Alessandra, and Simona Serban. 2019. "Industrial applications of immobilized enzymes—A review." *Molecular Catalysis* 479:110607.

Buenrostro-Figueroa, Juan, Marcela Miereles, Juan A Ascacio-Valdés, Antonio Aguilera-Carbo, Leonardo Sepúlveda, Juan Contreras-Esquivel, Raúl Rodríguez-Herrera, and Cristobal N Aguilar. 2020. "Enzymatic biotransformation of pomegranate ellagitannins: Initial approach to reaction conditions." *Iranian Journal of Biotechnology* 18 (2):30–36.

Cásedas, Guillermo, Francisco Les, Carmen Choya-Foces, Martín Hugo, and Víctor López. 2020. "The metabolite urolithin-A ameliorates oxidative stress in Neuro-2a Cells, becoming a potential neuroprotective agent." *Antioxidants* 9 (2):177.

Djedjibegovic, Jasmina, Aleksandra Marjanovic, Emiliano Panieri, and Luciano Saso. 2020. "Ellagic acid-derived urolithins as modulators of oxidative stress." *Oxidative Medicine and Cellular Longevity* 2020.

El-Hamamsy, S.M.A., and H.A.Z. El-khamissi. 2020. "Phytochemicals, antioxidant activity and identification of phenolic compounds by HPLC of pomegranate (*Punica granatum* L.) Peel Extracts." *Journal of Agricultural Chemistry and Biotechnology* 11 (4):79–84.

Farag, R., M. Abdelatif, S. Emam, and L. Tawfeek. 2014. "Phytochemical screening and polyphenol constituents of pomegranate peels and leave juices." *Agricultural Soil Science* 1 (6):86–93.

García-Villalba, Rocío, David Beltrán, María D. Frutos, María V. Selma, Juan C. Espín, and Francisco A. Tomás-Barberán. 2020. "Metabolism of different dietary phenolic compounds by the urolithin-producing human-gut bacteria *Gordonibacter urolithinfaciens* and *Ellagibacter isourolithinifaciens*." *Food & Function* 11 (8):7012–7022.

Gaya, Pilar, Ángela Peirotén, Margarita Medina, Inmaculada Álvarez, and José M. Landete. 2018. "Bifidobacterium pseudocatenulatum INIA P815: The first bacterium able to produce urolithins A and B from ellagic acid." *Journal of Functional Foods* 45:95–99. 10.1016/j.jff.2018.03.040.

Gayle, Engels, and Josef Brinckmann. 2013. Pomegranate. (100): 1–7. Accessed November 15, 2020.

Gligor, Octavia, Andrei Mocan, Cadmiel Moldovan, Marcello Locatelli, Gianina Crişan, and Isabel C. F. R. Ferreira. 2019. "Enzyme-assisted extractions of polyphenols – A comprehensive review." *Trends in Food Science & Technology* 88:302–315. 10.1016/j.tifs.2019.03.029.

Gmoser, Rebecca, Carissa Sintca, Mohammad J. Taherzadeh, and Patrik R. Lennartsson. 2019. "Combining submerged and solid state fermentation to convert waste bread into protein and pigment using the edible filamentous fungus *N. intermedia*." *Waste Management* 97:63–70.

Gumienna, Małgorzata, Artur Szwengiel, and Barbara Górna. 2016. "Bioactive components of pomegranate fruit and their transformation by fermentation processes." *European Food Research and Technology* 242 (5):631–640.

Javanmard, M. 2020. "Antimicrobial effects of grape and pomegranate waste extracts against two foodborne pathogens." *Journal of Food Biosciences and Technology* 10 (2):39–48.

Kamble, R.K., and D.U. Gawai. 2019. "Production and optimization of citric acid by *Aspergillus niger* ARU1 from pomegranate fruit waste." *Journal of the Gujarat Research Society* 21 (13):971–976.

Kandylis, Panagiotis, and Evangelos Kokkinomagoulos. 2020. "Food applications and potential health benefits of pomegranate and its derivatives." *Foods* 9 (2):122.

Khajebishak, Yaser, Laleh Payahoo, Mohammadreza Alivand, and Beitollah Alipour. 2019. "Punicic acid: A potential compound of pomegranate seed oil in type 2 diabetes mellitus management." *Journal of Cellular Physiology* 234 (3):2112–2120.

Mantzourani, Ioanna, Stavros Kazakos, Antonia Terpou, Athanasios Alexopoulos, Eugenia Bezirtzoglou, Argyro Bekatorou, and Stavros Plessas. 2019. "Potential of the probiotic *Lactobacillus plantarum* ATCC 14917 strain to produce functional fermented pomegranate juice." *Foods* 8 (1):4.

Mao, Ziling, Weibo Sun, Linyun Fu, Haiyu Luo, Daowan Lai, and Ligang Zhou. 2014. "Natural dibenzo-α-pyrones and their bioactivities." *Molecules* 19 (4):5088–5108.

Martini, Nataly. 2020. "Pomegranate." *Journal of Primary Health Care* 12 (3):293–294.

Marzolo, Gina. 2015. Pomegranates. AgMRC Agricultural Marketing Resource Center: AgMRC Agricultural Marketing Resource Center.

Moccia, Federica, Adriana C. Flores-Gallegos, Mónica L. Chávez-González, Leonardo Sepúlveda, Stefania Marzorati, Luisella Verotta, Lucia Panzella, Juan A. Ascacio-Valdes, Cristobal N. Aguilar, and Alessandra Napolitano. 2019. "Ellagic acid recovery by solid state fermentation of pomegranate wastes by *Aspergillus niger* and *Saccharomyces cerevisiae*: A comparison." *Molecules* 24 (20):3689.

Mohan, S. Venkata, G.N. Nikhil, P. Chiranjeevi, C. Nagendranatha Reddy, M.V. Rohit, A. Naresh Kumar, and Omprakash Sarkar. 2016. "Waste biorefinery models towards sustainable circular bioeconomy: critical review and future perspectives." *Bioresource Technology* 215:2–12.

Nadar, Shamraja S., Priyanka Rao, and Virendra K. Rathod. 2018. "Enzyme assisted extraction of biomolecules as an approach to novel extraction technology: A review." *Food Research International* 108:309–330.

Nazeam, Jilan A., Walaa A. AL-Shareef, Maged W. Helmy, and Alaadin E. El-Haddad. 2020. "Bioassay-guided isolation of potential bioactive constituents from pomegranate agrifood by-product." *Food Chemistry* 326.

Nigam, Poonam Singh-Nee, and Ashok Pandey. 2009. *Biotechnology for agro-industrial residues utilisation: utilisation of agro-residues*. Springer Science & Business Media.

Panth, Nisha, Bikash Manandhar, and Keshav Raj Paudel. 2017. "Anticancer activity of *Punica granatum* (pomegranate): A review." *Phytotherapy Research* 31 (4): 568–578.

Papaioannou, Emmanouil H., Soultana T. Mitrouli, Sotiris I. Patsios, Maria Kazakli, and Anastasios J. Karabelas. 2020. "Valorization of pomegranate husk – Integration of extraction with nanofiltration for concentrated polyphenols recovery." *Journal of Environmental Chemical Engineering* 8 (4):103951. 10.1016/j.jece.2020.103951.

Parashar, Surendra Kumar, S.K. Srivastava, and V.K. Garlapati. 2019. "Production of microbial enzyme triacylglycerol acylhydrolases by ASPERGILLUS SYDOWII JPG01 in submerged fermentation using agro-residues." *Asian Journal Microbiology Biotechnology Environmental Experimental Science* 21 (4):1076–1079.

Pathak, Pranav D., Sachin A. Mandavgane, and Bhaskar D. Kulkarni. 2017. "Valorization of pomegranate peels: A biorefinery approach." *Waste and Biomass Valorization* 8 (4): 1127–1137.

Pirzadeh, Maryam, Nicola Caporaso, Abdur Rauf, Mohammad Ali Shariati, Zhanibek Yessimbekov, Muhammad Usman Khan, Muhammad Imran, and Mohammad S Mubarak. 2020. "Pomegranate as a source of bioactive constituents: review on their characterization, properties and applications." *Critical Reviews in Food Science and Nutrition* 61: 982–999.

Reihani, Fatemeh S., and Kianoush Khosravi-Darani. 2019. "Influencing factors on single-cell protein production by submerged fermentation: A review." *Electronic Journal of Biotechnology* 37:34–40.

Robledo, Armando, Antonio F Aguilera-Carbo, Arely Prado-Barragan, Leonardo Sepulveda-Torre, Raul Rodríguez-Herrera, Juan C Contreras-Esquivel, and Cristobal N Aguilar. 2018. "Kinetics of ellagic acid accumulation by solid-state fermentation." *Theoretical Models and Experimental Approaches in Physical Chemistry: Research Methodology and Practical Methods* 1st edition, 267.

Robledo, Armando, Antonio Aguilera-Carbó, Raúl Rodriguez, José Luis Martinez, Yolanda Garza, and Cristobal N Aguilar. 2008. "Ellagic acid production by *Aspergillus niger* in solid state fermentation of pomegranate residues." *Journal of Industrial Microbiology & Biotechnology* 35 (6):507–513. 10.1007/s10295-008-0309-x.

Roukas, Triantafyllos, and Parthena Kotzekidou. 2020. "Pomegranate peel waste: A new substrate for citric acid production by *Aspergillus niger* in solid-state fermentation under non-aseptic conditions." *Environmental Science and Pollution Research* 27 (12): 13105–13113.https://doi.org/10.1007/s11356-020-07928-9

Ryu, Dongryeol, Laurent Mouchiroud, Pénélope A. Andreux, Elena Katsyuba, Norman Moullan, Amandine A. Nicolet-dit-Félix, Evan G. Williams, Pooja Jha, Giuseppe Lo Sasso, and Damien Huzard. 2016. "Urolithin A induces mitophagy and prolongs lifespan in *C. elegans* and increases muscle function in rodents." *Nature Medicine* 22 (8):879–888.

Salama, Amany A., Naglaa M. Ismael, and Magdy Bedewy. 2020. "The anti-inflammatory and antiatherogenic in vivo effects of pomegranate peel powder: From waste to medicinal food." *Journal of Medicinal Food* 24(2): 145–150. https://doi.org/10.1089/jmf.2019.0269

Sepúlveda, Leonardo, Jorge E. Wong-Paz, Juan Buenrostro-Figueroa, Juan A. Ascacio-Valdés, Antonio Aguilera-Carbó, and Cristóbal N. Aguilar. 2018. "Solid state fermentation of pomegranate husk: Recovery of ellagic acid by SEC and identification of ellagitannins by HPLC/ESI/MS." *Food Bioscience* 22:99–104. 10.1016/j.fbio.2018.01.006.

Sreekumar, Sreeja, Hima Sithul, Parvathy Muraleedharan, Juberiya Mohammed Azeez, and Sreeja Sreeharshan. 2014. "Pomegranate fruit as a rich source of biologically active compounds." *BioMed Research International* 2014.

Sundararajan, Aarthi, Radha Ganapathy, Lifang Huan, John R. Dunlap, Richard J. Webby, Girish J. Kotwal, and Mark Y. Sangster. 2010. "Influenza virus variation in susceptibility to inactivation by pomegranate polyphenols is determined by envelope glycoproteins." *Antiviral Research* 88 (1):1–9.

Talekar, Sachin, Antonio F. Patti, R. Vijayraghavan, and Amit Arora. 2019. "Rapid, enhanced and eco-friendly recovery of punicalagin from fresh waste pomegranate peels via aqueous ball milling." *Journal of Cleaner Production* 228:1238–1247.

Tavares, Walter S., Alberto G. Tavares-Júnior, Francisco J. Otero-Espinar, Manuel Martín-Pastor, and Francisco F.O. Sousa. 2020. "Design of ellagic acid-loaded chitosan/zein films for wound bandaging." *Journal of Drug Delivery Science and Technology* 59:101903. 10.1016/j.jddst.2020.101903.

Varadharajan, Venkatramanan, Sudhan Shanmuga Vadivel, Arulvel Ramaswamy, Venkatesaprabhu Sundharamurthy, and Priyadharshini Chandrasekar. 2017. "Modeling and verification of process parameters for the production of tannase by *Aspergillus oryzae* under submerged fermentation using agro-wastes." *Biotechnology and Applied Biochemistry* 64 (1):100–109.

Vilela, Carla, Ricardo J.B. Pinto, Joel Coelho, Maria R.M. Domingues, Sara Daina, Patrizia Sadocco, Sónia A.O. Santos, and Carmen S. R. Freire. 2017. "Bioactive chitosan/ellagic acid films with UV-light protection for active food packaging." *Food Hydrocolloids* 73:120–128. 10.1016/j.foodhyd.2017.06.037.

Yavari, N., M. Mazaheri-Assadi, Z.H. Mazhari, M.B. Moghadam, and K. Larijani. 2018. "Glucuronic acid rich kombucha-fermented pomegranate juice." *Journal of Food Research* 7 (1):61–69.

Zdarta, Jakub, Anne S. Meyer, Teofil Jesionowski, and Manuel Pinelo. 2018. "Developments in support materials for immobilization of oxidoreductases: A comprehensive review." *Advances in Colloid and Interface Science* 258:1–20.

Zheng, Dandan, Chanmei Lv, Xiao Sun, Jinglong Wang, and Zhongxi Zhao. 2019. "Preparation of a supersaturatable self-microemulsion as drug delivery system for ellagic acid and evaluation of its antioxidant activities." *Journal of Drug Delivery Science and Technology* 53:101209. 10.1016/j.jddst.2019.101209.

Zuccari, Guendalina, Sara Baldassari, Giorgia Ailuno, Federica Turrini, Silvana Alfei, and Gabriele Caviglioli. 2020. "Formulation strategies to improve oral bioavailability of ellagic acid." *Applied Sciences* 10 (10):3353.

4 Microbial Valorization of Agri-waste for Single Cell Protein: Current Status

Simranjeet Singh
Interdisciplinary Centre for Water Research (ICWaR), Indian Institute of Sciences, Bangalore, Karnataka, India

Dhriti Kapoor, Savita Bhardwaj, Dhriti Sharma, and Mamta Pujari
Department of Botany, School of Bioengineering and Biosciences, Lovely Professional University, Phagwara, Punjab, India

Praveen C Ramamurthy
Interdisciplinary Centre for Water Research (ICWaR), Indian Institute of Sciences, Bangalore, Karnataka, India

Joginder Singh
Department of Microbiology, School of Bioengineering and Biosciences, Lovely Professional University, Phagwara, Punjab, India

CONTENTS

DOI: 10.1201/9781003128977-4

4.1 INTRODUCTION

Addition of unwanted and discarded byproducts of a wide range of domestic, industrial, agricultural, plus commercial activities easily deteriorates the quality of our environment. Citing an example from the arena of agricultural production, in particular, the ever-increasing population and its impending demand of continuous food supply have escalated this process globally along with generating an enormous amount of agricultural wastes (Elijah and Edem 2017). Erroneous tackling of these agri-wastes causes pollution and diseases, thereby worsening the health of the environmental domain and its inhabitants as well (Yazid et al. 2017). However, profuse amount of organic substances found in these agri-wastes makes them a favorable option for bioconversion via microbial action into value-added products such as protein supplements, fodder, fuel and additives in some of the chemicals, and medicinal drugs (Ukaegbu-Obi 2016).

These products, more popularly known as single cell proteins (SCPs), subsequently offer certain advantages over conventional food products such as (i) enhancement in food security as there is round the year production owing to indifference to seasonal and climatic alterations; (ii) better waste management and reduction in pollution via reclamation of waste products into supportive growth material of micro-organisms; (iii) sustainable development because of comparatively very low water footprint than crops pertaining to the microbial growth inside closed bioreactors, this also makes them suitable for arid environments as well; (iv) noncompetitive nature in context to agricultural crops as the presence of fertile soil for cultivating microbes is not essential plus they do not pose a danger to existing biodiversity; and v) lesser emissions of greenhouse gases as compared to agriculture, thus not contributing to climate change (Ghosh et al. 2016).

In the early 1950s, SCP became an interesting topic in the field of agricultural research. SCP is basically the dried form of cells or micro-organisms, which has been used as an alternative protein source in animal feeds or as food supplements (Vermeulen et al. 2012). Micro-organisms such as microalgae, bacteria, yeast, and fungi are used in the production of SCP. The term SCP describes the production of protein from a cell that is totally different from animal and vegetable protein. The production of SCP also minimizes the impact of waste residues onto the environment.

Agricultural wastes such as juice, peels or pulp of fruits, sugar cane bagasse's, rice polishing, molasses, and corn stovers could be easily recycled as they act as cheap and natural substrate for the production of SCP. Diverse groups of micro-organisms ranging from algae, bacteria, fungi, and yeast are employed for the production of SCP, which can be then used to feed fish, cattle, and human beings (Azam et al. 2014; Yunus et al. 2015). High-priced protein supplements such as fish meal and soybean meal have been of late reported to be replaced by SCP (Goldberg 2013). The protein deficiency which the entire human race is facing can be given a plausible solution in the form of this unconventional food supplement. Besides yielding proteins, SCP biomass also acts as a source of carbohydrates, lipids, essential amino acids (such as lysine and methionine), nucleic acids, vitamins, and minerals (Jacob-Lopes et al. 2006). Furthermore, Ezejiofor et al. (2014) reported that the protein content and nutritional value of such microbial proteins are far higher as compared to those obtained from plant and animal sources.

Presently, several companies are working on the processing of biodegradable waste materials such as agri-wastes into SCPs, thereby managing these wastes in an eco-friendly manner and making the production of value-added products a profitable business enterprise. For instance, an Irish company Beltra is involved in four-nation (Ireland, Iceland, Germany, and France) development group initiative of setting up a pilot plant in Ireland whereby agri-wastes will be converted into such kind of relatively more valuable products. The advent of different modern waste management technologies has made this production possible which was earlier limited to only low-value products like bioenergy, biogas, and other biofuels (Browne et al. 2011). Various processes such as solid-state fermentation, semisolid fermentation, submerged fermentation, etc., are used for the microbial cultivation for SCP products. The basic step for the production of SCP includes the formation of medium with a carbon source, their mass propagation, and finally process of separation and processing.

This chapter intends to highlight agri-waste substrates and micro-organisms involved in their effective degradation in SCP production. Additionally, it also discusses the various fermentatio838n processes for SCP production. Efforts have also been made to summarize the various analytical methods for SCP analysis.

4.2 DIFFERENT AGRI-WASTE SUBSTRATES AND MICRO-ORGANISMS INVOLVED FOR THEIR EFFECTIVE DEGRADATION IN SCP PRODUCTION

Though different types of waste products can be utilized for the microbial degradation in SCP production, agri-wastes stand out apart in terms of their suitability and wide availability. These agri-wastes can be categorized into four broad groups after the classification of waste substrates used for single cell oil production proposed by Finco et al. in 2016. These groups are (i) monosaccharide- and disaccharide-rich waste products, (ii) starch-rich waste products, (iii) structural polysaccharide-rich waste products, and (iv) lignocellulose-rich waste products. Each of these waste types has its own merits and demerits for being used as a substrate in SCP production. SCP yield and all-inclusive digestibility of polysaccharides, protein, or lipid-rich waste substrates make their applicability as animal feed a notch higher than other groups. Still, monosaccharide- and disaccharide-rich wastes have a profitable and technological edge over them because of their bare minimum requirements of pretreatments.

Micro-organisms possess the potential to utilize the mineral elements such as carbon and nitrogen present in agri-wastes to transform themselves into high-quality SCPs. These agri-waste substrates and the micro-organisms carrying out their effective degradation through fermentation for the SCP production are enlisted in Table 4.1.

TABLE 4.1
Summary of Different Agri-wastes and Micro-organisms Involved in SCP Production

Type of Agri-Wastes Used for SCP Production	Monosaccharide- and Disaccharide-rich Waste	Starch-rich Waste	Structural Polysaccharide-rich Waste (%)	Lignocellulosic-rich Waste
Examples and protein content (%) by microbial action	Molasses (50.5), soy molasses (56.4), cheese whey (43.4) and its filtrate (34.2), sugarcane juice, spoiled date palm fruits (48.9)	Wheat (66.7) and rice (11.2) bran, starch (52.8), sorghum hydrolysate (47.5), leaf juice (45–55)	Corn cobs (30.4), deoiled rice bran (57), soybean hull (12.3), potato starch-rich processing waste (38–46), citrus pulp (31.9) and peel (5.7), poultry litter (29), Brewery' spent grains (31.8)	Stickwater (48–69), glutamic acid waste liquor (50.2), soybean meal, wheat capsicum powder (48.2), combined agricultural waste (33–38)
Micro-organisms involved in fermentation	*Kluyveromyces marxianus, Candida tropicalis, C. krusei, Trichoderma harzianum, Hanseniaspora uvarum*	*Aspergillus flavus, Candida krusei, Rhodopseudomonas gelatinosa, Schwanniomyces alluvius, S. occidentali, Saccharomyces cerevisiae, Torula utilis, Candida lipolytica*	*Aspergillus niger, A. oryzae, Bacillus licheniformis, B. pumilus, Candida utilis, Trichoderma viride, Debaryomyces hansenii*	*Aspergillus niger, Bacillus subtilis, Candida utilis, Lactobacillus acidophilus, Saccharomyces cerevisiae, Kluyveromyces marxianus*
References	Gao et al. (2012); Sisman et al. (2013); Yadav et al. (2014); Hashem et al. (2014)	Anupama and Ravindra (2000)	De Gregorio et al. (2002); Duarte et al. (2008); De Jalasutram et al. (2013); Liu et al. (2013, 2014); Wongputtisin et al. (2014)	Chiou et al. (2001); Zhao et al. (2010); Kam et al. (2012); Wongputtisin et al. (2012); Aggelopoulos et al. (2014)

A closer look at this table reveals that these micro-organisms indulged in SCP production span over different biological groups such as algae, bacteria, and fungi (especially yeast). However, out of all, most important ones belong to bacteria and fungi followed by some algal species owing to their rapid growth rates and comparatively higher protein concentrations (Voltolina et al. 2005; Gao et al. 2007). The preferences exhibited by these microbial groups for specific agri-waste substrates on an individual basis can be further elaborated as follows.

4.2.1 Bacteria

Bacteria have been used as SCP producers since a long time which is then utilized in a variety of ways, especially as animal feed. Though bacteria can grow on a wide array of substrates, yet the suitability of agri-wastes in context to their cost-effectiveness is very high. A number of bacterial species thrive on fruit processing wastes, but *Rhodopseudomonas gelatinosa* has been found to efficiently degrade wheat bran; *Rhodococcus opacus* strains particularly feed upon citrus and corn stovers; *Bacillus licheniformis* and *Bacillus pumilis* ferment potato wastes whereas *Cellulomonas* spp. prefers cellulose- and hemicellulose-rich agro-industrial wastes (Mahan et al. 2018).

4.2.2 Fungi

Protein yield from fungal species is favored over other source materials pertaining to its high content and distinct amino acid profile. As far as their biodegradative action on specific agri-wastes is concerned, the revelation of preference given by *Aspergillus flavus*, *A. ochraceus*, *A. oryzae*, *Cladosporiumcladosporioides,* and *Penicillium citrinum* to rice bran; *Kluyveromyces marxianus* and *Saccharomyces cerevisiae* to an orange pulp, molasses, brewer's spent grain; *Aspergillus niger* to apple pomace, banana waste, rice bran, and potato starch; *Trichoderma virideae* to citrus pulp has been found (Ravinder et al. 2003). Different agri-wastes for SCP production by yeast, bacterial and fungal strains have been mentioned in Table 4.2.

4.2.3 Algae

Some microalgae such as *Chlorella*, *Arthrospira platensis* (popularly *Spirulina*), and *Dulaniella* have been put under cultivation for extracting good-quality proteins in amounts as high as 70%. However, algal species for SCP production usually utilize CO_2 and sunlight as substrates, but simple sugar inputs from agricultural wastes such as corn stovers, sugar beets, and sugarcane can also be fermented via traditional methods for SCP production by them.

4.3 FERMENTATION PROCESS FOR SCP PRODUCTION

Fermentation progression needs an uncontaminated culture of the selected moieties, which require several aspects such as precise physical size, disinfection of the culture, and utilized for their growth. The construction fermenter is the apparatus utilized for

TABLE 4.2
Different Yeast and Fungal Strains Used for SCP Production Utilizing Agri-food Wastes

Agri-waste	Micro-organism	Results	References
Orange, lemon waste, and corn stover effluent	*Rhodococcus opacus*	Maximum protein amount was observed when microbe was cultivated on lemon agri-waste with a final protein amount of 62.6%	Mahan et al. (2018)
Sweet potato residues	*Saccharomyces* sp. IFO1426	Produced SCP which is used as an animal feed product, holding 16–21% protein	Panda and Ray (2015)
Cucumber and orange peel	*Saccharomyces cerevisiae*	Cucumber peel produces greater content of protein, i.e., 53.4%/100 g of the substrate in comparison to orange peel which produces only 30.5%/100 g of substrate	Mondal et al. (2012)
Orange, pineapple, banana, watermelon, and cucumber waste	*Aspergillus Niger*	Highest amount of *A. niger* biomass and protein amount of 2.29 ± 0.02 and 0.57 ± 0.01 g L^{-1}, respectively, were observed in case of banana waste as compared to other wastes	Oshoma and Eguakun-Owie (2018)
Potato, apple, carrot, and orange wastes	*Saccharomyces cerevisiae*	Maximum SCP biomass with a protein content of 49.29 ± 1.126% was found in case of potato peels followed by carrot peels	Bacha et al. (2011)
Banana skin, mango and apple waste, sweet orange peel, and rind of pomegranate	*Saccharomyces cerevisiae*	The highest protein content of 58.62% was observed in case of the banana skin, followed by a rind of pomegranate, apple waste, mango waste, and sweet orange peel with 54.28, 50.86, 39.98, and 26.26% protein content/100 g of substrate utilized, respectively	Khan et al. (2010)
Orange peel (*Citrus aurantium, C. sinensis* and *C. paradisi*)	*Aspergillus niger* and *Saccharomyces cerevisiae*	Maximum protein amount of 13.37% was found in *C. aurantium,* whereas in *C. paradisi* and *C. sinensis*, it was 11.9 and 11.53%, respectively	Azam et al. (2014)
Potato, orange pulp, whey, molasses	*Saccharomyces cerevisiae Kluyveromyces marxianus* and kefir	The maximum amount of protein, i.e., 38.5% w/w on a dry weight basis, was observed in case of a fermented product which was made with *S. cerevisiae*	Aggelopoulos et al. (2014)

Wheat bran	*Candida utilis* and *Rhizopus oligosporus*	The highest protein content of 41.02% was observed when fermented with *C. utilis* and *R. oligosporus* under the controlled circumstances as compared to nonfermented wheat bran which gave only 4.21% protein amount	Yunus et al. (2015)
Orange peel, beet pulp, and rice husk, whey	Yeasts (*Candida blankii, C. rugosa, Pichia anomala, Kluyveromyces lactis* and *Rhodotorula glutinis)*	The maximum protein content of was 6.78 g L^{-1} was found with *K. lactis* on whey waste and then 6.01 g L^{-1} in case of *Candida blankii* on an orange peel. Whereas, smaller protein amount was observed when rice husk and beet pulp were utilized for the growth of these strains	Khalil and Zaanon (2008)
Potato starch processing waste	*Aspergillus niger* and *Bacillus licheniformis*	Significantly produced SCP as high-protein feed	Liu et al. (2014)
Date's waste	*Aspergillus oryzae*	$(NH4)_2SO_4$ was the best nitrogen provider as the protein amount generated was expressively greater, i.e., 13.8%	Al-Farsi et al. (2019)
Rice bran	*Aspergillus oryzae*	The highest protein content of 57% was generated	Ravinder et al. (2003)
Mango waste	*Saccharomyces cerevisiae*	Protein amount of 79.14% was attained when inoculum and substrate amount was 8%, and sugar amount and highest cell dry weight obtained were 15.28 and 29.85% under optimal environments	Somda et al. (2017)
Corn cob	Endophytic fungi	Significantly improved the proximate constituent level and enhanced the total protein amount, which can be used as a substitute for costly animal feeds	Paynor et al. (2016)
Potato waste and corn cob	*Raoutella ornithinolytica*	SCP with 24.4% protein content was observed in potato peel through inoculation with *R. ornithinolytica*	Al-Hadithi et al. (2018)

taking the growth medium in the stable position, mass parting, supernatant gathering supernatant, refinement, and then by product isolation (Nasseri et al. 2011). SCP production occurs in a fermentation method (Chandrani-Wijeyaratne and Tayathilake 2000), in which chosen microbe species are cultured on appropriate growth medium in practical development progression leading to the microbe development and debris volume which is then separated by isolation method. Course progress starts with micro-organism identification, in which appropriate microbe species are attained from air, water, and soil or from minerals or organic resources and then augmented through the assortment, alteration, or additional genomic strategies. After this, the growth environment for the chosen species is regulated, and specific physiological aspects and cellular organization are specified. Simultaneously, the engineering method and organizational technique acclimatize the practical functioning of the progression and the instrument where the formation of micro-organism biomass takes place is regulated to prepare them for utilization on the huge industrial level.

Factors such as microbe availability, their generation, and energy requirement are the major assets for the entire procedure of biomass development. Different microbial strains must be examined for their appropriateness from cost-effective, commercial, and physiological view, and organized for the precise fermentation method. Safety requirement and queries about ecological defense appear during the formation of SCP in connection with the method and byproduct. Ultimately, protection and the defense of novelty lead to permissible and regulated facets, i.e., functioning authorizations, invention consents for specific submission, and the allowed guard of the novel procedure and species of microbe (Srividya et al. 2013). SCP can be formed through three different fermentation methods, i.e., submerged, semisolid, and solid state. In submerged progression, product utilized for fermentation is consistently in the fluid condition in which different nutrients necessary for growth are found. The apparatus holding the product is managed uninterruptedly, and the final obtained substrate biomass is unceasingly cultivated from the apparatus via practicing diverse methods. After this, the attained product is filtered and dried out. Aeration is a crucial practice in the production process, and heat is released during this process which is eliminated through a cooling instrument. The biomass of micro-organisms can be attained via different techniques to which microbes such as yeast and bacteria are isolated by centrifugation, whereas filamentous fungi are isolated by filtration (Suman et al. 2015).

Solid-state fermentation has been widely carried out via evaluating different kinds of bioreactors, course circumstances, and microbes for the formation of different important byproducts such as SCP, enzymes, C_2H_5OH, organic acids, vitamins, pigments, etc. (Singhania et al. 2009). This progression comprises adding a solid product like rice or wheat bran on flatbeds after combining it with microbes; the product is then allowed to stable in a temperature-regulated chamber for many days. The liquid medium is appropriate for the development of microbes like bacteria or yeasts. To accomplish this fluidic fermentation, it is essential to continuously provide the microbe with O_2, which is usually completed by moving the fermentation culture medium (Capalbo et al. 2001; Suman et al. 2015). In semisolid fermentation, the development of the substrate is unclear; however, mostly utilized in solid form. Fermentation through submerged growth medium demands high venture and has a large functioning price (Adedayo et al. 2011). Production comprises several procedures such as agitation and combining a multiphase apparatus, passage of $O_2$from the gas foams via the fluid system to the microbe, and then, progression of heat transport from fluid system to the adjacent environment (Andersen et al. 2005). U-loop fermenter, a distinct bioreactor, is made for categorizing biomass and energy transfer procedure (Jorgensen 2010). SCP formation includes normal stages in the making of appropriate standard, i.e., convenient C resource, inhibition of the culture and fermenter infection, the formation of the microbe with fascinating characteristics and isolation of generated biomass, and finally its processing (Soland 2005). The C resource utilized in this process can be gaseous hydrocarbons, n-alkenes, CH_3OH and C_2H_5OH, recyclable resources, for instance, carbohydrates, wastes of breweries, and other organic materials (Talebnia 2008).

4.4 MECHANISM OF SCP PRODUCTION

Particularly, microbes use the accessible wastes and practice them as a development medium to enhance their biomasses that are comprised of the SCP (Queiroz et al. 2007). Fermentation like submerged or solid in the form are the crucial aspects for SCP formation (Kadim et al. 2015). Whenever fermentation is achieved, then remaining biomass is collected that can be practiced as a protein moiety (Zepka et al. 2008). Afterward, this procedure experiences additional processing technologies such as decontamination, cell separation, cleaning, and then protein isolation (John et al. 2011) to provide normally greater manufacture proportion and maximum harvest and accomplishes the manufacture regulation comparatively easier (Figure 4.1). Agri-waste waste moieties have been considered as a virtuous candidate for SCP formation. SCP formed from these agri-wastes and micro-organisms rich in protein source which us of better superiority and efficient to be utilized for animal fodder. Extra processing makes it to be eatable even for the humans also (Yunus et al. 2015). There are several micro-organisms that are intricated in the transformation of distinct microbe biomass and agri-wastes into the commercial byproducts. Jacob-Lopes et al. (2006) practiced blue–green algae for the SCP

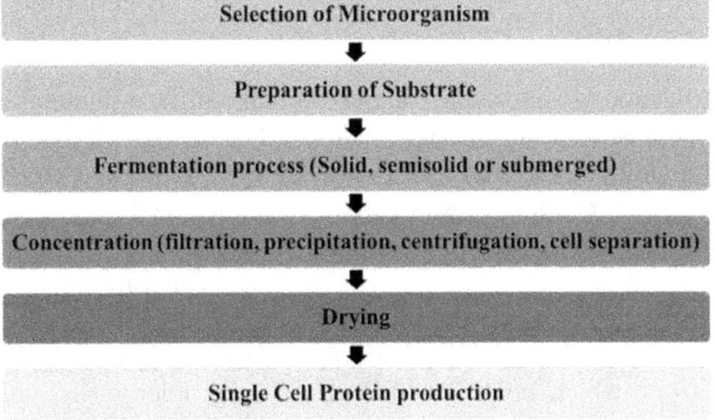

FIGURE 4.1 Stepwise representation of single-cell protein production.

generation. Nasseri et al. (2011) utilized various moieties for the SCP formation and evaluated that moieties with C structure were significant for this process. In addition to this, cellulose and hemicellulose also recognized as good substrates for SCP production. Fermentation of various moieties such as of plants cellulose and hemicellulose, and diverse N complexes from animals can be important moieties for SCP formation through distinct hydrolyzation techniques, i.e., physical, chemical, and enzymatic (Sharif et al. 2021).

4.5 ANALYTICAL METHODS FOR SCP ANALYSIS

4.5.1 Flow Cytometry

Flow cytometry is the utmost recognized and environment-friendly technique for SCP analysis. Flow cytometry efficacy is determined from the statistic that total cellular proteins' content is little, and the local protein amounts are high and quantifiable if the tissues are retained confined. In the present day time, this technique becomes more advanced due to improvement in both analytic techniques and accessibility of extremely precise antibodies. About 10–15 important proteins in signaling mechanisms can be evaluated together in single cells through the practice of several aspect evaluations by multicolor flow cytometers (De Rosa et al. 2001; Perfetto et al. 2004). The potential to accomplish interrelated extents of several proteins in single cells has converted cytometry into a commanding apparatus to semiquantitatively examine mechanisms causing numerous disorders (Irish et al. 2004; Sachs et al. 2005). DNA and RNA identification by *in situ* hybridization methods through Tyramide signal amplification has been incorporated to conventional antibody staining strategies for evaluation of small profusion proteins (Clutter et al. 2010). Whereas, the practice of multiparameter cytometry is still restricted as significant for drug screening, in spite of its greater amount of screening characteristics. However, this restriction was reduced by evolving a barcoding technique in which a complex of 3 dyes is used for tagging the distinct supplemented cells (Krutzik and Nolan 2006). A single dye allows around seven intensity phases; therefore, a complex of three dyes permits isolation of about 343 illustrations from a single cellular group.

4.5.2 Affinity Arrays

For the identification of single cell released proteins, surface-immobilized antibody methods can be practiced; for instance, industrially existing ELISpot technique that depends up on antibody-covered PVDF membrane to fix desired proteins and then finding with a 2nd antibody was improved to practice fluorescence-dependent evaluation of single cell released cytokines (Casey et al. 2010). But the major limitation is that it can take more time (days) prior cytokines are noticeable and complexing is restricted to one to three cytokines. To solve this problem, the high-throughput single cell barcoded chip is developed to modify complexing procedure (Ma et al. 2011), which holds 10,40 3-nL capacity microwells, holding DNA-barcoded antibodies to greater than 10 cytokines. Cytokine emissions of these cells are detected through fluorescence immune sandwich method, and around one to forty cells are identified in a single microwell. Barcode chip significantly utilized to detect cytokine release from cancer antigen-precise cytotoxic T cells and could be important for the detection of additional immune mechanisms. Affinity array technique is also modified by identifying antibodies in combination with covalently bind luminous oligomers in single cell released cytokines (Choi et al. 2011).

4.5.3 Mass Spectrometry

Single cell mass spectrometry exhibits the strong capacity to deliver label-free measurable investigation of the whole single cell proteomic structure, i.e., proteins, peptides, and post-translational modifications. The benefit of using mass spectroscopy is that it does not need any genomic labeling. Electrospray mass spectrometry, laser/desorption/ionization spectrometry, and secondary ion MS are the types of mass spectrometry which are practiced for SCP studies, whereas matrix-assisted laser desorption mass spectrometry is practiced for evaluation of peptides of the nervous system in individual neurons (Rubakhin and Sweedler 2008). However, mass spectrometry practice to examine SCP has certain restrictions, to which major one is the absence of sensitivity to sense small SCP content. Combining electrospray mass spectrometry with microfluidic cell lysis and capillary electrophoretic parting was helpful in greater recognition of hemoglobin in distinct red blood cells (Mellors et al. 2010). Combination of mass spectrometry with a microarray technique is a better method for improving the SCP analysis (Urban et al. 2010).

4.5.4 Separation-based SCP Analysis

Numerous expressed proteins are found in a single cell to which better quality isolation of the cellular moieties into separate proteins is carried out by electrophoresis or chromatography techniques to appropriately detect these proteins. The benefit of these separation techniques is that these permit balanced extent of whole proteins in a single trial. Separation methods like slab gel electrophoresis or high-performance liquid chromatography are not applicable for SCP analysis due to their incapability to isolate small proteins content. Capillary electrophoresis, executed in small vessels which impressionist the cellular size or nowadays majorly in microfluidic chips with μ extent pores, is highly capable due to the reason that it has the ability to isolate and evaluate SCP, particularly of huge mammalian tissues (Hu et al. 2003). 2-D isolation was performed to detect alterations in tumor cells in order to modify peak ability (Sobhani et al. 2007). A microfluidic instrument that combines seize, breakdown, capillary electrophoresis of single cells with individual fragment luminous estimate was practiced to detect unique protein moieties (Huang et al. 2007).

4.6 CONCLUDING REMARKS

SCPs could be formed from different inexpensive substrates, display significant features as a nutrient supplier for plants, humans, and animals, which can simply substitute their traditional protein resources with no harmful effect on them. To decline the manufacturing rate of SCP, the imperative aspect is the preference to economic and appropriate substrates or decomposable agro-trade wastes as a nutrient supplier for the microbe to cultivate and yield a greater amount of protein. Micro-organisms can be utilized to ferment several inexpensive agricultural wastes and transform them into valuable byproducts, and reuse of these agri-waste moieties can expressively diminish the SCP production expenses. Sustainable production of SCP will be highly relying upon decreasing manufacture charges and refining eminence, which can be accomplished with smaller feedstock expenses, better-quality fermentation, and development in the micro-organisms due to genomic applications.

ACKNOWLEDGEMENT

Dr. Simranjeet Singh is thankful to the Interdisciplinary Centre for Water and Applied Research (IcWaR), Indian Institute of Sciences, Bangalore for the financial assistance IOE-IISc Fellowship (Sr. No: IE/REAC-20–0134) and providing laboratory and library facilities.

REFERENCES

Adedayo, M. R., Ajiboye, E. A., Akintunde, J. K., and Odaibo, A. (2011). Single cell proteins: As nutritional enhancer. *Adv Appl Sci Res*, *2*(5), 396–409.

Aggelopoulos, T., Katsieris, K., Bekatorou, A., Pandey, A., Banat, I. M., and Koutinas, A. A. (2014). Solid state fermentation of food waste mixtures for single cell protein, aroma volatiles and fat production. *Food Chem*, *145*, 710–716.

Al-Farsi, M., Al Bakir, A., Al Marzouqi, H., and Thomas, R. (2019). Production of single cell protein from date waste. *By-products Palm Trees Appl*, *11*, 302.

Al-Hadithi, Q. A. H. A. J., Al-Rawi, D. F., and Al-Ani, M. Q. A. (2018). Single cell protein production by soil *Raoutella ornithinolytica* incubated on waste potato, paper and corn cob products. *Online J Vet Res*, *22*(12), 1137–1144.

Andersen, B. R., Andersen, J. B., and Jørgensen, S. B. (2005). U-loop reactor modelling for optimization–Part I: Estimation of heat loss. *J Environ Issues*, *9*: 88–90.

Anupama and Ravindra, P. (2000). Value-added food: single cell protein. *Biotech Adv 18*, 459–479.

Azam, S., Khan, Z., Ahmad, B., Khan, I., and Ali, J. (2014). Production of single cell protein from orange peels using *Aspergillus niger* and *Saccharomyces cerevisiae*. *Global J Biotechnol Biochem*, *9*(1), 14–18.

Bacha, U., Nasir, M., Khalique, A., Anjum, A. A., and Jabbar, M. A. (2011). Comparative assessment of various agro-industrial wastes for *Saccharomyces cerevisiae* biomass production and its quality evaluation as single cell protein. *J Anim Plant Sci*, *21*(4), 844–849.

Browne, J., Nizami, A. S., Thamsiriroj, T., and Murphy, J. D. (2011). Assessing the cost of biofuel production with increasing penetration of the transport fuel market: A case study of gaseous biomethane in Ireland. *Renew Sustain Energy Rev*, *15*, 4537–4547.

Casey, R., Blumenkrantz, D., Millington, K., Montamat-Sicotte, D., Kon, O. M., Wickremasinghe, M., … and Lalvani, A. (2010). Enumeration of functional T-cell subsets by fluorescence-immunospot defines signatures of pathogen burden in tuberculosis. *PLoS One*, *5*(12), e15619.

Chiou, P. W. S., Chiu, S. W., and Chen, C. R. (2001). Value of *Aspergillus niger* fermentation product as a dietary ingredient for broiler chickens. *Animal Feed Sci Technol*, *91*, 171–182.

Choi, J., Routenberg Love, K., Gong, Y., Gierahn, T. M., and Love, J. C. (2011). Immuno-hybridization chain reaction for enhancing detection of individual cytokine-secreting human peripheral mononuclear cells. *Anal Chem*, *83*(17), 6890–6895.

Clutter, M. R., Heffner, G. C., Krutzik, P. O., Sachen, K. L., and Nolan, G. P. (2010). Tyramide signal amplification for analysis of kinase activity by intracellular flow cytometry. *Cytometry Part A*, *77*(11), 1020–1031.

De Gregorio, A., Mandalari, G., Arena, N., Nucita F., Tripodo M. M., and Lo Curto, R. B. (2002). SCP and crude pectinase production by slurry-state fermentation of lemon pulps. *Bioresour Technol*, *83*, 89–94.

De Rosa, S. C., Herzenberg, L. A., Herzenberg, L. A., and Roederer, M. (2001). 11-color, 13-parameter flow cytometry: Identification of human naive T cells by phenotype, function, and T-cell receptor diversity. *Nature Med*, *7*(2), 245–248.

Duarte, L. C., Carvalheiro, F., Lopes, S., Neves, I., and Girio, F. M. (2008). Yeast biomass production in Appl. *Biochem Biotech, 148*, 119–129.

Elijah, A. I., and Edem, V. E. (2017). Value addition to food and agricultural wastes: A biotechnological approach. *Nig J Agric Food Environ, 13*(1), 139–154.

Ezejiofor, T. I. N., Enebaku, U. E., and Ogueke, C. (2014). Waste to wealth – Value recovery from agrofood processing wastes using biotechnology: A review. *Brit Biotechnol J.* *4*(4), 418–481.

Finco, A. M. O., Mamani, L. D. G., Carvalho, J. C., Pereira, G. V. M., Soccol, V. T., and Soccol, C. R. (2016). Technological trends and market perspectives for production of microbial oils rich in omega-3. *Crit Rev Biotechnol*, *37*(5), 656–671.

Fontana Capalbo, D. M., Valicente, F. H., Oliveira Moraes, I. D., and Pelizer, L. H. (2001). Solid-state fermentation of *Bacillus thuringiensis tolworthi* to control fall armyworm in maize. *Electron J Biotechnol*, *4*(2), 9–10.

Gao, L., Chi, Z., Sheng, J., Ni, X., and Wang, L. (2007). Single-cell protein production from Jerusalem artichoke extract by a recently isolated marine yeast *Cryptococcus aureus* G7a and its nutritive analysis. *Appl Microbiol Biotechnol*, *77*, 825–832.

Gao, Y., Li, D., and Liu, Y. (2012). Production of single cell protein from soy molasses using *Candida tropicalis*. *Ann Microbiol*, *62*, 1165–1172.

Ghosh, P. R., Fawcett, D., Sharma, S. B., and Poinern, G. E. J. (2016). Progress towards sustainable utilisation and management of food wastes in the global economy. *Int J Food Sci,* ID 3563478, 22 pages.

Goldberg, I. (2013). *Single Cell Protein*. Springer Science and Business Media.

Hashem, M., Hesham, A. E. L., Almari, S. A., and Alrumman, S. A. (2014). Production of single-cell protein from wasted date fruits by *Hanseniaspora uvarum* KKUY-0084 and *Zygosaccharomyces rouxii* KKUY-0157. *Ann Microbiol*, *64*, 1505–1511.

Huang, B., Wu, H., Bhaya, D., Grossman, A., Granier, S., Kobilka, B. K., and Zare, R. N. (2007). Counting low-copy number proteins in a single cell. *Science*, *315*(5808), 81–84.

Hu, S., Zhang, L., Newitt, R., Aebersold, R., Kraly, J. R., Jones, M., and Dovichi, N. J. (2003). Identification of proteins in single-cell capillary electrophoresis fingerprints based on comigration with standard proteins. *Anal Chem*, *75*(14), 3502–3505.

Irish, J. M., Hovland, R., Krutzik, P. O., Perez, O. D., Bruserud, Ø., Gjertsen, B. T., and Nolan, G. P. (2004). Single cell profiling of potentiated phospho-protein networks in cancer cells. *Cell*, *118*(2), 217–228.

Jacob-Lopes, E., Zepka, L. Q., Queiroz, M. I., and Netto, F. M. (2006). Protein characterisation of the Aphanothece Microscopica Nägeli cyanobacterium cultivated in parboiled rice effluent. *Food Sci Technol*, *26*(2), 482–488.

Jalasutram, V., Kataram, S., Gandu, B., and Anupoju, G. R. (2013). Single cell protein production from digested and undigested poultry litter by *Candida utilis*: Optimization of process parameters using response surface methodology. *Clean Technol Environ Policy*, *15*, 265–273.

John, R. P., Anisha, G. S., Nampoothiri, K. M., and Pandey, A. (2011). Micro and macroalgal biomass: A renewable source for bioethanol. *Bioresour Technol*, *102*(1), 186–193.

Jorgensen, J. B. 2010. Systematic model analysis for single cell protein (SCP), production in a U-loop reactor, 20th European symposium on computer aided process engineering escape. *Am.-Eur. J Agric Environ Sci*, *20*, 79–90.

Kadim, I. T., Mahgoub, O., Baqir, S., Faye, B., and Purchas, R. (2015). Cultured meat from muscle stem cells: A review of challenges and prospects. *J Integr Agricul*, *14*(2), 222–233.

Kam, S., Kenari, A. A., and Younesi, H. (2012). Production of single cell protein in stickwater by *Lactobacillus acidophilus* and *Aspergillus niger*. *J Aqua Food Prod Tech*, *21*, 403–417.

Khalil, A. I., and Zaanon, M. H. R. (2008). Utilization of food industries wastes for the production of single cell protein by yeasts. *Alexandria Sci Exchange J*, *29*(4), 315–324.

Khan, M., Khan, S. S., Ahmed, Z., and Tanveer, A. (2010). Production of single cell protein from *Saccharomyces cerevisiae* by utilizing fruit wastes. *Nanobiotechnica Universale*, *1*(2), 127–132.

Krutzik, P. O., and Nolan, G. P. (2006). Fluorescent cell barcoding in flow cytometry allows high-throughput drug screening and signaling profiling. *Nat Methods*, *3*(5), 361–368.

Liu, B., Li, Y., Song, J., Zhang, L., Dong, J., and Yang, Q. (2014). Production of single-cell protein with two-step fermentation for treatment of potato starch processing waste. *Cellulose*, *21*(5), 3637–3645.

Liu, B., Song, J., Li, Y., Niu, J., Wang, Z., and Yang, Q. (2013). Towards Industrially feasible treatment of potato starch processing waste by mixed cultures. *Appl Biochem Biotechnol*, *171*, 1001–1010.

Ma, C., Fan, R., Ahmad, H., Shi, Q., Comin-Anduix, B., Chodon, T., … and Ribas, A. (2011). A clinical microchip for evaluation of single immune cells reveals high functional heterogeneity in phenotypically similar T cells. *Nat Med*, *17*(6), 738–743.

Mahan, K. M., Le, R. K., Wells, T., Anderson, S., Yuan, J. S., Stoklosa, R. J., … and Ragauskas, A. J. (2018). Production of single cell protein from agro-waste using Rhodococcus opacus. *J Indust Microbiol Biotechnol*, *45*(9), 795–801.

Mellors, J. S., Jorabchi, K., Smith, L. M., and Ramsey, J. M. (2010). Integrated microfluidic device for automated single cell analysis using electrophoretic separation and electrospray ionization mass spectrometry. *Anal Chem*, *82*(3), 967–973.

Mondal, A. K., Sengupta, S., Bhowal, J., and Bhattacharya, D. K. (2012). Utilization of fruit wastes in producing single cell protein. *Int J Sci Environ Technol*, *1*(5), 430–438.

Nasseri, A. T., Rasoul-Amini, S., Morowvat, M. H., and Ghasemi, Y. (2011). Single cell protein: Production and process. *Am J Food Technol*, *6*(2), 103–116.

Oshoma, C. E., and Eguakun-Owie, S. O. (2018). Conversion of food waste to single cell protein using *Aspergillus niger*. *J Appl Sci Environ Manage*, *22*(3), 350–355.

Panda, S. K., and Ray, R. C. (2015). Microbial processing for valorization of horticultural wastes. In *Environmental Microbial Biotechnology* (pp. 203–221). Springer, Cham.

Paynor, K. A., David, E. S., and Valentino, M. J. G. (2016). Endophytic fungi associated with bamboo as possible sources of single cell protein using corn cob as a substrate. *Mycosphere*, *7*(2), 139–147.

Perfetto, S. P., Chattopadhyay, P. K., and Roederer, M. (2004). Seventeen-colour flow cytometry: Unravelling the immune system. *Nat Rev Immunol*, *4*(8), 648–655.

Queiroz, M. I., Lopes, E. J., Zepka, L. Q., Bastos, R. G., and Goldbeck, R. (2007). The kinetics of the removal of nitrogen and organic matter from parboiled rice effluent by cyanobacteria in a stirred batch reactor. *Bioresour Technol*, *98*(11), 2163–2169.

Ravinder, R., Venkateshwar Rao, L., and Ravindra, P. (2003). Studies on *Aspergillus oryzae* mutants for the production of single cell proteins from deoiled rice bran. *Food Technol Biotechnol*, *41*(3), 243–246.

Rubakhin, S. S., and Sweedler, J. V. (2008). Quantitative measurements of cell–cell signaling peptides with single-cell MALDI MS. *Anal Chem*, *80*(18), 7128–7136.

Sachs, K., Perez, O., Pe'er, D., Lauffenburger, D. A., and Nolan, G. P. (2005). Causal protein-signaling networks derived from multi-parameter single-cell data. *Science*, *308*(5721), 523–529.

Sharif, M., Zafar, M. H., Aqib, A. I., Saeed, M., Farag, M. R., and Alagawany, M. (2021). Single cell protein: Sources, mechanism of production, nutritional value and its uses in aquaculture nutrition. *Aquaculture*, *531*, 735885.

Singhania, R. R., Patel, A. K., Soccol, C. R., and Pandey, A. (2009). Recent advances in solid-state fermentation. *Biochem Engineer J*, *44*(1), 13–18.

Sisman, G., Dogan, M., and Algur, T. (2013). Single-cell protein as an alternative food for zebrafish, *Danio rerio*: A toxicological assessment. *Toxicol Ind Health*, *29*, 792–799.

Sobhani, K., Fink, S. L., Cookson, B. T., and Dovichi, N. J. (2007). Repeatability of chemical cytometry: 2-DE analysis of single RAW 264.7 macrophage cells. *Electrophoresis*, *28*(13), 2308–2313.

Soland, L. (2005). Characterization of liquid mixing and dispersion in a U-loop fermentor. *Am-Eur J Agric Environ Sci*, *67*, 99–109.

Somda, M. K., Ouattara, C. A. T., Mogmenga, I., Nikiema, M., Keita, I., Ouedraogo, N., … and Traore, A. S. (2017). Optimization of *Saccharomyces cerevisiae* SKM10 single cell protein production from mango (*Magnifera indica* L.) waste using response surface methodology. *African J Biotechnol*, *16*(45), 2127–2133.

Srividya, A. R., Vishnuvarthan, V. J., Murugappan, M., and Dahake, P. G. (2013). Single cell protein-a review. *Int J Pharmaceut Res Scholars*, *2*, 1–4.

Suman, G., Nupur, M., Anuradha, S., and Pradeep, B. (2015). Single cell protein production: A review. *Int J Curr Microbiol Appl Sci*, *4*(9), 251–262.

Talebnia, F. (2008). Ethanol production from cellulosic biomass by encapsulated *Saccharomyces cerevisiae*, PhD thesis. Chalmers Univ. Techno., Gothenburg (Sweden), *334*: 113 145.

Ukaegbu-Obi, K. M. (2016). Single cell protein: A resort to global protein challenge and waste management. *J Microbiol Microbial Technol, 1*(1), 5.

Urban, P. L., Jefimovs, K., Amantonico, A., Fagerer, S. R., Schmid, T., Mädler, S., … and Zenobi, R. (2010). High-density micro-arrays for mass spectrometry. *Lab on a Chip*, *10*(23), 3206–3209.

Vermeulen, S. J., Campbell, B. M., and Ingram, J. S. I. (2012). Climate change and food systems. *Ann Rev Environ Resour*, *37*, 195–222.

Voltolina, D., Gómez-Villa, H., and Correa, G. (2005). Nitrogen removal and recycling by *Scenedesmus obliquus* in semicontinuous cultures using artificial wastewater and a simulated light and temperature cycle. *Bioresour Technol*, *96*, 359–362.

Wijeyaratne, S. C., and Jayathilake, A. N. (2000). Characteristics of two yeast strains (*Candida tropicalis*) isolated from *Caryota urens* (Kithul) toddy for single cell protein production. *J Natl Sci Foundation Sri Lanka*, *28*(1), 79–86.

Wongputtisin, P., Khanongnuch, C., Kongbuntad, W., Niamsup, P., and Lumyong, S. (2012). Screening and selection of *Bacillus* spp. for fermented corticate soybean meal production. *J Appl Microbiol*, *113*, 798–806.

Wongputtisin, P., Khanongnuch, C., Kongbuntad, W., Niamsup, P., Lumyong, S., and Sarkar, P. K. (2014). Use of *Bacillus subtilis* isolates from Tea-nao towards nutritional improvement of soya bean hull for monogastric feed application. *Lett Appl Microbiol*, *59*, 328–333.

Yadav, J. S. S., Bezawada, J., Ajila, C. M., Yan, S., Tyagi, R. D., and Surampalli, R. Y. (2014). Mixed culture of *Kluyveromyces marxianus* and *Candida krusei* for single-cell protein production and organic load removal from whey. *Bioresour Technol*, *164*, 119–127.

Yazid, N. A., Barrena, R., Komilis, D., and Sánchez, A. (2017). Solid-state fermentation as a novel paradigm for organic waste valorization: A review. *Sustainability*, *9*: 224.

Yunus, F. U. N., Nadeem, M., and Rashid, F. (2015). Single-cell protein production through microbial conversion of lignocellulosic residue (wheat bran) for animal feed. *J Instit Brew*, *121*(4), 553–557.

Zepka, L. Q., Jacob-Lopes, E., Goldbeck, R., and Queiroz, M. I. (2008). Production and biochemical profile of the microalgae *Aphanothece microscopica* Nägeli submitted to different drying conditions. *Chem Engineer Process: Process Intensif*, *47*(8), 1305–1310.

Zhao, G., Zhang, W., and Zhang, G. (2010). Production of single cell protein using waste capsicum powder produced during capsanthin extraction. *Lett Appl Microbiol*, *50*, 187–191.

5 Biovalorization of Agri-waste for the Production of Polyphenols

Marco Túlio Pardini Gontijo
Genetics, Evolution, Microbiology and Imunology Department, Biology Institute, Universidade Estadual de Campinas (UNICAMP), Campinas, São Paulo, Brazil

José Guilherme Prado Martin
Microbiology of Fermented Products Lab (FERMICRO), Microbiology Department, Universidade Federal de Viçosa, Avenida Peter Henry Rolfs, s/n, Campus Universitário, Viçosa, Minas Gerais, Brazil

CONTENTS

5.1 INTRODUCTION

The losses related to the disposal of food represent direct financial disadvantages and water, agrarian, and input resources. Agro-industrial wastes, rich in biomass and organic matter, represent a significant source of biological compounds of interest for commercial exploitation (Bibra et al. 2020). Recent studies have demonstrated the benefits of reuse of wastes from the food industry with greater efficiency, as lignocellulosic compounds used for renewable energy generation

DOI: 10.1201/9781003128977-5

and production of additives for food and cosmetics products (Rani et al. 2020), nutraceuticals (Yang, Fu, and Yang 2020), pharmaceutical properties to diseases treatment (Mokochinski et al. 2015), functional food (Mirab et al. 2020; Zhu et al. 2020), among others.

Besides that, the modern industry has employed biotechnological processes to comply with stricter requirements, reducing costs and damages to the environment (Sepúlveda et al. 2020). It is according to the concept of "circular bioeconomy", an intersection between circular economy and bioeconomy (Álvarez-Cao, Becerra, and González-Siso 2020). Considering the global population growth and urbanization, and novel agricultural practices, it is essential to rethink how to produce food, feed, and fuel to reduce the large amounts of organic wastes or manage its discard (Surendra et al. 2020).

The biovalorization of wastes with high content of bioactive compounds could improve the reuse of natural resources, adding value and generating income for the industry (Usmani et al. 2020). Considering the wide range of substances with biological activity, antioxidant activity is among the most researched (Nayak and Bhushan 2019; Portilla Rivera et al. 2021). In this context, the class of polyphenols represents an essential target to the development of novel products to be exploited on an industrial scale.

Ultrasound-assisted extraction, microwave-assisted extraction, pressurized liquid extraction, supercritical fluid extraction, gas expanded liquids, and pulsed electric field are new methods used to extract polyphenols from agro-industrial wastes. Nevertheless, the use of solvents is one of the main methods used for this purpose (Portilla Rivera et al. 2021). It has disadvantages, especially regarding the degradation of bioactive compounds related to high temperatures for prolonged extraction periods (Pimentel-Moral et al. 2020). Therefore, to improve the obtaining of polyphenols from industrial wastes, strategies in line with bioeconomy practices represent a significant way.

In this sense, this chapter presents the main results from scientific studies about the biovalorization of wastes from the agri-food industry to produce polyphenols, considering the role of different groups of micro-organisms in this process.

5.2 POLYPHENOLS IN NATURAL SOURCES

5.2.1 Main Types of Polyphenols

Raw materials of plant origin, including fruits, vegetables, and grains, are a source of phenolic compounds (Cheynier 2012). Phenolic compounds are ubiquitous and among the most abundant secondary metabolites found in plants and plant-derived products, including agro-industrial wastes (Abbas et al. 2017; Cheynier 2012). Phenol and its chemical derivatives' immediate action is related to the plant's defense against biotic and abiotic stresses (Böttger et al. 2018). Phenolic compounds comprise at least one phenol unit in their structures (Tsao 2010; Gan et al. 2019). Polyphenolic content in plants varies depending on the species, climatic conditions, plant part (roots, stems, leaves, flowers, fruits, and seeds), cultivation conditions, and degree of maturation (Gharras 2009). Phenolics classification is divided into phenolic acids and polyphenols (Figure 5.1). Currently, more than 8,000 phenolic structures are known (Tsao 2010).

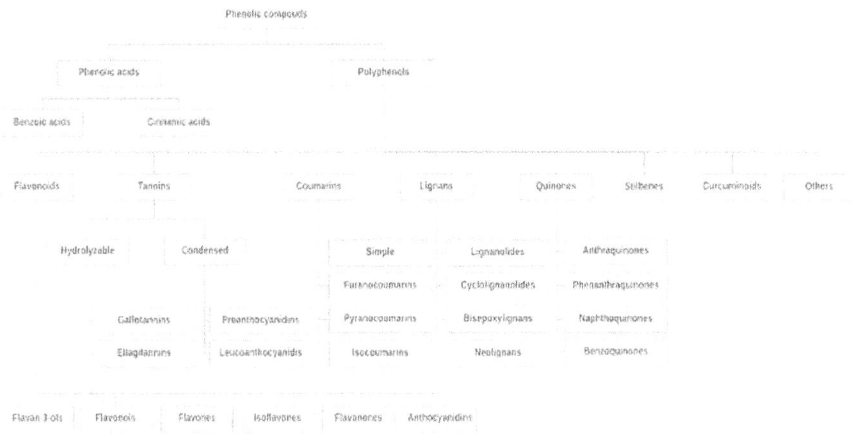

FIGURE 5.1 Categories of phenolic compounds.

5.2.1.1 Phenolic Acids

5.2.1.1.1 Benzoic Acids

Phenolic acids can be divided into derivatives of benzoic and cinnamic acids. Benzoic acid derivatives (Figure 5.2A) are based on C1–C6 structure, consisting of a C6 ring and a C1 carboxyl group (Tsao 2010). Fruits and vegetables contain majorly free phenolic acids, and grains and seeds contain phenolic acids mostly bound to macromolecules (Tsao 2010; Abbas et al. 2017). The most common benzoic acid derivative found in plants is gallic acid. However, p-hydroxybenzoic acid and p-hydroxybenzoic aldehyde, vanillic acid and vanillic aldehyde, protocatechuic acid, ethyl gallate, syringic acid, salicylic acid, and 2,3,4-trihydroxybenzoic acid are also found in some plants (Shirahigue et al. 2020; de Araújo et al. 2021; Gharras 2009; Tresserra-Rimbau, Lamuela-Raventos, and Moreno 2018).

5.2.1.1.2 Cinnamic Acids

Cinnamic acid derivatives (Figure 5.2B) are based on C3–C6 structure, consisting of a C6 ring and a three-carbon backbone comprising a carboxyl group (Tsao 2010). Similarly, cinnamic acids are majorly found as bound phenolics (Abbas et al. 2017; Gan et al. 2019). The most common cinnamic acid derivatives are caffeic acid, chlorogenic acid, o-coumaric, m-coumaric, p-coumaric acids, ferulic acid, isoferulic acid, and sinapic acid (Erukainure, Sanni, and Islam 2018; Gan et al. 2019; Gharras 2009).

5.2.1.2 Polyphenols

5.2.1.2.1 Flavonoids

Flavonoids have the C6–C3–C6 general structure, consisting of two C6 rings of phenolic nature (rings A and B) and one heterocyclic three-carbon ring (ring C) comprising an oxygen atom (Tsao 2010). The variations in the heterocyclic ring divide flavonoids into flavan-3-ols (Figure 5.2C), flavonols (Figure 5.2D), flavones (Figure 5.2E), isoflavones (Figure 5.2F), flavanones (Figure 5.2G), and anthocyanidins (Figure 5.2H). Currently, more than 4,000 flavonoids have been reported in

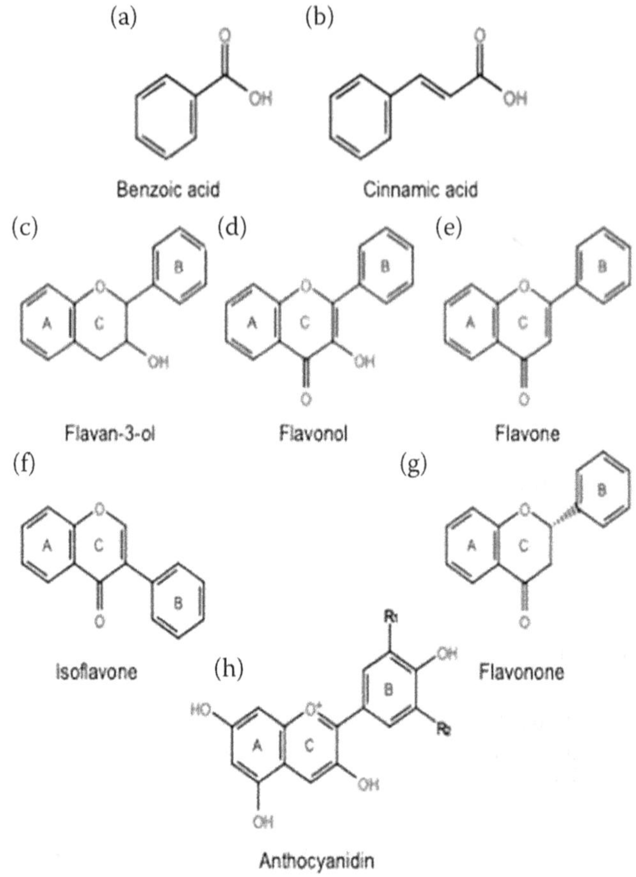

FIGURE 5.2 Benzoic (a) and cinnamic (b) acids, precursors of the primary phenolic acids found in plants. Flavan-3-ol (c), flavonol (d), flavone (e), isoflavone (f), flavone (g), and anthocyanidin (h), precursors of the primary flavonoids found in plants.

various raw materials of plant origin (Shirahigue et al. 2020; de Araújo et al. 2021). The major representants in each of the six subclasses of flavonoids are: (i) flavan-3-ols: catechin and epicatechin; (ii) flavonols: quercetin, kaempferol, and myricetin; (iii) flavones: apigenin and luteolin; (iv) isoflavones: daidzein, genistein, and glycitein; (v) flavanones: hesperetin and naringenin; and (vi) anthocyanidins: cyanidin, pelargonidin, peonidin, delphinidin, and malvidin (Tsao 2010; de Araújo et al. 2021).

5.2.1.2.2 Tannins

Tannins are a class of water-soluble polyphenols comprising many hydroxyl and other functional groups. The presence of these chemical groups makes tannins a molecule capable of binding to macromolecules, especially proteins and carbohydrates (Le Bourvellec and Renard 2012). Tannins are classified as hydrolyzable (Figure 5.3A and 5.3B) and condensed tannins (Figure 5.3C and 5.3D). Hydrolyzable tannins are composed of ellagic and gallic acids with a small sugar core (Das et al. 2020). On the

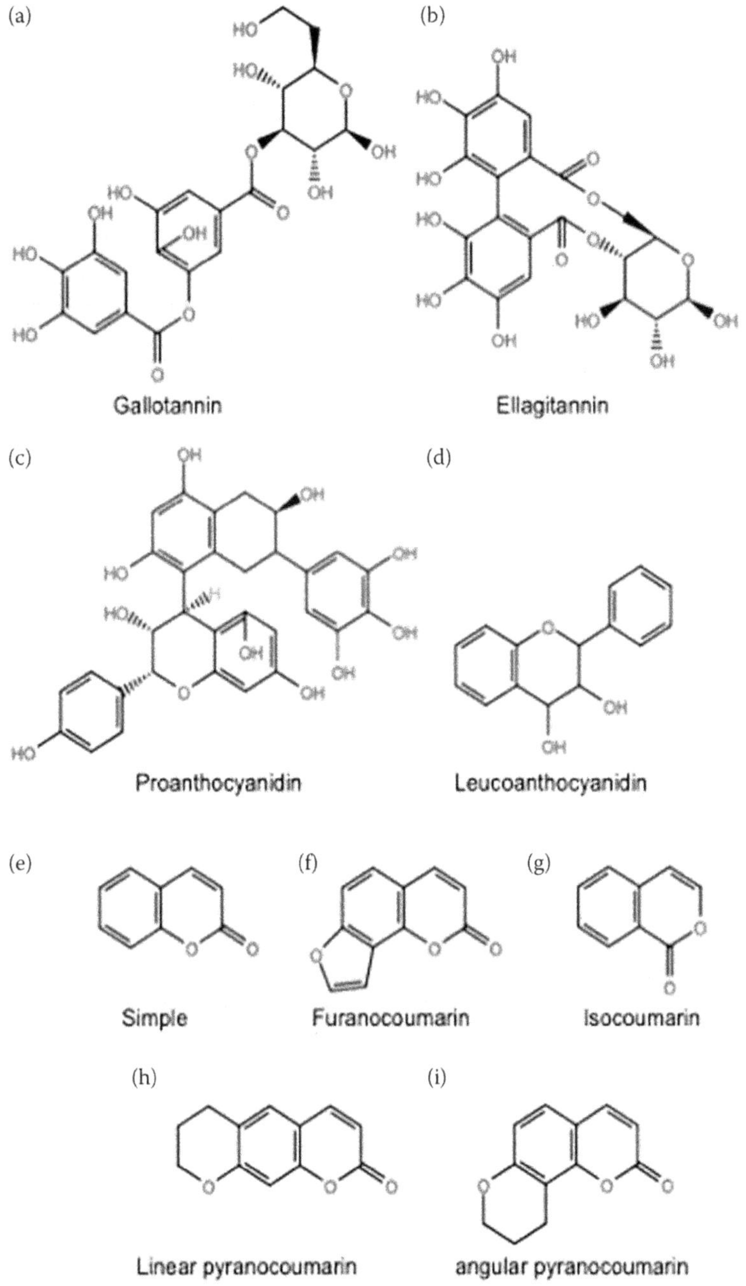

(a) Gallotannin

(b) Ellagitannin

(c) Proanthocyanidin

(d) Leucoanthocyanidin

(e) Simple

(f) Furanocoumarin

(g) Isocoumarin

(h) Linear pyranocoumarin

(i) angular pyranocoumarin

FIGURE 5.3 Gallotannin (a), ellagitannin (b), proanthocyanidin (c), and leucoanthocyanidin (d), precursors of the primary tannins found in plants. Simple coumarin (e), furanocoumarin (f), isocoumarin (g), linear (h), and angular (i) pyranocoumarin, precursors of the mains coumarins found in plants.

other hand, condensed tannins are formed of either flavan-3-ol or flavan-3,4-diol without a sugar core (de Araújo et al. 2021; Cheynier 2012). Tannins correspond from 5 to 10% (dry weight) of barks, stems, seeds, roots, buds, and leaves and, in smaller amounts, some fruits, such as grapes, blackberries, and strawberries (Das et al. 2020).

5.2.1.2.3 Coumarins

Coumarins are a class of lactone molecules comprising a benzene ring fused to an α-pyrone ring (Hussain et al. 2018; Venugopala, Rashmi, and Odhav 2013). Based on their chemical structure, coumarins are classified into simple hydroxycoumarins (Figure 5.3E), furocoumarins (Figure 5.3F), isocoumarins (Figure 5.3G), and linear or angular pyranocoumarins (Figure 5.3H and 5.3I). About 1,300 coumarins have already been identified in plants. Coumarins are reported in more than 150 different species and over 30 families of plants, such as *Rutaceae*, *Umbelliferae*, *Clusiaceae*, *Guttiferae*, *Caprifoliaceae*, *Oleaceae*, *Nyctaginaceae*, and *Apiaceae* (Matos et al. 2015; Venugopala, Rashmi, and Odhav 2013).

5.2.1.2.4 Lignans

Lignans consist of two propyl-benzene belonging to diphenolic compounds (Soleymani et al. 2020; Rodríguez-García et al. 2019). Lignans are characterized into four main subgroups based on oxidation and cyclization patterns: lignanolides (Figure 5.4A), cyclolignanolide (Figure 5.4B), bisepoxylignan (Figure 5.4C), and neolignane (Figure 5.4D). Lignans are found in several vegetables, fruits, legumes, whole grain cereals, and oilseeds (Soleymani et al. 2020). Similar to benzoic and cinnamic acids, lignans can be found in plants either free or bound to sugars (Rodríguez-García et al. 2019). The lignans widespread among plants are lariciresinol, matairesinol, pinoresinol, and secoisolariciresinol (Peterson et al. 2010).

5.2.1.2.5 Quinones

Quinones are cyclohexadienediones comprising two carbonyls (Ouellette and Rawn 2018; El-Najjar et al. 2011). Depending on the number of cyclic structures and the position of the carbonyl group, quinones can be classified into anthraquinones (Figure 5.4E), phenanthraquinone (Figure 5.4F), naphthoquinone (Figure 5.4G), and benzoquinone (Figure 5.4H). Quinones are compounds widely distributed in nature, comprising more than 1,200 structures identified (El-Najjar et al. 2011). Quinones are found in several plant families, such as *Ranunculaceae*, *Aphodelaceae*, *Fabaceae*, *Ebenaceae*, and *Rhamnaceae* (El-Najjar et al. 2011; Kishikawa and Kuroda 2014).

5.2.1.2.6 Stilbenes

The stilbene structure (Figure 5.5A) is based on two aromatic rings linked by an ethylene bridge (El Khawand et al. 2018). Stilbene structures range from monomers to octamers and carry functional groups (glycosyl, hydroxyl, methyl, or isopropyl) at different positions (Chong, Poutaraud, and Hugueney 2009; El Khawand et al. 2018). This class of compounds is synthesized in lower quantities than other polyphenols found in plants (Reinisalo et al. 2015; El Khawand et al. 2018).

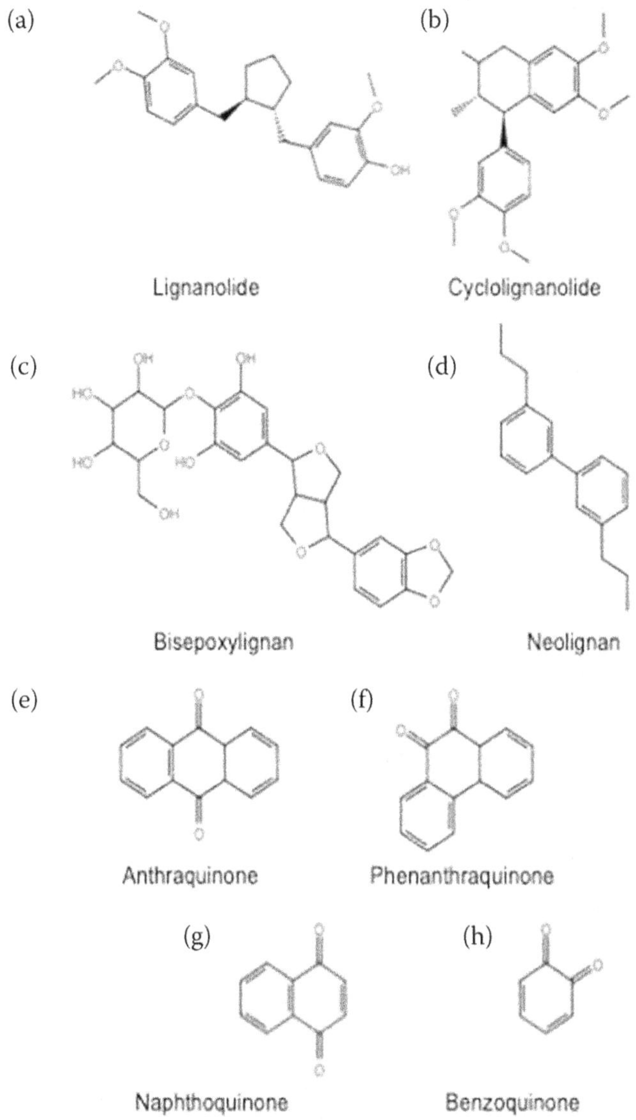

(a) Lignanolide (b) Cyclolignanolide (c) Bisepoxylignan (d) Neolignan (e) Anthraquinone (f) Phenanthraquinone (g) Naphthoquinone (h) Benzoquinone

FIGURE 5.4 Lignanolide (a), cyclolignanolide (b), bisepoxylignan (c), and neolignan (d), precursors of the main lignans found in plants. Anthraquinone (e), phenanthraquinone (f), naphthoquinone (g), and benzoquinone (h), precursors of the mains quinones found in plants.

Stilbenes are naturally found in cocoa, grapes, hops, peanuts, pines, rhubarbs, strawberries, sugar cane, tomatoes, and others (Reinisalo et al. 2015). The main stilbenes found in plants are resveratrol, piceatannol, isorhapontigenin, pinosylvin, and rhapontigenin (Reinisalo et al. 2015; El Khawand et al. 2018). However, more than 400 stilbenes are currently known (El Khawand et al. 2018).

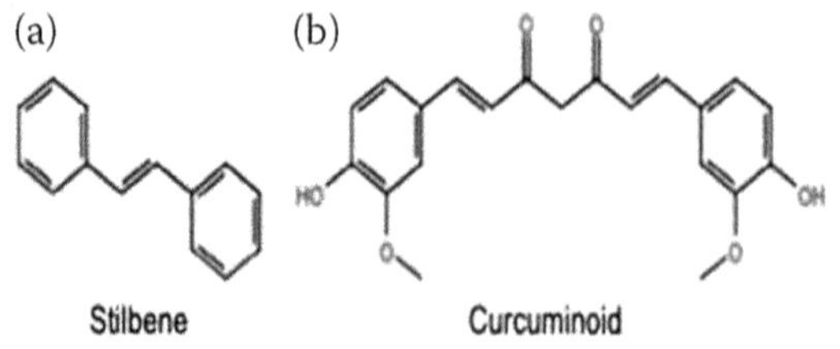

(a) (b)

Stilbene Curcuminoid

FIGURE 5.5 Precursors of the stilbenes (a) and curcuminoids (b).

5.2.1.2.7 Curcuminoids

Curcuminoids are a subclass of polyphenols derived from curcumin (Figure 5.5B), the primary polyphenol found in the rhizomes of *Curcuma longa* (turmeric) and other plants belonging to *Curcuma* genera (Hewlings and Kalman 2017; Priyadarsini 2014). The main curcuminoids found in plants are curcumin, demethoxycurcumin, bis-demethoxycurcumin, and cyclic curcumin (Amalraj et al. 2017; Priyadarsini 2014). Depending on several cultivation factors, turmeric contains from 2 to 9% of curcuminoids, being curcumin the significant component, and cyclic curcumin is the minor (Priyadarsini 2014).

5.2.2 Polyphenols in Agro-industrial Wastes

Polyphenols are compounds found naturally in fruits and vegetables and, consequently, in their byproducts (Dzah et al. 2020). Table 5.1 summarizes some wastes of vegetable origin whose polyphenol composition was identified.

5.3 EXTRACTION OF POLYPHENOLS IN AGRO-INDUSTRIAL WASTES USING TRADITIONAL METHODS

Several studies about the reuse of organic wastes to obtain compounds with biological activities have been developed. The most part involves the extraction using solid-liquid methods and different types of solvents and their solutions, as ethanol, methanol, ethyl acetate, chloroform, and hexane, among others (Azabou et al. 2017; Martin et al. 2012; Sánchez-Gómez et al. 2014; Zagklis and Paraskeva 2018). The principle of this method is based on lixiviation. In general, the sample is placed in a Soxhlet apparatus containing an organic solvent. Cycles of volatilization and condensation of the solvent are repeated several times until the extraction of the compounds. The process is easily operable and does not require professional expertise (Suwal and Marciniak 2018).

On the other hand, novel technologies have been evaluated to improve the extraction yield and reduce the use of chemical compounds. The extraction procedures should be optimized for the maximum quality of the extracts, besides the most economical production. For this, it is essential to consider the polarity of the target molecules since some extractions work better than others. In this sense, if the target comprises polar molecules, supercritical CO_2 extraction represents an interesting option; for less polar molecules, accelerated solvent extraction or pressurized hot

TABLE 5.1
Polyphenolic Compounds found in Agro-industrial Wastes

Source	Compound	References
Coffee pulp waste	Gallic acid, Vanillin, Ferulic acid, Coumaric acid, Catechin, Epicatechin, Epicatechol, Ethyl catechol, Caffeic acid, Myristicin, Cinnamic acid, Rutin, and Kampherol	Manasa, Padmanabhan, and Anu Appaiah (2021)
Tomatoes peel	Caffeic acid, Chlorogenic acid, p-coumaric acid, and Quercetin	Ninčević Grassino et al. (2020)
Pomegranate peels	Eugenic acid, Digalloyl, Gallic acid, Ferulic acid, Cyanidin-3-O-pentoside, Kaempferol 3-O-glucoside, p-hydroxybenzoic, Cyanidin-3-O-rutinoside, Brevifolin carboxylic acid, Galloyl-bis-HHDP, Vanillic acid, Coutaric acid, Quercetin, Dihydroxy gallocatechin, Ellagic acid, Cyanidin-3-O-glucoside, Caffeic acid, Caffeic acid phenethyl ester, Tetra galloyl glucopyranose, Monogalloyl, Gallocatechin, Protocatechuic acid	Selvakumar et al. (2021)
Citrus peels	Gallic acid, p-Coumaric acid, trans-Ferulic acid, Naringin, Hesperidin, Rutin, Myricetin, and Resveratrol	Gómez-Mejía et al. (2019)
Red cabbage and Brussels sprout waste	Quinic acid, Progoitrin, Sinigrin, Gluconapin, Hydroxyglucobrassicin, Glucobrassicin, Quercetin, Chlorogenic acid, Kaempferol, Ferulic acid, and Sinapic acid	Gonzales et al. (2015)
Onion solid wastes	Protocatechuic acid, Quercetin, Cyanidin, and Isorhamnetin	Katsampa et al. (2015)
Chestnut solid wastes	Gallic acid, Vescalagin, Castalagin, Ellagic acid, (−) Epigallocatechin, (+) Catechin, and (−) Epicatechin	Aires, Carvalho, and Saavedra (2016)
Olive mill wastes	Gallic acid, Protocatechuic acid, p-hydroxybenzoic acid, Caffeic acid, Vanillic acid, p-coumaric acid, Tyrosol, and Vanillin	Michailof, Manesiotis, and Panayiotou (2008)
Spent coffee grounds	Gallic acid, 3,4-dihydroxybenzoic acid., Chlorogenic acid, Caffeine, p-Coumaric acid, trans-Ferulic acid, Rutin, Naringin, Quercetin, and Kaempferol	Ramón-Gonçalves et al. (2019)
Black mulberry pomace	Chlorogenic acid, Gallic acid, Syringic acid, Neochlorogenic acid, Caffeic acid, Rutin, Quercetin, Cyanidin-3-O-glucoside, and Cyanidin-3-O-rutinoside	Khodaiyan and Parastouei (2020)

water extraction could be a viable alternative. In some situations, ultrasound-assisted extraction represents an interesting pre-treatment procedure (Putnik et al. 2018).

Supercritical fluid extraction, a method widely used to extract natural compounds from plant materials (Suwal and Marciniak 2018), is frequently applied to obtaining substances with biological activities from agri-food wastes (Makris 2018; Manna, Bugnone, and Banchero 2015). The method includes the use of supercritical fluids for substrate processing, where the fluid does not condense during the process, alternating between a gas-like and liquid-like condition, resulting in a specific property able to extract substances with biological value. Furthermore, it does not require organic solvents or uses it in few quantities and shorter time, providing higher yields and making it an interesting option to extract compounds on an industrial scale (Vidović et al. 2021).

Da Porto and Natolino (2017) used supercritical carbon dioxide to extract polyphenols from grape seeds. The experiment was performed using an ethanol–water mixture (57% v/v) as cosolvent at 40°C and specific conditions. The parameters that allowed the highest polyphenol extraction (7,132 mg GAE/100 g) were the pressure of 80 bar, a CO_2 flow rate of 6 kg/h, and 20% of cosolvent. In another study, the authors evaluated a combination of supercritical CO_2 extraction and ultrasound-assisted extraction to recover polyphenols from grape marc waste. The polyphenol yield obtained (3493 mg GAE/100 g) and the antioxidant activity (7503 mg α-tocopherol/100 g) were significantly increased (Da Porto and Natolino 2017).

Ultrasound-assisted extraction techniques have been used to improve the extraction of polyphenols from different types of wastes. Punzi et al. (2014) evaluated a procedure to obtain solvent-free extracts (water) with considerable bioactive compounds contents from artichoke wastes. The use of ultrasounds for 1 h resulted in a better recovering of phenolic compounds than the other treatments. 5-O-caffeoylquinic and 1,5-di-O-caffeoylquinic acids were identified among the major phenols using chromatographic techniques. The artichoke wastes were considered a good source of nutraceuticals compounds, and the procedure evaluated, a powerful eco-friendly approach to be explored by the food industry.

Escobar-Avello et al. (2019) identified 75 compounds (including phenolic acids, flavanols, flavanonas, and stilbenoids) in grape canes extracts obtained using an ultrasound-assisted procedure. The study identified 17 polyphenols never described before in this type of agri-waste, demonstrating its viability as a source of polyphenols. Boukroufa et al. (2015) extracted polyphenols from orange peel using an optimized ultrasound-assisted procedure by response surface methodology. A higher polyphenol yield (50.02 mg GA/100 g dm) was obtained under 0.956W/cm^2 and 59.83°C conditions. A combination of ultrasound and microwave techniques allowed to add value to target compounds in a shorter time, using only natural resources, contributing to energy saving and reducing the discard of wastes.

Ninčević Grassino et al. (2020) extracted pectin using high hydrostatic pressure (HHP) from dried tomato peel samples using nitric acid (c = 0.1 mol/L) at 80°C with a pressure of 300 MPa and extraction times of 10, 20, 30, and 45 min. Then, the authors extracted polyphenols tomato peels with and without pectin using ultrasound-assisted extraction (UAE) at a power of 400 W, frequency of 30 kHz, using ethanol (70 and 96%) at a solvent ratio of 1:50 (m/V) for 5, 10, and 15 min with temperature ranging from 20.8 to 59.7°C. They observed that depectinized samples contained lower mass fractions of polyphenols when compared to pectinized samples. However, their quantities are still significant, and the combination of HHP and UAE can be used to recover residual polyphenols from industrial depectinized wastes.

Some studies have accounted for high-pressure homogenization (HPH) as a pretreatment to optimize the extraction of polyphenols and other bioactive compounds from agro-industrial wastes due to the disruption of plant tissues (Velázquez-Estrada et al. 2013). However, HPH was carried out as a pretreatment to ultrasound (US) extraction of polyphenols of lettuce wastes, resulting in lower phenolic yield than ultrasound without pretreatment. The authors accounted that HPH also releases oxidative enzymes from plant tissues, which could reduce the total phenolic content of the waste. Alternative treatments that improve polyphenol release and inactivate oxidative enzymes, such as blanching, pulsed electric fields, and high hydrostatic pressures, might improve polyphenol extraction yield (Stella Plazzotta and Manzocco 2018).

Conversely, a study accounting for the total phenolic content of blanched lettuce wastes was smaller than those observed in not-blanched feces (S. Plazzotta and Manzocco 2019). Despite the deactivation of oxidative enzymes, the phenolic content is also reduced in blanching. On the other hand, pulsed electric field treatment of citrus fruits at 3 kV/cm yielded 16 mg/g of phenolics from skins, improving the extraction of polyphenols by 50% by a synergistic effect (El Kantar et al. 2018). The yield increment was probably due to oxidative enzymes deactivation.

Rajha et al. (2019) evaluated the extraction of polyphenols from pomegranate peels assisted by infrared (IR), ultrasound (US), pulsed electric fields (PEF), and high-voltage electrical discharges (HVED) compared to the classical extraction method using a water bath (WB). The higher concentration of polyphenols (46 ± 0.5 mg GAE/g) was obtained with HVED, followed by PEF (39 ± 2 mg GAE/g), US (14.5 ± 0.8 mg GAE/g), IR (8 ± 0.2 mg GAE/g), and WB (5 ± 0.1 mg GAE/g). In general, all of the improved methods enhanced the extraction yield of polyphenols.

Zhang et al. (2013) extracted polyphenols from peanut shells using an optimized microwave-assisted enzymatic extraction (MAEE) method. The yield in polyphenols reached $1.75 \pm 0.06\%$ using the optimum conditions (irradiation time of 2.6 min, cellulase of 0.81 wt.%, pH of 5.5, and incubation at 66°C for 2 h). The result ($1.75 \pm 0.06\%$) using the MAEE method was significantly higher when compared to heat reflux extraction ($1.53 \pm 0.03\%$), ultrasonic-assisted extraction ($1.56 \pm 0.02\%$), and enzyme-assisted extraction (1.62 ± 0.04). Furthermore, Gonzalez-Rivera et al. (2021) evaluated an *in situ* microwave-assisted extraction (ISMAE) of polyphenols from clove buds applying the method for 10 min using 150 W and stirring 250 rpm. The authors identified gallic acid, chlorogenic acid, eugenol, tyrosol, catechin, (–) epicatechin, acetylsalycilic acid, oleuropein, and pinoresinol. The proposed method yielded a higher level of total polyphenols (ranging from 5.5 ± 0.8 to $16 \pm 1.5\%$) compared to conventional hydrodistillation ($7.8 \pm 0.8\%$) even at higher extraction times. The results highlight the ISMAE as a faster and greener approach than traditional methods, even without enzyme-assisted extraction.

Another microwave-assisted extraction method was evaluated by Khodaiyan and Parastouei (2020) using dried black mulberry pomace powder in acidified distilled water at different pH levels (1, 2, and 3), different liquid/solid ratios (20, 30, and

40 mL/g), irradiation time (60, 180, and 300 s), and microwave power (200, 500, and 800 W). The MAE optimum condition (700 W, 300 s, pH of 1.42, and liquid/solid ratio of 20 mL/g) resulted in a pectin and phenolics yields of 10.95% and 12.11%, respectively. The results highlight the simultaneous biovalorization of pectin and polyphenols from agro-industrial wastes.

Another method to improve the extraction of polyphenol consists of the use of an accelerated solvent extractor. It has demonstrated exciting results in the improvement of substances with biological activity. Suárez et al. (2009) observed that this procedure resulted in final extracts rich in phenolic compounds from olive cake bioproduct generated after olive oil extraction. Great antioxidant activities were obtained, ranging from 3,450 to 17,900 µmol of Trolox equivalents/g of extract. According to the authors, the accelerated solvent extractor method showed potential to be applied on an industrial scale for the phenolic compound's exploitation from this kind of waste.

Drosou et al. (2015) evaluated different methods for the polyphenol extraction from grape pomace dehydrated by air drying and accelerated solar drying. Microwave-assisted, ultrasound-assisted, and conventional Soxhlet extraction (SE) were evaluated, considering extracts obtained using water, water:ethanol (1:1), and ethanol as solvents. Ultrasound in combination with water:ethanol extraction showed the highest phenolic compound content among the parameters tested. Besides that, air-dried grape pomace extracts showed the best performance to total anthocyanin and flavonol content.

At long last, the polyphenol extraction using nonorganic solvents could be an interesting alternative for the valorization of agri-food wastes. Eutectic solvents considered a "green option" produced using natural materials such as glycerol and organic acid salts represent a novel strategy to extract solvent-free polyphenols to application in the food industry (Ben-Othman, Jõudu, and Bhat 2020).

Fernández et al. (2018) developed an environmentally friendly method to extract phenolic compounds from onion, olive, tomato, and pear industrial byproducts. A eutectic solvent based on lactic acid, glucose, and 15% water was designed and compared with organic solvents. The ultrasound-assisted extraction resulted in 14 phenols from the wastes analyzed, being considered a simple, nonexpensive, eco-friendly, and robust system, with potential use to exploit polyphenols by food and pharmaceutical industries.

Cabezudo et al. (2021) proposed a green and environmentally friendly method as an alternative to solvent use to improve the extraction of polyphenols from soybean hulls. The study evaluated the effect of an optimized free-solvent pre-treatment using enzymatic digestion (α-amylase). The results demonstrated a significant increase in the antioxidant activity related to different polyphenols, mainly phenolic acids and isoflavones. The protocol could contribute to the exploitation of phytochemicals to be used by food and pharmaceutical industries.

5.4 BIOVALORIZATION OF POLYPHENOLS FROM AGRI-FOOD WASTES

In general, the polyphenol extraction from plant material or vegetal wastes comprises solid–liquid extraction, as discussed above. The ease of use is its main advantage.

However, the requirement of large amounts of chemical solvents, besides a long time for the extraction and, mainly, the polyphenol losses due to the extraction procedures, represents the several disadvantages of traditional methods employed for that. Even if novel technologies have been developed, as discussed above, a considerable fraction of bound polyphenols cannot be extracted using them (Bhanja Dey et al. 2016).

In this context, the development of processes involving bioconversion has improved the bound polyphenols extraction. These processes comprise reactions catalyzed by biological systems commonly constituted by enzymes (amylases, cellulases, xylanases, and esterases) or micro-organisms responsible for the breakdown of the linkages on the cell wall, making possible the release and recovery of desired polyphenol molecules (Khosravi and Razavi 2020; Madeira and Macedo 2015). The different methods for the biovalorization of polyphenols from agro-industrial wastes are summarized in Table 5.2.

The biovalorization of agri-food wastes includes fermentation processes, including submerged (SmF) or solid-state fermentation (SSF) based on wild micro-organisms' strain, as well as the use of modified micro-organisms (by heterologous expression, for example). In recent years, the first has received more attention; in submerged fermentation, micro-organisms are cultivated in liquid substrate containing essential nutrients under specific conditions such as aeration, pH, temperature, final volume, nutritional requirements, and inoculum amount. In solid-state fermentation, micro-organisms grow on a solid substrate with low quantities of water and produce extracellular enzymes necessary to release the phenolic compounds from the matrix. At the same time, that novel is generated by the microbial secondary metabolism (Bhanja Dey et al. 2016).

The SSF has been used in many applications, including enzyme, antibiotic, and bioactive compounds production (Lizardi-Jiménez and Hernández-Martínez 2017). It is considered a clean technology that can be used to produce and extract compounds with biological activities from natural sources. For this purpose, factors as micro-organism strains and solid support must be considered. Fungi, yeasts, and bacteria could be used in two processes; however, considering the specific conditions in this type of method (e.g., moisture content and nutritional availability), fungi and yeasts represent the most commonly used micro-organisms (Martins et al. 2011).

Pomegranate wastes were used as a substrate for ellagic acid production using a continuous system based on enzymes produced by SSF attached to polyurethane foam particles (Buenrostro-Figueroa et al. 2014). The reactions were performed in a bioreactor fed with pomegranate ellagitannins solution under specific conditions (0.1% ellagitannin w/v, 22.22 units/g dry solid, flow rate 0.27 mL/min at 60°C). Significant productivity of ellagic acid yield was obtained (1.09 g/L/h and 235.89 mg/g of pomegranate ellagitannins in the first 70 min of hydrolysis). The system demonstrated potential for hydrolysis of ellagitannin from pomegranate, considering the operational conditions and mechanical stability.

The same waste (pomegranate husks) was used to grow two *Aspergillus niger* strains to obtain ellagic acid using SSF. *A. niger* GH1 and PSH strains were responsible for yields of around 6.3 and 4.6 mg/g of dried husks, respectively. The degradation of the polyphenols in the samples was performed after 72 h of culture. According to the authors, the husks showed an excellent substrate for the

TABLE 5.2
Studies about Biovalorization of Polyphenols from Agri-food Wastes

Compounds	Agri-food Waste	Micro-organisms	Processes*	Use of Enzymes	References
Catechin, ellagic acid, quercetin	Orange peel	*Aspergillus fumigatus* MUM 1603	SmF	–	Sepúlveda et al. (2020)
Phenolic acids Anthocyanis Stilbens and isoflavones	Soybean hulls	*Aspergillus oryzae*	SSF	α-Amylase	Cabezudo et al. (2021)
Ellagic acid	Pomegranate seeds and husks	*Aspergillus niger* GH1	SSF	Ellagin tannin hydrolase	Robledo et al. (2008)
Gama-oryzanol	Rice bran	*Rhizopus oryzae*	SSF	–	Massarolo et al. (2017)
Total phenolic content	Fruit and vegetable	*Rhizopus* sp.	SSF and SmF	–	Ibarruri, Cebrián, and Hernández (2021)
Total polyphenol	Orange residue	*Diaporthe* sp.	SSF	–	Bier et al. (2019)
Protocatechuic acid and caffeic acid-3-glucoside	Poplar bark sawdust	*Ceriporiopsis subvermispora, Cantharellus cinereus,* and *Pleurotus ostreatus*	SSF	–	Xie et al. (2020)
Phenolic compounds, flavonoids	Grape, apple, and pineapple pomaces, pineapple peel	*Agaricus brasiliensis*	SSF and SmF	–	Mokochinski et al. (2015)
Total polyphenol	*Okara* (soybean byproduct)	*Lentinus edodes* *Ganoderma lucidum*	SSF	–	Yang, Fu, and Yang (2020)
Epigallocatechin-3-gallate, epicatechin-3-gallate, phelligridin G, davallialactone, and inoscavin B	Peanut shell	*Inonotus obliquus*	SmF	–	Xu, Hu, and Zhu (2014)

Hesperetin Naringenin	Citrus juice	–	Enzymatic extraction and biotransformation	Pectinase, cellulase, tannase, and β-glucosidase	Ruviaro, Barbosa, and Macedo (2019)
Hesperetin, naringenin, ellagic acid	Citrus residue	–	Enzymatic extraction and biotransformation	Cellulase, pectinase, tannase	Madeira and Macedo (2015)
Hydroxyphenylpropionic, hydroxyphenylacetic and hydroxybenzoic acids	Mango peel	Human gut microbiota	*In vitro* human colon model	Pepsin and pancreatin	Sáyago-Ayerdi et al. (2021)
Total phenolic, quercetin-3-glucoside	Plum pomace	*Aspergillus niger* *Rhizopus oligosporus*	SSF	–	Dulf, Vodnar, and Socaciu (2016)

Note

* SmF, submerged fermentation; SSF, solid-state fermentation.

production of polyphenol molecules; however, according to the authors, it was not possible to relate the role of tannin acyl hydrolases with ellagic acid production (Robledo et al. 2008).

A. niger was compared to *Rhizopus oligosporus* regarding the biovalorization of polyphenol content from plum pomaces generated by the juice industry (Dulf, Vodnar, and Socaciu 2016). After SSF processes, total phenolic amounts increased around 30% for *R. oligosporus* and 21% for *A. niger*. The antioxidant activity of the waste was improved; quercetin-3-glucoside was the primary phenolic compound in both fermented plum wastes.

Rice bran was used to cultivate *Rhizopus oryzae* and evaluate γ-oryzanol production and its antioxidant potential (Massarolo et al. 2017). After the SSF process, the compound was extracted using organic solvents and analyzed by chromatographic techniques. Considering the increase of γ-oryzanol after 48 h of cultivation and the high levels of oxidation inhibition (90.5%) related to its activity, the SSF was deemed efficient in improving the functional properties of the rice bran.

This filamentous fungus was also used for the valorization of fruit and vegetable wastes. SSF and SmF were performed to improve the nutritional and biological properties of the wastes in feeding application. *Rhizopus* sp. was responsible for increasing biomass, amino acids, and phenolic compounds content, besides changes in fatty acid profile. The phenolic compound concentration in the final product was doubled (up to 8.9 mg GAE/g), and the antioxidant activity was higher in the fermented substrate (DPPH reduction of 81.3 and TEAC of 3.6 mg/g) (Ibarruri, Cebrián, and Hernández 2021).

Soybean hull was used to extract polyphenols with antioxidant activity by *Aspergillus oryzae* fermentation using SSF (Cabezudo et al. 2021). An alkaline hydrolysis treatment optimized by response surface methodology was applied to soybean hulls. Polyphenols responsible for the antioxidant capacity were identified as phenolic acids, anthocyanins, stilbenes, and isoflavones. The fermentation by *A. oryzae* for 120 h was accountable for an increase of around 160% for polyphenols and 270% for the antioxidant activity compared to the control treatment. The authors related this improvement to amylase production by the filamentous fungi, resulting in a higher liberation of bound phenolics by degrading the starch fibers from the substrate evaluated.

The endophytic fungus *Diaporthe* sp. was responsible for a considerable increase of polyphenol content in orange waste (Bier et al. 2019). After an SSF process, its levels (271.33 mg GAE/g extract) were more than 8 times more significant than the nonfermented extract (36.39 mg/g). The polyphenol content was related to the high antioxidant activity evaluated using CUPRAC, ORAC, and DPPH methods. The fungus was considered an excellent option to improve the antioxidant potential of the orange wastes.

Rot fungi have also been used in fermentation processes for polyphenol extraction improvement. *Ceriporiopsis subvermispora, Cantharellus cinereus,* and *Pleurotus ostreatus* were utilized in a cofungal treatment of poplar bark sawdust polyphenol content sixfold in comparison to control treatment was observed after 12 days of fermentation (Xie et al. 2020). The authors identified chromatographic peaks corresponding to protocatechuic acid and caffeic acid-3-glucoside. The results were obtained due to degradation of lignin by the mix-fungi inoculum enzymes (laccase, lignin peroxidase, and manganase-dependent peroxidase, as well as β-glycosidases). The synergistic effect of the fungi species evaluated was highlighted as a promisor strategy to polyphenol enrichment in this type of waste.

Mushrooms also can be used in fermentation procedures involving agro-industrial wastes as substrate. Yang, Fu, and Yang (2020) evaluated the effects of the fermentation by *Lentinus edodes* and *Ganoderma lucidum* in the polyphenol content of *okara*, a soybean processing byproduct. The results showed that SSF improved the antioxidant activity more than threefold and the content of the bioactive compound. Flavonoids and phenolic acids were identified as significant polyphenol molecules. The total polyphenols in *okara* fermented by *G. lucidum* (12.0 mg/g) and *L. edodes* (7.0 mg/g) presented an increase more than ninefold in comparison with the substrate not fermented (0.8 mg/g). Furthermore, the experiments about biological effects in the animal model showed the potential of the waste in treating postmenopausal osteoporosis in humans.

Other agri-food wastes (grape, apple and, pineapple pomaces, pineapple peel) were used as substrates for SSF and SmF by *Agaricus bisporus*, a medicinal mushroom (Mokochinski et al. 2015). The myceliated substrate showed phenolic contents ranging from 18.57 to 70.46 mg/g and flavonoids from 0.83 to 4.51 mg/g for SSF and ranging from 27.19 to 66.99 mg/g of phenolics and 0.75 to 5.34 mg/g of flavonoids for SmF. The mushroom cultivated on biological wastes was considered an essential source of nutraceutical compounds for incorporation in food supplements.

Inonotus obliquus, a medicinal mushroom, was used in an SmF process to investigate the enhancement of antioxidant activity and polyphenol production of peanut shell biomass (Xu, Hu, and Zhu 2014). The study demonstrated many active polyphenols (epigallocatechin-3-gallate, epicatechin-3-gallate, phelligridin G, davallialactone, and inoscavin B) after lignocellulose degradation after 12 days of fermentation. According to the authors, the peanut shell can be considered a cost-effective source of compounds with biological activities; the improvement in antioxidant activity and polyphenol content was related to the fermentation processes by *I. obliquus* because of its capacity to degrade lignocellulose in the substrate. However, studies about enzyme activity and production are necessary to explain better the mechanisms involved in these results.

Sepúlveda et al. (2020) evaluated the production of molecules of polyphenolic origin using orange wastes as substrates to fermentation by *Aspergillus fumigatus* MUM 1603 in an SmF system. The results showed a high concentration of polyphenols (28% condensed, 27% ellagitannins, 25% flavonoids, and 20% gallotannins). Catechin was the primary polyphenol compound, followed by ellagic acid and quercetin. The authors appointed the best conditions for the maximum production of ellagic acid: temperature of 30°C, inoculum around 2×10^7 *A. fumigatus* spores/g, and orange-peel polyphenols of 6.2 g/L. The fermentation process was considered an effective method for the biotransformation of wastes generating polyphenol molecules with antioxidant and antibacterial agents.

Madeira and Macedo (2015) combined extraction (cellulase and pectinase) and biotransformation (tannase) enzymes to extract polyphenols from Brazilian citrus residues. The best results for hesperetin (120 ug/g), naringenin (80 ug/g), and ellagic acid (1,125 ug/g) production were obtained using 5.0 U/mL of cellulase and 7.0 U/mL of tannase at 40°C and 200 rpm. The authors concluded that there was a significant increase of 77% in the total antioxidant capacity of the citrus residue extracts, which represents an exciting alternative to be considered a green technology since the aims were achieved without the employ of organic solvents.

A similar experiment using an enzyme-assisted extraction combined with the biotransformation process was used to recover polyphenols from citrus juice bioproducts (Ruviaro, Barbosa, and Macedo 2019). The authors used pectinase, cellulase, tannase, and β-glucosidase alone or combined to evaluate the recovery of polyphenols. A better extraction yield was observed after the enzyme application, resulting in higher antioxidant activity. Best results were reached after a 24 h reaction with β-glucosidase (20 U/g), yielding high amounts of hesperetin and naringenin. The citrus bioproduct was considered a potential natural source of polyphenol compounds for food and pharmaceutical products.

Considering the use of modified micro-organisms for the biovalorization of polyphenols from vegetal material, Zeng et al. (2019) demonstrated the enzymatic production of theaflavins, bioactive molecules found in black tea, using recombinant polyphenol oxidases (PPOs) and tea polyphenols as substrate. Genes for the PPO production were cloned from nine plant species and expressed in *E. coli*. The higher enzymatic activities were observed for genes extracted from apple, pear, and loquat. The authors demonstrated that the PPO immobilization on mesoporous silica improves the activity and the theaflavin production. Even though the authors have used tea leaves as a substrate, this method could be applied to tea wastes that contain a considerable number of bioactive molecules (Sanz et al. 2020; Sui et al. 2019).

It is essential to highlight novel studies about the bioconversion of polyphenols in agro-industrial wastes by the human microbiota. Sáyago-Ayerdi et al. (2021) studied the improvement of polyphenol content in predigested mango peel using an *in vitro* human colon model with fecal microbial inoculum. After enzymatic treatment, the mango wastes were fermented for 72 h under specific conditions. Derivatives of hydroxyphenylpropionic, hydroxyphenylacetic, and hydroxybenzoic acids were identified by the HPLC-QToF method. The results suggest that these compounds could be generated after the biotransformation of flavonoids, gallate, and gallotannin found in the waste evaluated. This approach could indicate a direction to further studies to understand the role of human microbiota on the nutritional improvement of potential ingredients produced from agri-food wastes.

5.5 CONCLUDING REMARKS

In this chapter, the most recent results of scientific studies about the biovalorization of polyphenols from agro-industrial wastes were summarized. Green procedures have replaced traditional methods involving chemical solvents to avoid environmental damages and add value to wastes generated by the agri-food industry. In this context, besides the novel methodologies involving supercritical CO_2, ultrasound-assisted and pressurized hot water extraction, the use of enzymes and microbial fermentation represent an important alternative. Several studies have demonstrated the importance of bacteria and fungi in SSF and SmF processes on fermentation of different agro-industrial wastes; even if traditional filamentous fungi are considered the most important group of micro-organisms used for this purpose, such as *Aspergillus* and *Rhizopus*, endophytic and rot fungi as well as mushrooms have demonstrated a significant potential of use as inoculum to improve the extraction of polyphenols from wastes. A different approach related to the biotransformation of polyphenols by human gut microbiota could indicate a future direction for the studies in the area. Novel research focused on the chemical changes during the fermentation process. Its impacts on the polyphenol content are necessary, mainly considering the costs involved and the application potential on an industrial scale.

REFERENCES

Abbas, Munawar, Farhan Saeed, Faqir Muhammad Anjum, Muhammad Afzaal, Tabussam Tufail, Muhammad Shakeel Bashir, Adnan Ishtiaq, Shahzad Hussain, and Hafiz Ansar Rasul Suleria. 2017. "Natural Polyphenols: An Overview". *International Journal of Food Properties* 20 (8): 1689–1699. 10.1080/10942912.2016.1220393

Aires, Alfredo, Rosa Carvalho, and Maria José Saavedra. 2016. "Valorization of Solid Wastes from Chestnut Industry Processing: Extraction and Optimization of Polyphenols, Tannins and Ellagitannins and Its Potential for Adhesives, Cosmetic and Pharmaceutical Industry". *Waste Management* 48 (February): 457–464. 10.1016/j.wasman.2015.11.019

Álvarez-Cao, María-Efigenia, Manuel Becerra, and María-Isabel González-Siso. 2020. "Chapter 8 - Biovalorization of Cheese Whey and Molasses Wastes to Galactosidases by Recombinant Yeasts". In *Biovalorisation of Wastes to Renewable Chemicals and Biofuels*, edited by Navanietha Krishnaraj Rathinam and Rajesh K. Sani, 149–161. Elsevier. 10.1016/B978-0-12-817951-2.00008-0

Amalraj, Augustine, Anitha Pius, Sreerag Gopi, and Sreeraj Gopi. 2017. "Biological Activities of Curcuminoids, Other Biomolecules from Turmeric and Their Derivatives – A Review". *Journal of Traditional and Complementary Medicine* 7 (2): 205–233. 10.1016/j.jtcme.2016.05.005

Araújo, Fábio Fernandes de, David de Paulo Farias, Iramaia Angélica Neri-Numa, and Glaucia Maria Pastore. 2021. "Polyphenols and Their Applications: An Approach in Food Chemistry and Innovation Potential". *Food Chemistry* 338 (February): 127535. 10.1016/j.foodchem.2020.127535

Azabou, Samia, Fadia Ben Taheur, Mourad Jridi, Mohamed Bouaziz, and Moncef Nasri. 2017. "Discarded Seeds from Red Pepper (*Capsicum annum*) Processing Industry as a Sustainable Source of High Added-Value Compounds and Edible Oil". *Environmental Science and Pollution Research* 24 (28): 22196–22203. 10.1007/s11356-017-9857-9

Ben-Othman, Sana, Ivi Jõudu, and Rajeev Bhat. 2020. "Bioactives from Agri-Food Wastes: Present Insights and Future Challenges". *Molecules* 25 (3). 10.3390/molecules25030510

Bhanja Dey, Tapati, Subhojit Chakraborty, Kavish Kr. Jain, Abha Sharma, and Ramesh Chander Kuhad. 2016. "Antioxidant Phenolics and Their Microbial Production by Submerged and Solid State Fermentation Process: A Review". *Trends in Food Science & Technology* 53 (July): 60–74. 10.1016/j.tifs.2016.04.007

Bibra, Mohit, Navanietha K. Rathinam, Glenn R. Johnson, and Rajesh K. Sani. 2020. "Single Pot Biovalorization of Food Waste to Ethanol by *Geobacillus* and *Thermoanaerobacter* spp." *Renewable Energy* 155 (August): 1032–1041. 10.1016/j.renene.2020.02.093

Bier, Mário César Jucoski, Adriane Bianchi Pedroni Medeiros, Norbert De Kimpe, and Carlos Ricardo Soccol. 2019. "Evaluation of Antioxidant Activity of the Fermented Product from the Biotransformation of R-(+)-Limonene in Solid-State Fermentation of Orange Waste by *Diaporthe* Sp." *Biotechnology Research and Innovation* 3 (1): 168–176. 10.1016/j.biori.2019.01.002

Böttger, Angelika, Ute Vothknecht, Cordelia Bolle, and Alexander Wolf. 2018. "Plant Secondary Metabolites and Their General Function in Plants". In *Lessons on Caffeine, Cannabis & Co: Plant-Derived Drugs and Their Interaction with Human Receptors*, edited by Angelika Böttger, Ute Vothknecht, Cordelia Bolle, and Alexander Wolf, 3–17. Learning Materials in Biosciences. Cham: Springer International Publishing. 10.1007/978-3-319-99546-5_1

Boukroufa, Meryem, Chahrazed Boutekedjiret, Loïc Petigny, Njara Rakotomanomana, and Farid Chemat. 2015. "Bio-Refinery of Orange Peels Waste: A New Concept Based on Integrated Green and Solvent Free Extraction Processes Using Ultrasound and Microwave Techniques to Obtain Essential Oil, Polyphenols and Pectin". *Ultrasonics Sonochemistry* 24 (May): 72–79. 10.1016/j.ultsonch.2014.11.015

Buenrostro-Figueroa, Juan, Sergio Huerta-Ochoa, Arely Prado-Barragán, Juan Ascacio-Valdés, Leonardo Sepúlveda, Raúl Rodríguez, Antonio Aguilera-Carbó, and Cristóbal N. Aguilar. 2014. "Continuous Production of Ellagic Acid in a Packed-Bed Reactor". *Process Biochemistry* 49 (10): 1595–1600. 10.1016/j.procbio.2014.06.005

Cabezudo, Ignacio, María-Rocío Meini, Carla C. Di Ponte, Natasha Melnichuk, Carlos E. Boschetti, and Diana Romanini. 2021. "Soybean (Glycine Max) Hull Valorization through the Extraction of Polyphenols by Green Alternative Methods". *Food Chemistry* 338 (February): 128131. 10.1016/j.foodchem.2020.128131

Cheynier, Véronique. 2012. "Phenolic Compounds: From Plants to Foods". *Phytochemistry Reviews* 11 (2): 153–177. 10.1007/s11101-012-9242-8

Chong, Julie, Anne Poutaraud, and Philippe Hugueney. 2009. "Metabolism and Roles of Stilbenes in Plants". *Plant Science* 177 (3): 143–155. 10.1016/j.plantsci.2009.05.012

Da Porto, Carla, and Andrea Natolino. 2017. "Supercritical Fluid Extraction of Polyphenols from Grape Seed (*Vitis vinifera*): Study on Process Variables and Kinetics". *The Journal of Supercritical Fluids* 130 (December): 239–245. 10.1016/j.supflu.2017.02.013

Das, Atanu Kumar, Md. Nazrul Islam, Md. Omar Faruk, Md. Ashaduzzaman, and Rudi Dungani. 2020. "Review on Tannins: Extraction Processes, Applications and Possibilities". *South African Journal of Botany* 135 (Decemeber): 58–70. 10.1016/j.sajb.2020.08.008

Drosou, Christina, Konstantina Kyriakopoulou, Andreas Bimpilas, Dimitrios Tsimogiannis, and Magdalini Krokida. 2015. "A Comparative Study on Different Extraction Techniques to Recover Red Grape Pomace Polyphenols from Vinification Byproducts". *Industrial Crops and Products, Advances in Industrial Crops and Products Worldwide: AAIC 2014 International Conference*, 75 (November): 141–149. 10.1016/j.indcrop.2015.05.063

Dulf, Francisc Vasile, Dan Cristian Vodnar, and Carmen Socaciu. 2016. "Effects of Solid-State Fermentation with Two Filamentous Fungi on the Total Phenolic Contents, Flavonoids, Antioxidant Activities and Lipid Fractions of Plum Fruit (*Prunus domestica* L.) by-Products". *Food Chemistry* 209 (October): 27–36. 10.1016/j.foodchem.2016.04.016

Dzah, Courage Sedem, Yuqing Duan, Haihui Zhang, Nana Adwoa Serwah Boateng, and Haile Ma. 2020. "Latest Developments in Polyphenol Recovery and Purification from Plant By-Products: A Review". *Trends in Food Science & Technology* 99 (May): 375–388. 10.1016/j.tifs.2020.03.003

El Kantar, Sally, Nadia Boussetta, Nikolai Lebovka, Felix Foucart, Hiba N. Rajha, Richard G. Maroun, Nicolas Louka, and Eugene Vorobiev. 2018. "Pulsed Electric Field Treatment of Citrus Fruits: Improvement of Juice and Polyphenols Extraction". *Innovative Food Science & Emerging Technologies, Food Science and Technology in France: INRA's Contribution to This Area*, 46 (April): 153–161. 10.1016/j.ifset.2017.09.024

El Khawand, Toni, Arnaud Courtois, Josep Valls, Tristan Richard, and Stéphanie Krisa. 2018. "A Review of Dietary Stilbenes: Sources and Bioavailability". *Phytochemistry Reviews* 17 (5): 1007–1029. 10.1007/s11101-018-9578-9

El-Najjar, Nahed, Hala Gali-Muhtasib, Raimo A. Ketola, Pia Vuorela, Arto Urtti, and Heikki Vuorela. 2011. "The Chemical and Biological Activities of Quinones: Overview and Implications in Analytical Detection". *Phytochemistry Reviews* 10 (3): 353. 10.1007/s11101-011-9209-1

Erukainure, Ochuko L., Olakunle Sanni, and Md. Shahidul Islam. 2018. "Chapter 6 – Clerodendrum Volubile: Phenolics and Applications to Health". In *Polyphenols: Mechanisms of Action in Human Health and Disease (Second Edition)*, edited by Ronald Ross Watson, Victor R. Preedy, and Sherma Zibadi, 53–68. Academic Press. 10.1016/B978-0-12-813006-3.00006-4

Escobar-Avello, Danilo, Julián Lozano-Castellón, Claudia Mardones, Andy J. Pérez, Vania Saéz, Sebastián Riquelme, Dietrich von Baer, and Anna Vallverdú-Queralt. 2019. "Phenolic Profile of Grape Canes: Novel Compounds Identified by LC-ESI-LTQ-Orbitrap-MS". *Molecules* 24 (20). 10.3390/molecules24203763

Fernández, María de Los Ángeles, Magdalena Espino, Federico J. V. Gomez, and María Fernanda Silva. 2018. "Novel Approaches Mediated by Tailor-Made Green Solvents for the Extraction of Phenolic Compounds from Agro-Food Industrial by-Products". *Food Chemistry* 239 (January): 671–678. 10.1016/j.foodchem.2017.06.150

Gan, Ren-You, Chak-Lun Chan, Qiong-Qiong Yang, Hua-Bin Li, Dan Zhang, Ying-Ying Ge, Anil Gunaratne, Jiao Ge, and Harold Corke. 2019. "9 - Bioactive Compounds and Beneficial Functions of Sprouted Grains". In *Sprouted Grains*, edited by Hao Feng, Boris Nemzer, and Jonathan W. DeVries, 191–246. AACC International Press. 10.1016/B978-0-12-811525-1.00009-9

Gharras, Hasna El. 2009. "Polyphenols: Food Sources, Properties and Applications – A Review". *International Journal of Food Science & Technology* 44 (12): 2512–2518. 10.1111/j.1365-2621.2009.02077.x

Gómez-Mejía, Esther, Noelia Rosales-Conrado, María Eugenia León-González, and Yolanda Madrid. 2019. "Citrus Peels Waste as a Source of Value-Added Compounds: Extraction and Quantification of Bioactive Polyphenols". *Food Chemistry* 295 (October): 289–299. 10.1016/j.foodchem.2019.05.136

Gonzales, Gerard Bryan, Katleen Raes, Hanne Vanhoutte, Sofie Coelus, Guy Smagghe, and John Van Camp. 2015. "Liquid Chromatography–Mass Spectrometry Coupled with Multivariate Analysis for the Characterization and Discrimination of Extractable and Nonextractable Polyphenols and Glucosinolates from Red Cabbage and Brussels Sprout Waste Streams". *Journal of Chromatography A* 1402 (July): 60–70. 10.1016/j.chroma.2015.05.009

Gonzalez-Rivera, José, Celia Duce, Beatrice Campanella, Luca Bernazzani, Carlo Ferrari, Eleonora Tanzini, Massimo Onor, et al. 2021. "In Situ Microwave Assisted Extraction of Clove Buds to Isolate Essential Oil, Polyphenols, and Lignocellulosic Compounds". *Industrial Crops and Products* 161 (March): 113203. 10.1016/j.indcrop.2020.113203

Hewlings, Susan J., and Douglas S. Kalman. 2017. "Curcumin: A Review of Its' Effects on Human Health". *Foods* 6 (10). 10.3390/foods6100092

Hussain, M. I., S. Qamar Abbas, M. J. Reigosa, M. I. Hussain, S. Qamar Abbas, and M. J. Reigosa. 2018. "Activities and Novel Applications of Secondary Metabolite Coumarins". *Planta Daninha* 36. 10.1590/s0100-83582018360100016

Ibarruri, Jone, Marta Cebrián, and Igor Hernández. 2021. "Valorisation of Fruit and Vegetable Discards by Fungal Submerged and Solid-State Fermentation for Alternative Feed Ingredients Production". *Journal of Environmental Management* 281 (March): 111901. 10.1016/j.jenvman.2020.111901

Katsampa, Photene, Evdokea Valsamedou, Spyros Grigorakis, and Dimitris P. Makris. 2015. "A Green Ultrasound-Assisted Extraction Process for the Recovery of Antioxidant Polyphenols and Pigments from Onion Solid Wastes Using Box–Behnken Experimental Design and Kinetics". *Industrial Crops and Products* 77 (December): 535–543. 10.1016/j.indcrop.2015.09.039

Khodaiyan, Faramarz, and Karim Parastouei. 2020. "Co-Optimization of Pectin and Polyphenols Extraction from Black Mulberry Pomace Using an Eco-Friendly Technique: Simultaneous Recovery and Characterization of Products". *International Journal of Biological Macromolecules* 164 (December): 1025–1036. 10.1016/j.ijbiomac.2020.07.107

Khosravi, Azin, and Seyed Hadi Razavi. 2020. "The Role of Bioconversion Processes to Enhance Bioaccessibility of Polyphenols in Rice". *Food Bioscience* 35 (June): 100605. 10.1016/j.fbio.2020.100605

Kishikawa, Naoya, and Naotaka Kuroda. 2014. "Analytical Techniques for the Determination of Biologically Active Quinones in Biological and Environmental Samples". *Journal of Pharmaceutical and Biomedical Analysis*, Review Papers on Pharmaceutical and Biomedical Analysis 2013, 87 (January): 261–270. 10.1016/j.jpba.2013.05.035

Le Bourvellec, C., and C. M. G. C. Renard. 2012. "Interactions between Polyphenols and Macromolecules: Quantification Methods and Mechanisms". *Critical Reviews in Food Science and Nutrition* 52 (3): 213–248. 10.1080/10408398.2010.499808

Lizardi-Jiménez, M. A., and R. Hernández-Martínez. 2017. "Solid State Fermentation (SSF): Diversity of Applications to Valorize Waste and Biomass". *3 Biotech* 7 (1): 44. 10.1007/s13205-017-0692-y

Madeira, Jose Valdo, and Gabriela Alves Macedo. 2015. "Simultaneous Extraction and Biotransformation Process to Obtain High Bioactivity Phenolic Compounds from Brazilian Citrus Residues". *Biotechnology Progress* 31 (5): 1273–1279. 10.1002/btpr.2126

Makris, Dimitris P. 2018. "Green Extraction Processes for the Efficient Recovery of Bioactive Polyphenols from Wine Industry Solid Wastes – Recent Progress". *Current Opinion in Green and Sustainable Chemistry*, Reuse and Recycling/UN SGDs: How can Sustainable Chemistry Contribute?/Green Chemistry in Education, 13 (October): 50–55. 10.1016/j.cogsc.2018.03.013

Manasa, Vallamkondu, Aparna Padmanabhan, and K. A. Anu Appaiah. 2021. "Utilization of Coffee Pulp Waste for Rapid Recovery of Pectin and Polyphenols for Sustainable Material Recycle". *Waste Management* 120 (February): 762–771. 10.1016/j.wasman.2020.10.045

Manna, Luigi, Cristiano Agostino Bugnone, and Mauro Banchero. 2015. "Valorization of Hazelnut, Coffee and Grape Wastes through Supercritical Fluid Extraction of Triglycerides and Polyphenols". *The Journal of Supercritical Fluids* 104 (September): 204–211. 10.1016/j.supflu.2015.06.012

Martin, José Guilherme Prado, Ernani Porto, Severino Matias Alencar, Eduardo Micotti Gloria, Ingridy Simone Ribeiro Cabral, and Ligia Maria Aquino. 2012. "Antimicrobial Potential and Chemical Composition of Agro-Industrial Wastes." *Journal of Natural Products* 5: 27–36.

Martins, Silvia, Solange I. Mussatto, Guillermo Martínez-Avila, Julio Montañez-Saenz, Cristóbal N. Aguilar, and Jose A. Teixeira. 2011. "Bioactive Phenolic Compounds: Production and Extraction by Solid-State Fermentation. A Review". *Biotechnology Advances* 29 (3): 365–373. 10.1016/j.biotechadv.2011.01.008

Massarolo, Kelly Cristina, Taiana Denardi de Souza, Carolina Carvalho Collazzo, Eliana Badiale Furlong, and Leonor Almeida de Souza Soares. 2017. "The Impact of Rhizopus Oryzae Cultivation on Rice Bran: Gamma-Oryzanol Recovery and Its Antioxidant Properties". *Food Chemistry* 228 (August): 43–49. 10.1016/j.foodchem.2017.01.127

Matos, Maria João, Lourdes Santana, Eugenio Uriarte, Orlando A. Abreu, Enrique Molina, and Estela Guardado Yordi. 2015. "Coumarins — An Important Class of Phytochemicals". *Phytochemicals – Isolation, Characterisation and Role in Human Health*, September. 10.5772/59982

Michailof, Chrysa, Panagiotis Manesiotis, and Costas Panayiotou. 2008. "Synthesis of Caffeic Acid and P-Hydroxybenzoic Acid Molecularly Imprinted Polymers and Their Application for the Selective Extraction of Polyphenols from Olive Mill Waste Waters". *Journal of Chromatography A* 1182 (1): 25–33. 10.1016/j.chroma.2008.01.001

Mirab, Bijan, Hassan Ahmadi Gavlighi, Roghayeh Amini Sarteshnizi, Mohammad Hossein Azizi, and Chibuike C. Udenigwe. 2020. "Production of Low Glycemic Potential Sponge Cake by Pomegranate Peel Extract (PPE) as Natural Enriched Polyphenol Extract: Textural, Color and Consumer Acceptability". *LWT* 134 (December): 109973. 10.1016/j.lwt.2020.109973

Mokochinski, João B., Begoña G. C. López, Vanessa Sovrani, Herta S. Dalla Santa, Pedro Pablo González-Borrero, Alexandra Christine Helena F. Sawaya, Eduardo M. Schmidt, Marcos N. Eberlin, and Yohandra R. Torres. 2015. "Production of Agaricus *Brasiliensis mycelium* from Food Industry Residues as a Source of Antioxidants and Essential Fatty Acids". *International Journal of Food Science & Technology* 50 (9): 2052–2058. 10.1111/ijfs.12861

Nayak, A., and Brij Bhushan. 2019. "An Overview of the Recent Trends on the Waste Valorization Techniques for Food Wastes". *Journal of Environmental Management* 233 (March): 352–370. 10.1016/j.jenvman.2018.12.041

Ninčević Grassino, Antonela, Jelena Ostojić, Vicenzia Miletić, Senka Djaković, Tomislav Bosiljkov, Zoran Zorić, Damir Ježek, Suzana Rimac Brnčić, and Mladen Brnčić. 2020. "Application of High Hydrostatic Pressure and Ultrasound-Assisted Extractions as a Novel Approach for Pectin and Polyphenols Recovery from Tomato Peel Waste". *Innovative Food Science & Emerging Technologies* 64 (August): 102424. 10.1016/j.ifset.2020.102424

Ouellette, Robert J., and J. David Rawn. 2018. "25 - Aryl Halides, Phenols, and Anilines". In *Organic Chemistry (Second Edition)*, edited by Robert J. Ouellette and J. David Rawn, 801–828. Academic Press. 10.1016/B978-0-12-812838-1.50025-6

Peterson, Julia, Johanna Dwyer, Herman Adlercreutz, Augustin Scalbert, Paul Jacques, and Marjorie L. McCullough. 2010. "Dietary Lignans: Physiology and Potential for Cardiovascular Disease Risk Reduction". *Nutrition Reviews* 68 (10): 571–603. 10.1111/j.1753-4887.2010.00319.x

Pimentel-Moral, Sandra, María de la Luz Cádiz-Gurrea, Celia Rodríguez-Pérez, and Antonio Segura-Carretero. 2020. "7 - Recent Advances in Extraction Technologies of Phytochemicals Applied for the Revaluation of Agri-Food by-Products". In *Functional and Preservative Properties of Phytochemicals*, edited by Bhanu Prakash, 209–239. Academic Press. 10.1016/B978-0-12-818593-3.00007-5

Plazzotta, S., and L. Manzocco. 2019. "High-Pressure Homogenisation Combined with Blanching to Turn Lettuce Waste into a Physically Stable Juice". *Innovative Food Science & Emerging Technologies* 52 (March): 136–144. 10.1016/j.ifset.2018.11.008

Plazzotta, Stella, and Lara Manzocco. 2018. "Effect of Ultrasounds and High Pressure Homogenization on the Extraction of Antioxidant Polyphenols from Lettuce Waste". *Innovative Food Science & Emerging Technologies* 50 (December): 11–19. 10.1016/j.ifset.2018.10.004

Portilla Rivera, Oscar Manuel, María Dolores Saavedra Leos, Vicente Espinosa Solis, and José Manuel Domínguez. 2021. "Recent Trends on the Valorization of Winemaking Industry Wastes". *Current Opinion in Green and Sustainable Chemistry* 27 (February): 100415. 10.1016/j.cogsc.2020.100415

Priyadarsini, Kavirayani Indira. 2014. "The Chemistry of Curcumin: From Extraction to Therapeutic Agent". *Molecules (Basel, Switzerland)* 19 (12): 20091–20112. 10.3390/molecules191220091

Punzi, Rossana, Annalisa Paradiso, Cristina Fasciano, Antonio Trani, Michele Faccia, Maria Concetta de Pinto, and Giuseppe Gambacorta. 2014. "Phenols and Antioxidant Activity in Vitro and in Vivo of Aqueous Extracts Obtained by Ultrasound-Assisted Extraction from Artichoke by-Products". *Natural Product Communications* 9 (9): 1315–1318.

Putnik, Predrag, Jose M. Lorenzo, Francisco J. Barba, Shahin Roohinejad, Anet Režek Jambrak, Daniel Granato, Domenico Montesano, and Danijela Bursać Kovačević. 2018. "Novel Food Processing and Extraction Technologies of High-Added Value Compounds from Plant Materials". *Foods (Basel, Switzerland)* 7 (7). 10.3390/foods7070106

Rajha, Hiba N., Anna-Maria Abi-Khattar, Sally El Kantar, Nadia Boussetta, Nikolai Lebovka, Richard G. Maroun, Nicolas Louka, and Eugene Vorobiev. 2019. "Comparison of Aqueous Extraction Efficiency and Biological Activities of Polyphenols from Pomegranate Peels Assisted by Infrared, Ultrasound, Pulsed Electric Fields and High-Voltage Electrical Discharges". *Innovative Food Science & Emerging Technologies* 58 (December): 102212. 10.1016/j.ifset.2019.102212

Ramón-Gonçalves, Marina, Esther Gómez-Mejía, Noelia Rosales-Conrado, María Eugenia León-González, and Yolanda Madrid. 2019. "Extraction, Identification and Quantification of Polyphenols from Spent Coffee Grounds by Chromatographic Methods and Chemometric Analyses". *Waste Management* 96 (August): 15–24. 10.1 016/j.wasman.2019.07.009

Rani, Jyoti, Indrajeet, Akhil Rautela, and Sanjay Kumar. 2020. "Chapter 4 - Biovalorization of Winery Industry Waste to Produce Value-Added Products". Em *Biovalorisation of Wastes to Renewable Chemicals and Biofuels*, edited by Navanietha Krishnaraj Rathinam and Rajesh K. Sani, 63–85. Elsevier. 10.1016/B978-0-12-817951-2.00004-3

Reinisalo, Mika, Anna Kårlund, Ali Koskela, Kai Kaarniranta, and Reijo O. Karjalainen. 2015. "Polyphenol Stilbenes: Molecular Mechanisms of Defence against Oxidative Stress and Aging-Related Diseases". *Oxidative Medicine and Cellular Longevity* 2015: 340520. 10.1155/2015/340520

Robledo, Armando, Antonio Aguilera-Carbó, Raúl Rodriguez, José Luis Martinez, Yolanda Garza, and Cristobal N. Aguilar. 2008. "Ellagic Acid Production by *Aspergillus niger* in Solid State Fermentation of Pomegranate Residues". *Journal of Industrial Microbiology & Biotechnology* 35 (6): 507–513. 10.1007/s10295-008-0309-x

Rodríguez-García, Carmen, Cristina Sánchez-Quesada, Estefanía Toledo, Miguel Delgado-Rodríguez, and José J. Gaforio. 2019. "Naturally Lignan-Rich Foods: A Dietary Tool for Health Promotion?" *Molecules* 24 (5). 10.3390/molecules24050917

Ruviaro, Amanda Roggia, Paula de Paula Menezes Barbosa, and Gabriela Alves Macedo. 2019. "Enzyme-Assisted Biotransformation Increases Hesperetin Content in Citrus Juice by-Products". *Food Research International*, *SLACA 2017: Food Science and Its Role in a Changing World*, 124 (October): 213–221. 10.1016/j.foodres.2018.05.004

Sánchez-Gómez, Rosario, Amaya Zalacain, Gonzalo L. Alonso, and M. Rosario Salinas. 2014. "Vine-Shoot Waste Aqueous Extracts for Re-Use in Agriculture Obtained by Different Extraction Techniques: Phenolic, Volatile, and Mineral Compounds". *Journal of Agricultural and Food Chemistry* 62 (45): 10861–10872. 10.1021/jf503929v

Sanz, Vanesa, Noelia Flórez-Fernández, Herminia Domínguez, and María Dolores Torres. 2020. "Clean Technologies Applied to the Recovery of Bioactive Extracts from *Camellia sinensis* Leaves Agricultural Wastes". *Food and Bioproducts Processing* 122 (July): 214–221. 10.1016/j.fbp.2020.05.007

Sáyago-Ayerdi, Sonia G., Koen Venema, Maria Tabernero, Beatriz Sarriá, L. Laura Bravo, and Raquel Mateos. 2021. "Bioconversion by Gut Microbiota of Predigested Mango (*Mangifera indica* L.) 'Ataulfo' Peel Polyphenols Assessed in a Dynamic (TIM-2) in Vitro Model of the Human Colon". *Food Research International* 139 (January): 109963. 10.1016/j.foodres.2020.109963

Selvakumar, P., V. Karthik, P. Senthil Kumar, P. Asaithambi, S. Kavitha, and P. Sivashanmugam. 2021. "Enhancement of Ultrasound Assisted Aqueous Extraction of Polyphenols from Waste Fruit Peel Using Dimethyl Sulfoxide as Surfactant: Assessment of Kinetic Models". *Chemosphere* 263 (January): 128071. 10.1016/j.chemosphere.2020.128071

Sepúlveda, Leonardo, Elan Laredo-Alcalá, José Juan Buenrostro-Figueroa, Juan Alberto Ascacio-Valdés, Zlatina Genisheva, Cristobal Aguilar, and José Teixeira. 2020. "Ellagic Acid Production Using Polyphenols from Orange Peel Waste by Submerged Fermentation". *Electronic Journal of Biotechnology* 43 (January): 1–7. 10.1016/j.ejbt.2019.11.002

Shirahigue, Ligianne Din, Sandra Regina Ceccato-Antonini, Ligianne Din Shirahigue, and Sandra Regina Ceccato-Antonini. 2020. "Agro-Industrial Wastes as Sources of Bioactive Compounds for Food and Fermentation Industries". *Ciência Rural* 50 (4). 10.1590/0103-8478cr20190857

Soleymani, Samaneh, Solomon Habtemariam, Roja Rahimi, and Seyed Mohammad Nabavi. 2020. "The What and Who of Dietary Lignans in Human Health: Special Focus on Prooxidant and Antioxidant Effects". *Trends in Food Science & Technology* 106 (December): 382–390. 10.1016/j.tifs.2020.10.015

Suárez, Manuel, Maria-Paz Romero, Tomás Ramo, Alba Macià, and Maria-José Motilva. 2009. "Methods for Preparing Phenolic Extracts from Olive Cake for Potential Application as Food Antioxidants". *Journal of Agricultural and Food Chemistry* 57 (4): 1463–1472. 10.1021/jf8032254

Sui, Wenjie, Ying Xiao, Rui Liu, Tao Wu, and Min Zhang. 2019. "Steam Explosion Modification on Tea Waste to Enhance Bioactive Compounds' Extractability and Antioxidant Capacity of Extracts". *Journal of Food Engineering* 261 (November): 51–59. 10.1016/j.jfoodeng.2019.03.015

Surendra, K. C., Jeffery K. Tomberlin, Arnold van Huis, Jonathan A. Cammack, Lars-Henrik L. Heckmann, and Samir Kumar Khanal. 2020. "Rethinking Organic Wastes Bioconversion: Evaluating the Potential of the Black Soldier Fly (*Hermetia illucens* (L.)) (Diptera: Stratiomyidae) (BSF)". *Waste Management* 117 (November): 58–80. 10.1016/j.wasman.2020.07.050

Suwal, Shyam, and Alice Marciniak. 2018. "Technologies for the Extraction, Separation and Purification of Polyphenols – A Review". *Nepal Journal of Biotechnology* 6 (1): 74–91. 10.3126/njb.v6i1.22341

Tresserra-Rimbau, Anna, Rosa M. Lamuela-Raventos, and Juan J. Moreno. 2018. "Polyphenols, Food and Pharma. Current Knowledge and Directions for Future Research". *Biochemical Pharmacology* 156 (October): 186–195. 10.1016/j.bcp.2018.07.050

Tsao, Rong. 2010. "Chemistry and Biochemistry of Dietary Polyphenols". *Nutrients* 2 (12): 1231–1246. 10.3390/nu2121231

Usmani, Zeba, Minaxi Sharma, Yevgen Karpichev, Ashok Pandey, Ramesh Chander Kuhad, Rajeev Bhat, Rajesh Punia, Mortaza Aghbashlo, Meisam Tabatabaei, and Vijai Kumar Gupta. 2020. "Advancement in Valorization Technologies to Improve Utilization of Bio-Based Waste in Bioeconomy Context". *Renewable and Sustainable Energy Reviews* 131 (October): 109965. 10.1016/j.rser.2020.109965

Velázquez-Estrada, R. M., M. M. Hernández-Herrero, C. E. Rüfer, B. Guamis-López, and A. X. Roig-Sagués. 2013. "Influence of Ultra High Pressure Homogenization Processing on Bioactive Compounds and Antioxidant Activity of Orange Juice". *Innovative Food Science & Emerging Technologies* 18 (April): 89–94. 10.1016/j.ifset.2013.02.005

Venugopala, K. N., V. Rashmi, and B. Odhav. 2013. "Review on Natural Coumarin Lead Compounds for Their Pharmacological Activity". *BioMed Research International* 2013: 963248. 10.1155/2013/963248

Vidović, Senka, Jelena Vladić, Nataša Nastić, and Stela Jokić. 2021. "2.47 – Subcritical and Supercritical Extraction in Food By-Product and Food Waste Valorization". In *Innovative Food Processing Technologies*, edited by Kai Knoerzer and Kasiviswanathan Muthukumarappan, 705–721. Oxford: Elsevier. 10.1016/B978-0-08-100596-5.23014-X

Xie, Pujun, Linlin Fan, Lixin Huang, and Caihong Zhang. 2020. "An Innovative Co-Fungal Treatment to Poplar Bark Sawdust for Delignification and Polyphenol Enrichment". *Industrial Crops and Products* 157 (December): 112896. 10.1016/j.indcrop.2020.112896

Xu, Xiang-qun, Yan Hu, and Ling-hui Zhu. 2014. "The Capability of *Inonotus obliquus* for Lignocellulosic Biomass Degradation in Peanut Shell and for Simultaneous Production of Bioactive Polysaccharides and Polyphenols in Submerged Fermentation". *Journal of the Taiwan Institute of Chemical Engineers* 45 (6): 2851–2858. 10.1016/j.jtice.2014.08.029

Yang, Li-Chan, Tzu-Jung Fu, and Fan-Chiang Yang. 2020. "Biovalorization of Soybean Residue (Okara) via Fermentation with *Ganoderma lucidum* and *Lentinus edodes* to Attain Products with High Anti-Osteoporotic Effects". *Journal of Bioscience and Bioengineering* 129 (4): 514–518. 10.1016/j.jbiosc.2019.10.003

Zagklis, Dimitris P., and Christakis A. Paraskeva. 2018. "Isolation of Organic Compounds with High Added Values from Agro-Industrial Solid Wastes". *Journal of Environmental Management* 216 (June): 183–191. 10.1016/j.jenvman.2017.04.083

Zeng, Jun, Gang Du, Xue Shao, Ke-Na Feng, and Ying Zeng. 2019. "Recombinant Polyphenol Oxidases for Production of Theaflavins from Tea Polyphenols". *International Journal of Biological Macromolecules* 134 (August): 139–145. 10.1016/j.ijbiomac.2019.04.142

Zhang, Guowen, Mingming Hu, Li He, Peng Fu, Lin Wang, and Jia Zhou. 2013. "Optimization of Microwave-Assisted Enzymatic Extraction of Polyphenols from Waste Peanut Shells and Evaluation of Its Antioxidant and Antibacterial Activities in Vitro". *Food and Bioproducts Processing* 91 (2): 158–168. 10.1016/j.fbp.2012.09.003

Zhu, Y. F., J. P. Wang, X. M. Ding, S. P. Bai, S. R. N. Qi, Q. F. Zeng, Y. Xuan, Z. W. Su, and K. Y. Zhang. 2020. "Effect of Different Tea Polyphenol Products on Egg Production Performance, Egg Quality and Antioxidative Status of Laying Hens". *Animal Feed Science and Technology* 267 (September): 114544. 10.1016/j.anifeedsci.2020.114544

6 Valorization of Food Waste via Fermentation: Ethanol and Biogas Production from Agriculture Food Waste

Hiba N. Abu Tayeh
Institute of Agricultural Engineering, ARO, Volcani Institute, Rishon LeZion, Israel

Institute of Applied Research, The Galilee Society, Shefa-Amr, Israel

Department of Natural Resources & Environmental Management, University of Haifa, Haifa, Israel

Manal Haj-Zaroubi
Institute of Applied Research, The Galilee Society, Shefa-Amr, Israel

Department of Natural Resources & Environmental Management, University of Haifa, Haifa, Israel

Hassan Azaizeh
Institute of Applied Research, The Galilee Society, Shefa-Amr, Israel

Department of Natural Resources & Environmental Management, University of Haifa, Haifa, Israel

Department of Environmental Science, Tel Hai College, Upper Galilee, Israel

CONTENTS

DOI: 10.1201/9781003128977-6

6.1 INTRODUCTION

A rapid increase in food production every year to support a growing global population plays a major role in generating food waste (FW). FW is a major issue faced by food producers, retailers, consumers, and society as a whole. The Food and Agriculture Organization (FAO) of the United Nations reported in 2015 that an estimated one-third of all food produced globally is lost or goes to waste, which is equivalent to 1.3 billion tons of FW annually (Gustavsson et al. 2011) with a cost of around 0.65 trillion USD. FW generated in Asian countries alone was found to be around 278 million tons in 2020 and is expected to increase to 416 million tons by 2025 (Paritosh et al. 2017; Melikoglu et al. 2013). Based on FUSION, 2016; around 88 million tons of FW are generated every year in the EU, with costs estimated at 143 billion euros. In the United States, FW is estimated between 30 and 40% of the food supply. This is based on the US Department of Agriculture (USDA) estimates that 31% of food is lost at the retail and consumer levels.

FW is generated from food processing industries and domestic sectors and is composed of protein, lipid, organic acids, and carbohydrate polymers such as starch, cellulose, hemicelluloses, and a small fraction of lignin and inorganic mineral components (Banu et al. 2019; Kavitha et al. 2019; Yukesh Kannah et al. 2020). Typical FW in several Asia-Pacific countries and around the globe are cereal (95,245 kilotons (KT)), rice (26,738 KT), sugar (459.9 KT), Oil crops (18,424 KT), vegetables (81,441 KT), milk (16,560 KT), potatoes (62,229 KT), fruits (53,796 KT), coffee (105.0 KT), animal fats (174.1 KT), meat (1,184 KT), offal (63.0 KT), banana (13,532 KT) and coconut (3,038 KT) (Kiran et al. 2014). FW can be categorized into (1) edible FW, which can be managed easily to minimize its production, and (2) nonedible FW, which needs a special management system and can be used for the production of value-added chemicals and fuels (Cho et al. 2018, 2019).

The majority of FW generated worldwide are currently managed by landfills, which means FW decomposes to form methane, a greenhouse gas with a global warming potential of 25 times greater than CO_2 on a 100-year time scale (Solomon et al. 2007). Furthermore, landfill management forms a highly contaminated leachate with dangerous and poisonous properties to the surrounding habitats (Abu Amr et al. 2013; Zhang et al. 2013). Another FW management is incineration with other

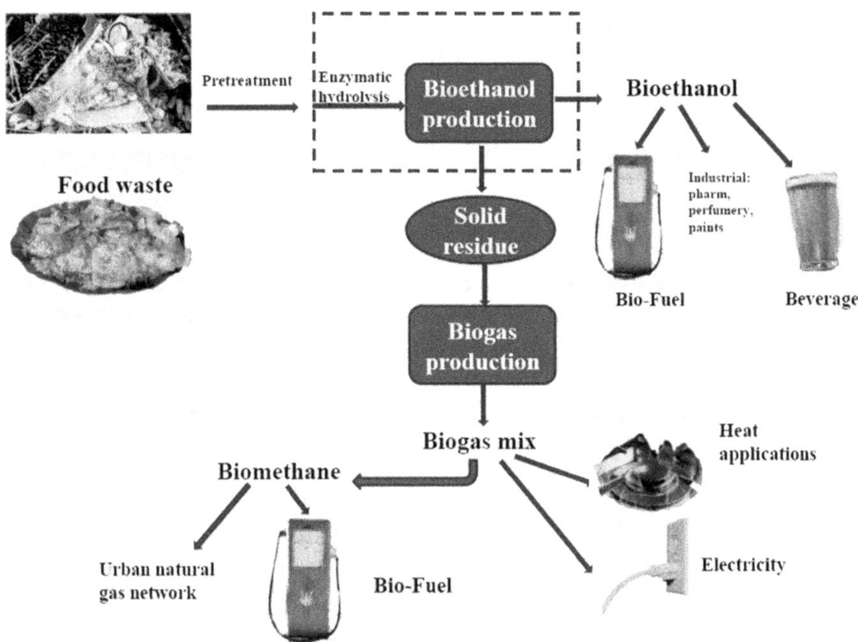

FIGURE 6.1 Stages of ethanol and biogas production from food wastes and their uses.

combustible municipal wastes for the generation of heat or energy. One should note that FW contains a high level of moisture, which may lead to the production of dioxins during its combustion and other wastes of low humidity and high calorific value (Katami et al. 2004). Furthermore, incineration of FW causes air pollution and loss of chemical values of FW. Therefore, different other methods have been studied for FW management such as composting, gasification, waste to energy, and recovery of value-added products. It turned out that waste to energy and recovery of value-added product from FW is the most profitable method of FW management.

FW has high organic content, approximately 60% carbohydrates in its total solids, although composition varies from source to source, where FW is recently intensely studied worldwide as a potential substrate for biogas production, liquid biofuel production, commodity chemicals, biohydrogen, and bioelectricity (Dahiya et al. 2018) via biological processes (fermentation). Beyond the environmental impact of treatment and disposal of the FW, the fermentation process for producing renewable resources from FW and other crop residues (i.e. FW to alcohol) also has great potential to enhance the economic impact (Kim et al. 2018). This review highlights the ethanol and biogas production from FW (Figure 6.1).

6.2 BIOETHANOL PRODUCTION

The term biofuel refers to solid, liquid, or gaseous fuels that are derived from biomass. Bioethanol is one of the liquid biofuels; it can be produced by the alcoholic fermentation of simple sugars or sucrose, which are derived from plants, from sugar crops such as sugar beet and sugar cane, starchy crops such as grain, potatoes, and corn, or cellulosic crops such as wood, lignocellulose such as agricultural residues and even household wastes such as FW. While a large amount of synthetic ethanol facility uses ethylene, a petroleum byproduct, in a simple one-step process of catalytic hydration process of ethylene to produce ethanol (Equation 6.1):

$$C_2H_4 + H_2O \rightarrow CH_3CH_2OH \tag{6.1}$$

Saccharomyces ceveresiae is a commercial yeast used for sugar fermentation. The conversion of sucrose to ethanol by the yeast *S. ceveresiae* is consisted of catalyzing the breakdown of sucrose into glucose and fructose by the sucrase/invertase enzyme present in the yeast (Equation 6.2), then converting the glucose and the fructose into ethanol and carbon dioxide by another enzyme in the yeast called zymase (Equation 6.3):

$$C_{12}H_{22}O_{11} + H_2O = C_6H_{12}O_6 + C_6H_{12}O_6 \tag{6.2}$$

$$C_6H_{12}O_6 \rightarrow 2C_2H_5OH + 2CO_2 \tag{6.3}$$

Starch feedstock such as corn (60–70% starch) is first converted into glucose with enzymatic hydrolysis by gluco-amylase enzymes, which is followed by the fermentation process into bioethanol. Carbohydrates (hemicelluloses and cellulose) found in the lignocellulosic biomass could be converted into bioethanol after three main processing steps of pretreatment, enzymatic hydrolysis, and fermentation (Figure 6.1).

Renewable fuels and biofuels have gained increasing importance because of the combustion of fossil fuel resources resulting in environmental pollution and concerns associated with the produced gases (Hegde et al. 2018). One of the applications of bioethanol is as a petrol additive/alternative. Ethanol is used widely as an alternative fuel in the transportation sector. The octane number of ethanol is higher than this in gasoline. The presence of about 35% oxygen in the bioethanol is responsible for more effective combustion and resulting in a decrease in NO_x, hydrocarbons, carbon monoxide, and particle emission which is considered a less polluting agent compared to fossil fuel burning. Ethanol burns less polluting than gasoline fuel and it produces less CO, CO_2, and NO_X (Ilhak et al. 2019; Ozdingis and Kocar 2018; Tye et al. 2011).

Based on the report of the International Energy Agency (IEA), in May 2020, 24% of direct CO_2 emissions are from fuel burning, at the transportation sector.

The international awareness against fossil fuels and the attempting to reduce greenhouse gas emissions are also criteria that increase the importance of bioethanol. Besides that, bioethanol can be used in the pharmaceutical sector and in cosmetic and industrial processes (Sa´nchez and Cardona 2008). Therefore there is an increase in the amount of global scale of bioethanol production.

The widespread interest of bioethanol depends on three main elements; First, increasing in technological developments and using cheap raw material and the possibility to obtain more than one product at the same process. Second, the development of the agricultural sector is expected to increase employment and production in the agricultural sector and improve the local economy. Third, global climate change; comparing the fossil fuel with the bioethanol, using the latter can reduce carbon dioxide (CO_2) emission that is related to the greenhouse gases (Ozdingis and Kocar 2018).

Several limitations of technical and economic challenges can face the second generation bioethanol. First limitation is the formation of fermentation inhibitors in the pretreatment process of the lignocellulosic biomass (Shaohua et al. 2014). The second limitation is using the enzyme to convert the cellulose polymer into monomeric sugars, where the enzyme cost increasing the overall cost of bioethanol production (Takei et al. 2011). The third limitation interests when using the bioethanol as an alternative fuel, so it needs the dehydration process to obtain at least 99% ethanol purity. This process needs more energy and costly equipment and chemicals which may increase the production cost of bioethanol.

6.3 PRETREATMENT OF FW

FW is containing various components including starch, fat, and cellulose. A pretreatment step is required in order to change the structure of FW to increase the accessibility of enzymes in the sequential step of hydrolyzing into monomeric sugars. The pretreatment can be done using physical, chemical, biological, and combination pretreatment (Kondusamy and Kalamdhad 2014). The strategy of selecting the right pretreatment method for FW should take into consideration several points; improving the sugar formation during the enzymatic hydrolysis process, lesser formation of inhibitor compounds, increasing the total sugar yield by the availability of biomass, and lower energy demand (Table 6.1).

According to previous studies there is no need for severe pretreatment for converting organic FW into fermentable sugars (Kiran et al. 2014).

6.4 PHYSICAL/MECHANICAL PRETREATMENT

Physical pretreatment such as grinding or milling techniques aims to reduce the size of the particles and crystalline fibers by breaking them into small particles. Reducing particle size increases the surface area of cellulose available for enzymatic hydrolysis/saccharification. Izumi et al. 2010 investigated the effects of size reduction of FW as pretreatment on biogas production. These results showed that size reduction pretreatment promoted solubilization and increased methane yield in the anaerobic digestion process.

Microwave (MW) pretreatment has been commonly studied as a way of increasing the substrate biodegradability (Ortigueira et al. 2019). Liu et al. 2020 investigated the effect of MW pretreatment on anaerobic codigestion of sludge and FW. The MW pretreatment of both sludge and FW had the highest nutrient dissolution and the highest methane production (higher than the control), in addition to yield the highest energy recovery rate (76.25 kJ/g Fed VS).

TABLE 6.1
Pretreatment Methods: Advantages vs Disadvantages

Pretreatment method	Advantages	Disadvantages
Physical – grinding/ milling	Increasing the surface area	Energy demand
Microwave	Energy-saving and effective, reducing the treatment time (Liu et al. 2020)	
Ionizing irradiation	Low energy consumption, and excellent environmental benefits (Fei et al. 2020)	
Autoclave	No chemicals being required and the possibility of preserving the nutrition present within the waste (Banu et al. 2020)	
Chemical –	Easy operative models and lack of requirements for additional equipment	Not suitable for easily biodegradable substrate containing high amount of carbohydrate (Ariunbaatar et al. 2014)
Dilute acid	Hydrolyzes hemicellulose to xylose and other sugars, alters lignin structure (Mosier et al. 2005) high sugar recovery and very high reaction rate.	Cost of chemicals, recovery and neutralization, equipment corrosion, formation of toxic inhibitor by products (Yu et al. 2008)
Alkaline	Remove hemicelluloses and lignin (Kumar et al. 2009), High reaction rate (Mosier et al. 2005).	Low sugar yield, sugar decomposition by alkali attack (Yu et al. 2008)
Biological	Do not involve exceedingly high temperatures or pressure, and they also do not demand chemicals such as acids, alkalis. Does not yield any unwanted products. Lower generation of inhibitory substance and eco-friendliness. Demands low cost and low energy.	Highly restricted control is required over the entire pretreatment process (Banu et al. 2020). A slow process that requires longer retention time.

Ionizing irradiation pretreatment is used to improve the microbial availability of FW and it can oxidize solid organic matters and transforming them into liquid then increase the amount of dissolved organic matter in the FW. Fei et al. 2020 investigated the effects of ionizing irradiation pretreatment on the dissolution of organics and on the biogas product yield of FW in anaerobic fermentation. The results showed a yield of 504 mL of biogas/g VS, with an increase of 14.3% over the nonirradiated FW.

6.5 CHEMICAL PRETREATMENT

Chemical treatment relies on the presence of alkaline or acid, oxidizing agents and organic solvents, to change the physico-chemical and biological properties of FW. The main acids used in the acid pretreatment are sulfuric acid (H_2SO_4), hydrochloric acid (HCl), hydrogen peroxide (H_2O_2); whereas the main chemicals used in an alkali pretreatment included sodium hydroxide (NaOH)(Kavitha et al. 2016), potassium hydroxide (KOH), calcium hydroxide ($Ca(OH)_2$), and magnesium hydroxide ($Mg(OH)_2$). Acid pretreatment is known to be efficient in solubilizing carbohydrates, while alkali pretreatment is efficient in solubilizing proteins and lignin, as well as lipid saponification (Galbe and Wallberg 2019; Kim et al. 2016).

Strong acidic pretreatment may produce inhibitory byproducts, such as furfural and hydroxymethylfurfural (HMF). Hence, strong acidic pretreatment is avoided and pretreatment with dilute acids is coupled with thermal methods. Other disadvantages related to acid pretreatment are the loss of fermentable sugar due to increasing degradation of complex substrates, a high cost of acids, and the extra cost for acid neutralizing prior to the fermentation process (Loow et al. 2016).

Kim et al. 2018 optimized the hydrolysis conditions with the dilute acid fractionation of H_2SO_4 concentration (0–0.8% w/v), temperature (130–190°C), and residence time of (1–128 min) where the highest concentration of glucose obtained experimentally from FW was 24.89 g/L at 0.4% H_2SO_4, 160°C, and 64.5 min.

Hafid et al. 2017 used two different pretreatments for FW, hydrothermal and dilute acid pretreatment. Dilute acid pretreatment is conducted at a temperatures range (80°C–100°C) and varied acid concentrations of 0.5–2.0% v/v of HCl and H_2SO_4. Pretreatment of both HCl and H_2SO_4 yields higher sugar content. A modified sequential acid-enzymatic hydrolysis process was developed to produce a high concentration of fermentable sugars including glucose, sucrose, fructose and maltose. The process was based on hydrothermal and dilute acid pretreatment using HCl and H_2SO_4 which is sufficient to degrade larger molecules of polysaccharide followed by enzymatic hydrolysis using glucoamylase. The integrated pretreatment successfully increased sugar production by 2.04 folds (Hafid et al. 2017).

Deheri and Acharya (2021) investigated the effect of sodium hydroxide (NaOH) and calcium peroxide (CaO_2) addition with the FW, cow dung, and sludge solution for anaerobic codigestion. The results showed that the highest methane concentration was increased up to 10% with NaOH, and the hydrogen (H_2) concentration increased up to 7% by adding CaO_2 with the feedstock.

6.6 BIOLOGICAL PRETREATMENT

Biological pretreatment is employed using bacteria, enzymes, and fungi for solubilizing biomass. Biological pretreatment is a usually slow technique that requires increased retention time and the micro-organisms utilize the free accessible sugars as the main carbon source through the pretreatment process. Usually, it is difficult to cultivate and optimize pure cultures of microbes for an FW pretreatment, since the microbes compete with the indigenous micro-organisms through the pretreatment process (Banu et al. 2020).

The pure microbial enzymes for FW pretreatment were investigated in a few studies, due to enzymatic pretreatment needs higher enzyme concentrations to achieve an effective pretreatment which cause increasing in the costs. As an alternative, for cost reduction, the option of using the crude enzymes produced from biomass hydrolysis directly was suggested for the pretreatment for reducing the costs (Kiran et al. 2014; Yin et al. 2016). Since the study cases of the biological pretreatment so far are limited, the conclusion that this is a practical pretreatment could be too early.

6.7 HYDROLYSIS AND FERMENTATION OF AGRICULTURAL FW

Hydrolysis or saccharification is the second step in the process of bioethanol production. After the pretreatment process, this often takes place to hydrolyze the complex of polysaccharides (cellulose and hemicellulose) into simple monomers prior to bioethanol production (Abu Tayeh et al. 2014, 2016). Usually, hydrolysis of complex polysaccharides can be done using thermal, acid, or enzyme treatments. As a result, the cellulose and hemicellulose are hydrolyzed into hexoses and pentose. A small amount of hexoses may also be derived from hemicellulose during acid hydrolysis. In addition, some inhibitory compounds are also generated, including phenol (lignin deviates), furfural, and 5-hydroxyl-methyl-furfural (5-HMF) compounds (Abu Tayeh et al. 2014, 2016). Similarly, thermal hydrolysis also generates the above-listed inhibitory compounds and denatures the sugar monomers. In the case of enzyme hydrolysis (EH), the amount of phenolic compounds generated is insignificant. Several factors can influence EH, including biomass composition, crystalline structure complexity, and particle size. In addition, lignin composition in the biomass plays a major role in the performance of EH (Kennes et al. 2016). In order to overcome these issues, pretreatment is mandatory before the EH process. During pretreatment, the hemicellulose fraction is broken down and undigested. In that case, the undigested hemicellulose can be hydrolyzed using different hemicellulose enzymes. The structure of hemicellulose (a nonlinear chain of glucose) is more complicated than cellulose (a linear chain of glucose) but usually hydrolyzed easily.

The conversion efficiency of FW to ethanol depends on the extent of carbohydrate saccharification since the yeast cells cannot ferment starch or cellulose directly into bioethanol. Many researchers have suggested that EH is the best option for FW hydrolysis (Abu Tayeh et al. 2014, 2016). During EH, the cellulose content in the FW is converted into simple sugars with the help of the following cellulolytic enzymes: endocellulases, exocellulases, and β-glucosidases, which break down the internal structure of cellulose. Exocellulases convert the product of endocellulases into fine sugar molecules referred to as cellobiose. Similarly, β-glucosidase cleaves fine cellobiose into glucose monomers. Usually, cellulase employed for EH is from *Clostridium* sp. (anaerobic bacteria) or *Aspergillus* sp. (aerobic fungi). A mixture of α-amylase, β-amylase, and glucoamylase of various origins is more effective for substrates with higher molecular weight. When both enzymes are combined in a single final product such as glucose is produced. The α-amylase breaks down 1,4-α-glucosidic linear connections in the starch into small compounds such as maltose, glucose, and maltotriose. Glucoamylase breaks 1,6-α-glucosidic connections at the branching point of amylopectin along with 1,4-α-glucosidic (Kiran et al. 2014).

In the bioethanol fermentation process, glucose produced during enzyme hydrolyses is converted into ethanol by the metabolic activity of *S. cerevisiae* (yeast), *Zymomonas mobilis* (bacteria), or *Mucoralean* (fungi). A series of enzymatic reactions take place to convert glucose into bioethanol and carbon dioxide. Theoretically during this process, one mole of 100% fermentable sugar ($C_6H_{12}O_6$) can be fermented into 51% bioethanol (C_2H_5OH) and 49% carbon dioxide (CO_2) (Balat and Balat 2009). Actually, the micro-organisms consume part of fermentable sugar for growth and the actual yield is less than 100%.

Four types of the fermentation process are in general employed in bioethanol production including separate hydrolysis and fermentation (SHF), simultaneous saccharification and fermentation (SSF), simultaneous saccharification and co-fermentation (SSCF), and consolidated bioprocessing (CBP). SHF is a two-step process, in which EH and fermentation are performed separately. EH and fermentation have different optimal conditions. Therefore, SHF has the drawbacks of a long processing time, high energy demand, and high processing charge. SSF is a single-step process, in which enzymatic hydrolysis is combined with fermentation to attain a higher bioethanol yield. In this process, hydrolysis of FW produces glucose, and accumulation of glucose in the medium affects the enzyme activity and decreases the hydrolysis efficiency. In order to overcome this issue, simultaneous production of glucose and its utilization is encouraged in SSF. Rapid utilization of the produced glucose improves the EH efficiency and leads to higher bioethanol production. In addition, the processing cost of SSF is cheaper than SHF. It requires a single reactor for hydrolysis and fermentation. It also reduces the processing time, energy demand, and cost of alcohol synthesis (Hafid et al. 2017)

SSCF is an improved SSF and has benefits such as a shorter fermentation time, continuous conversion of glucose into bioethanol, and less generation of recalcitrant compounds. In this process, the bioethanol yield is based on the cofermentation of both hexose and pentose in a single-step process (Liu and Chen 2016). CBP combines three biological methods that aim to achieve cellulase production, enzymatic hydrolysis, and microbial fermentation in a single operation process (Fan 2014). CBP is recognized as a potential process for the conversion of complex carbohydrates into bio alcohol. Recently, CBP has been widely explored using different waste biomasses as a substrate.

Yan et al. 2011 investigated the use of FW for ethanol production by using glucoamylase for enzymatic hydrolysis, and *S. cerevisiae* H058 for fermentation. The maximum reducing sugar production of 164.8 g/L was at the condition of 142.2 u/g glucoamylase loud, pH of 4.82, 55°C, for 2.48 h. Hong and Yoon 2011 investigated ethanol fermentation of FW by simultaneous saccharification with an amylolytic enzyme complex (a mixture of amyloglucosidase, a-amylase, and protease), and fermentation (SSF) with the yeast, *S. cerevisiae*. Hafid et al. 2017 developed a modified sequential acid-enzymatic hydrolysis process to produce a high concentration of fermentable sugars from FW, using glucoamylase enzyme and *S. cerevisiae* for fermentation (SHF), in a batch reactor. The results showed 86.8% fermentable sugar conversion efficiency, 10.92 g/L ethanol yield, 0.46 g/L/h ethanol productivity, and 85.38% conversion efficiency. Aruwajoye et al. 2020 investigated the enzymatic saccharification of soaking assisted and thermal pretreated cassava peels waste, with the effects of substrate loading, α-amylase, amyloglucosidase, and cellulase at different concentrations. *S. cerevisiae* BY4743 was used for the fermentation process (SHF), in a batch reactor. The results showed 0.53 g/g and 0.29 g/L/h of maximum ethanol yield and productivity, respectively. Chohan et al. 2020 investigated bioethanol production from soaking assisted and thermal pretreated potato peel. In a batch reactor, glucose (Amyloglucosidase) and S. cerevisiae BY4743 were used at an SSF process. α-Amylase enzyme was used in the previous liquefaction process. The results showed 22.5 g/L ethanol, and 0.32 g/g ethanol yield. Jeong et al. 2012 investigated the SSCF for FW using *S. coreanus* and *Pichia stipites* in batch reactor, using Spirizyme Plus FG and Viscozyme L for saccharification. The results showed 48.63 g/L ethanol and 2.03 g/L/h ethanol productivity. Hossain et al. 2018 investigated potato peel waste as a substrate for ethanol production using the CBP method. The maximum bioethanol yield of 30.4 g/L was achieved utilizing 100 g/L of starch in 4 days.

6.8 LARGE-SCALE BIOETHANOL PRODUCTION

Corn and sugarcane molasses are the main feedstock used for commercially large-scale ethanol production, whereas other feedstocks have not been widely used (Abu Tayeh et al. 2014, 2016). The viability of operative an ethanol plant using FW is not totally understood. The production capability of corn ethanol exceeds all other technologies including lignocelluloses, municipal wastes, and FW. Yan et al. 2013 investigated the bioprocess to produce ethanol from FW at pilot scales comparing to laboratory and semi-pilot scales. The pilot-scale was with a production capacity of 80,000 L ethanol (95 %) per year, where the results showed a maximum ethanol concentration of 96.46 ± 1.12 g/L.

The commercial ethanol production from FW depends on the availability of FW, storage and transportation, the effectiveness of hydrolysis and fermentation process, the amount of lipid and carbohydrate found in FW (varying waste composition), clear policy and incentives, the dominance of corn ethanol and others (Hegde et al. 2018; Karmee and Lin 2014). Therefore, the potential barriers of each new raw material such as FW in the production of bioethanol or other products require different considerations and systematic planning to be a commercial process.

Ethanol can be produced by four main types of bioprocess operational modes: batch, continuous, fed-batch, and semi-continuous. In batch fermentation, substrate and yeast culture are charged into the bioreactor together with nutrients and the product is discharged after this period of processing (Bušić et al. 2018). The reactor is well stirred. The investment costs of this reactor are low, do not require much control. Complete sterilization and management of feedstock are easier than in the other processes. The other advantage of batch operation is the greater flexibility of using a bioreactor for various product specifications.

In the continuous process, feed, which contains substrate, culture medium and others required nutrients, is pumped continuously into an agitated vessel where the micro-organisms are active (Azaizeh and Jadoun 2010; Sánchez and Cardona 2008). The product, which is taken from the top of the bioreactor, contains ethanol, cells, and residual sugar. In continuous systems of bioethanol production increasing air supply can improve yeast cell viability, yield, and concentration. Comparison between continuous and batch bioprocesses for bioethanol production shows the following advantages of continuous bioprocesses: reduced costs of bioreactor constructions, lower plant maintenance and operation costs, better bioprocess control, and higher productivities (Sa´nchez and Cardona 2008). Most bioethanol production plants in Brazil are still employing the fed-batch operational mode because of its practical advantages on an industrial scale. However, 30% of industrial facilities for bioethanol production in Brazil are using continuous bioprocess systems due to their advantages related to the higher yeast cell concentrations (Sánchez and Cardona 2008).

The fed-batch process, which may be considered as a combination of batch and continuous operations, is very popular in the ethanol industry (Aylak and Sukan 1998). In this operation, the feed solution, which contains substrate, yeast culture, and the required minerals and vitamins, is fed at constant intervals until filling the reactor and then the process will continue as batch fermentation. The main advantage of the fed-batch method is that the discontinuous feeding of the substrate prevents inhibition and catabolite repression. If the substrate has an inhibitory effect, discontinuous addition improves fermentation productivity by maintaining a low substrate concentration (Aylak and Sukan 1998). It is essential to keep the culture volume constant in continuous operation, whereas there is volume variation in the fed-batch processes. This approach is the most common industrial technology in Brazil for bioethanol production because it can achieve the highest bioprocess volumetric productivity (Sánchez and Cardona 2008). In this bioprocess operational mode, the optimization of the feeding process plays a critical role in increasing ethanol yield and productivity.

In semi-continuous processes, a portion of the culture is withdrawn at intervals and a fresh medium is added into the system. In the continuous processes, it is essential to maintain a constant culture volume and keep the microbes in the system and not wash them out, whereas there is volume variation in semi-continuous processes. This method has some of the advantages of continuous and batch operations. There is no need for a separate inoculum tank, except at the initial startup. Time is also not wasted in nonproductive idle time for cleaning and resterilization of the reactors. Another advantage of this operation is that not much control is required. However, there is a high risk of contamination and mutation of the microbes due to long cultivation periods and periodic handling. Furthermore, since larger reactor volumes are needed, slightly higher investment costs are required especially in the initial phase of production.

6.9 STRATEGIES REQUIRED FOR IMPROVING BIOETHANOL YIELD

Various strategies were explored for improving the ethanol yield, including engineering the conventional *S. cerevisiae* yeast for xylose utilization, by using genetic materials of *P. stipites* and *S. passalidarum*, the main native yeasts for xylose fermentation. Xylose is the major pentose sugar in the hemicellulose of the plant cell wall and the second follow/after glucose, therefore the effective metabolic rate and fermentation of xylose are important for economical bioconversion of hexose and pentose sugars in hemicellulosic hydrolysates on industrial scales; nevertheless, the *S. cerevisiae* cannot convert xylose into bioethanol (Cadete et al. 2018; Jeffries 2006; Jeffries et al., 2009; Van Vleet and Jeffries 2009). Recombination of strains with the amylase-producing gene applied in order to hydrolyze starch completely with improved ethanol tolerance has been investigated (He et al., 2009; Wang et al. 2012). Cell recycling for improving the performance of continuous fermentation was investigated (Wang and Lin 2010). Mixing different feedstocks for ethanol production, even though few laboratory studies assessments have been demonstrated that multiple biomasses give similar or higher ethanol yields compared to single feedstock, therefore, there is need of more studies to establish the possibility of cost effective process (Nielsen et al. 2020). Coculture strategy for fermenting mixed sugar in a single process is important as well. The coculture strategy where *S. cerevisiae* ferments glucose, and the remaining xylose can be fermented by xylose-fermenting species, e.g., *S. stipites*, *Scheffersomyces stipitis*, and *Spathaspora passalidarum* (Ashoor et al. 2015; De Bari et al., 2013; Farias and Filho 2019; Gutiérrez-Rivera et al. 2015; Sreemahadevan et al. 2018; Taniguchi et al. 1997).

6.10 BIOGAS PRODUCTION FROM AGRICULTURAL FW

The food industry is a large sector that includes fruits and vegetables, edible oil, dairy products, seafood, meat product, sugar, brewing and different beverages (Chen et al. 2008). During processing and use, high amounts of FWs are produced, consisting of organic material rich with fats, proteins, and carbohydrates, especially in dairy production industries. The amount is high and suitable for the generation of biogas through the anaerobic digestion (AD) process. FW shows variations in biomass content and organic structure depending on the region and collection season. The characteristic of FW affects the principles of AD, where the process parameters during the digestion in addition to the pretreatment and codigestion for enhancing the AD of the FW are important for biogas production (Zhang C. et al. 2014). Different studies have shown that the total solid (TS) and volatile solid (VS) contents of different FW were in the ranges of 18.1–30.9% and 17.1–26.4%, respectively, indicating that water accounts for 70–80% in FW which is important for AD process (Zhang et al. 2014). The hydrolysis of the biomass is considered to be the rate-limiting step in AD of FW. To enhance the performance of AD, different physical treatments including mechanical grinding, thermo-chemical, biological, or combined pretreatments were used (Gonzales et al. 2005; Kumar et al. 2009; Ma et al. 2011; Vavouraki et al. 2013). The codigestion of FW with other organic substrates is a promising way to improve the AD performance, where a higher buffer capacity and an optimum nutrient balance enhance the codigestion system's biogas/methane yields.

FW'S different AD processes have shown 0.440–0.480 m^3/kg VS methane yield (Zhang C. et al. 2014). The FW has some important benefits such as low cost for collection and transportation, high biodegradability, and high methane yields (Wang et al. 2013). However, some works using a single FW substrate system for an AD process might cause failure of the system because of acid accumulation if the organic load rate (OLR) is not fixed in order to enhance biodegradation or to include codigestion with other organic wastes to correct the pH (Dhar et al. 2016). In addition, vegetable fruit waste also has low biodegradability due to its high lignocellulosic content.

The advantages of AD from organic waste are reducing fossil fuel combustion, enabling a sustainable renewable energy provision, and lowering the emission of the greenhouse gas. Biogas principally contained methane (CH_4) (50–75%), carbon dioxide (CO_2) (25–50%), other gases such as hydrogen sulfides (H_2S), hydrogen (H_2), ammonia (NH_3), water (H_2O), and traces of other gases such as oxygen (O_2) and nitrogen (N_2) (Atelge et al. 2018). Methane can be used instead of fossil fuels for several purposes for instance heating and power generation and transportation. The most effective commercial-scale using FW is the AD process, where nutrient-rich digestate and biofuel including biogas and bioethanol that can be produced and converted into electrical or thermal energy (Figure 6.1). FW is considered an inexpensive substrate for AD and meaningfully increasing the biogas production compared to systems that convert other organic biomass or sewage sludge alone or as the codigestion process (Ebner et al. 2016). Some challenges are associated with using FW-only as a single substrate for digestion since without any cosubstrates, we may get instability that could cause from lacking in trace elements that adjust enzymes reactions; for that reason, digesting of FW in many cases is conducted at a comparatively small portion, and the main substrates are sewage sludge or animal manure used for codigestion. The use of cosubstrates in AD of FW helps maintain AD process stability by balancing the C/N ratio or by supplyng trace minerals and buffering capacity to the process. FW is rich in nitrogen, which is attributed to the high content of proteins. In several cases, FW digestion causes formation of ammonia, which is toxic to methanogenic bacteria (WRAP 2012). The accumulation of volatile fatty acids (VFA) and ammonia formation are major challenges in FW digestion, which needs decreasing the ammonia or trace element supplementation through codigestion. Therefore, several research groups dealing with FW digestion recommended the needing for intense processing monitoring and microbial management to control instability cases (Li et al. 2018), especially most of the researches used a secondary FW stream as a cosubstrate and different studies were recommended to enhance AD of FW, through codigestion, the addition of micronutrients, control of foaming, and process design (Xu et al. 2018).

Thermal pretreatment is used to enhance the solubilization of organic matter and its subsequent fermentation of FW. Ariunbaatar et al. 2015 investigated the possibility to enhance the AD of FW at mesophilic conditions by applying a thermophilic (heating the full reactor content prior to mesophilic digestion) or conventional thermal pretreatment (heating only the substrate) was studied through a series of batch experiments. The highest enhancement of the biomethane production (higher than 40%) was obtained with a thermophilic pretreatment at 50°C for 6–12 h or thermal pretreatment at 80°C for 1.5 h. El Gnaoui et al. 2020 investigated the effects of thermal pretreatment in terms of solubilization rate, CODs and methane yield improvement. The results showed that the CODs increased with temperature increase reaching the maximum value of 68.54 ± 2.4 mg/L at a treatment temperature of 100°C for 30 min. The solubilization rates in terms of CODs increased by 43.41%. Methane yield of FW pretreated at 100°C for 30 min was 382.82 mLSTP CH4/g VS showing an improvement of 23.68% compared to the control.

AD includes a synergistic metabolism between several types of microbes: hydrolyzing bacteria, acetogens, acidogens, and methanogens. These microbial groups vary in their optimum pH conditions for growth and product formation, and sensitiveness to microenvironments changes which needs monitoring of different process parameters to keep a satisfactory balance between microbial groups and get a stable process. Biogas production using AD of organic matter is one of the preferred technologies for treating organic municipal solid waste (MSW), including FW. Biogas production, mainly, methane using anaerobic procedure is a suitable solution for waste management because of its low cost, low production of residual waste, and its utilization as a renewable energy source (Morita and Sasaki 2012; Nasir et al. 2012). The AD process has four main stages: pretreatment, waste digestion, gas recovery, and residue treatment. Pretreatment of waste, especially MSW or FW, is very important to obtain homogeneous feedstock for AD. The preprocessing involves the separation of nondigestible materials and shredding of the organic material to be ready for AD where the organic waste is converted into biogas through biodegradation that involves four stages: hydrolysis, acidogenesis, acetogenesis, and methanogenesis processes (Molino et al. 2013).

The AD process can be defined as a net zero-waste process due to the production of energy from biogas and utilizing the digested organic material (digestate) as an organic fertilizer. In addition, the AD system releases equal or less CO_2 to the atmosphere than biomass does due to fertilizers recovered and land applied, which contains carbon. Therefore, it has a positive impact on reducing global warming (Abbasi et al. 2011). Capturing CH_4 in biogas can reduce greenhouse gas emissions (GHGs), production of renewable energy, and management of waste disposal. Using biogas disposal as an organic fertilizer closes the life cycle. Using FW to produce biogas still has conflicting viewpoints around the world. However, using waste and inedible energy crops as a substrate is a promising alternative to produce energy via AD. Mixing of FW with wood chips in the AD process increased the production of methane and hydrogen, where the FW to wood chip ratio of 0.5, 20 mL/g of methane and 13.9 mL/g of hydrogen were produced during 15 days at 35°C (Oh et al. 2018). Codigestion of FW with other organic substrates could enhance the biodegradation of long-chain fatty acids as well as the methane yield. In addition, codigestion could also improve the buffer capacity and increase acceptable organic loadings compared to single digestion. The AD process includes 4 stages that could work simultaneously in the same digester (Single-stage AD) or in a two-stage AD in two different reactors (Diagram 2).

6.11 SINGLE-STAGE AD FOR BIOGAS PRODUCTION FROM FW

Single-stage AD processes have been widely employed for different organic wastes where all processes (hydrolysis, acidogenesis, acetogenesis, and methanogenesis) occur simultaneously in a single reactor (Figure 6.2). The system faces less repeated technical failures and has a lower investment cost (Forster-Carneiro et al. 2008). The waste can be treated as is in the AD, then call it wet AD. While in the dry AD, the waste can be treated after lowering the water content (12% of total solid) (Nasir et al. 2012). Compared to wet AD, dry AD provides lower methane production and VS reduction due to the VFA transport limitation (Nagao et al. 2012). Several works reported that the VFA accumulation and low pH causes unstable digester treating FW and low biogas production (El-Mashad et al. 2008; Fisgativa et al. 2016). High rate of biogas production was obtained when the pH of the FW was 7, while a pH of 5 and 9 in the reactor resulted in a major decrease in producing biogas (Jayaraj and Deepanraj 2014). Carbohydrates, lipids, and proteins are important to maintain C/N ratio. Proteins are the nitrogen source obtained during degradation into ammonia, where high levels of proteins in FW such as meats cause a reduction in C/N ratio and lead to instabilities by producing excessive ammonia. A balanced metabolism of acetogens and methanogens is required to maintain the pH of a digester within the optimum range. Methanogenic microbes need a pH of 6.8–7.5 for optimum growth for manure and codigestion (Shah et al. 2015); therefore for AD digestion, it is recommended to maintain a pH of 7.2–7.8. Different strategies were suggested to decrease the negative effect of VFAs, including codigestion, the addition of certain metal ions such as Ca^{2+} (Kumar and Singh 2016), and increasing alkalinity using different minerals and compounds (Zhang J. et al. 2014), and discontinuous feeding of the reactors was also suggested to avoid VFA accumulation in the digester (Cavaleiro et al. 2008). Using eggshells and lime from different sources has been

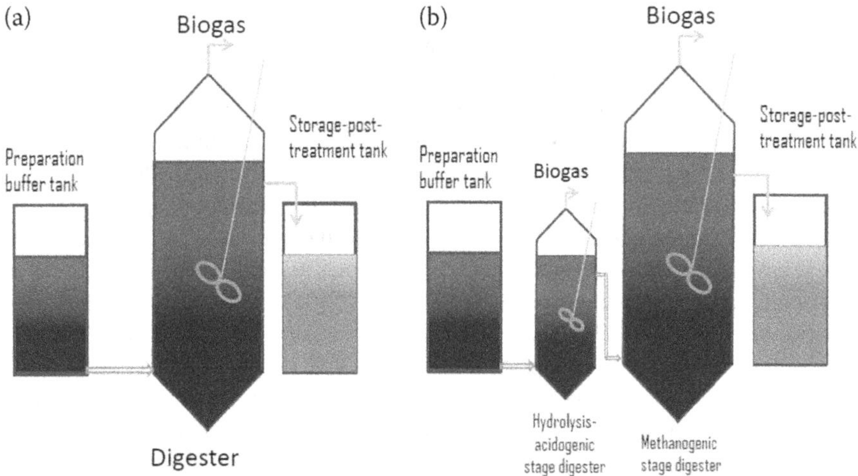

FIGURE 6.2 The single stage (a, left) versus the two stages (b, right) of anaerobic digestion systems for biogas production.

proposed for maintaining digester alkalinity (Zhang J. et al. 2014). During the AD process, the methane level of the biogas varies between 55 and 65%, and in a continuously fed digester at a steady state, the daily biogas composition should remain constant over time since levels below 55%, where CO_2 content could reach above 35–40%, in addition to VFA accumulation will cause inhibition in the activity of methanogenic bacteria and failure of the single-stage reactors.

Based on 102 samples which were obtained from 70 different published papers, Fisgativa et al. 2016 reported that the average pH, dry matter (DM), and VSs of FW were 5.1, 88.2% DM, and 22.8% weight/weight, respectively. They concluded that the high carbohydrate contents (36.4%VS) and the low pH (5.1) might cause inhibitions by the rapid acidification of the digesters. Moreover, the characteristics of FW (including lignocellulosic content) show differentiation between summer and winter collection periods, long holidays, and regular working days, and regions (Edwiges et al. 2018; Fisgativa et al. 2016). The best biochemical methane potential (BMP) calculation was achieved via the statistical model that contained lipid, protein, cellulose, lignin, and high calorific value (HCV), with R^2 of 92.5%; lignin was negatively correlated to methane production (Edwiges et al. 2018). BMP was most intensely correlated to lipid content and HCV. Because HCV and lipids are strongly correlated, and because HCV can be determined more rapidly compared to the overall chemical component, HCV was proposed to be a helpful parameter for calculating BMP.

Fruit and vegetable waste has more biodegradability and higher methane yields than other agriculture products due to its lower lignin content. The biodegradability of fruit and vegetable waste is between 50 and 70% (Dahunsi, et al. 2016; Gil et al. 2018). Principal components analysis provides information about the feasibility of optimal waste mixtures to be selected for biological processes. It was found that physicochemical characterization combined with statistical tools could be an innovative alternative to simplify the selection of mixtures and hence the valorization of the organic wastes for the AD process (Gil et al. 2018). Beside high methane yields and biodegradability, the FW in some cases has low pH values such as 3.88 for strawberry (Serrano et al. 2015), and 3.5 for onion waste (Pereira et al. 2017). Codigestion of sewage sludge and strawberry, both without pretreatment, enhanced the stability inside the digesters and the most suitable combination was found to be the codigestion of pre-treated sewage sludge and raw strawberry which showed a stable process and resulted in synergy in methane production (Serrano et al. 2015).

The AD of mixed cafeteria food waste (CFW) as the main substrate was combined in a semi-continuous mode with other wastes such as acid whey, waste bread as cosubstrates where the digestion of CFW without any cosubstrates, the maximum specific methane yield (SMY) was 363 mL/gVS/d at organic loading rate (OLR) of 2.8/gVSL/d, and the reactor failed at OLR of 3.5/gVSL/d (Hegde and Trabold 2019). Cosubstrates resulted in maximum SMY of 455, 453, and 479 mL/gVS/d, respectively, and it was possible to achieve stable digestion at OLR as high as 4.4/gVSL/d which offers a potential approach to high organic loading rate digestion of FW without using animal manure (Hegde and Trabold 2019). Therefore, FW for AD is advantageous using the codigestion process.

The single-stage wet low-rate digestion systems are described by low biogas yield and low OLR (0.5–1.6 kg-VS/m^3/d), needing long retention time (from 30 up to 60 days) and larger reactors, and depending on the outside temperature during the process (Van et al. 2020). Their advantages include easy to operate and inexpensive investment. These systems are suitable without any pretreatment process for small biodegradable wastes such as animal manures. For that reason, the wet low-rate system is a suitable solution for warm-climate rural areas, where farming land is available, and the leftover digestive products after the treatment can be of use to agricultural performance. The wet low-rate systems with OLR up to 4–8 kg-VS/m^3/d are much less effective than the single-stage wet high-rate digestion systems are much more effective than (Van et al. 2020). In spite of this, the procedure is more complicated than the low-rate systems because of the involved equipment and pretreatment step required. Since the low pH value may affect the AD process because of acidification; therefore, the waste should be used with a cosubstrate for AD or to be digested using a two-stage AD process.

6.12 TWO-STAGE AD FOR BIOGAS PRODUCTION FROM FW

Most full-scale AD systems are of a single-stage standard design of reactors used to produce methane (De Baere 2006). However, there are other variations on this standard configuration, one of which is the two stages of AD systems (Figure 6.2), where an additional acidogenic stage is used in order to produce the organic acids from the biomass and then to be used as a substrate for the next methanogenic stage for methane production (Khalid et al. 2011). The VFA/alkalinity ratio should be held for 0.10–0.30 to avoid acidification of the AD process (Barampouti et al. 2005), whereas the single-stage AD process could be an imbalance due to the fast change in pH for the low buffered and fast acidified high moisture of FW (Zhang et al. 2016).

The two-stage AD is used to produce hydrogen and methane in two single reactors to decrease the inhibition of the AD process caused by the pH level of the first two stages, especially the acidogenesis and acetogenesis processes (Chen et al. 2008). In this system, the fast-growing of acidogens and hydrogen-producing micro-organisms are improved for the hydrogen production and VFAs in the first stage in order to be utilized in the next AD stages especially the methanogenesis microbes to produce methane. In the second stage, slow-growing acetogens and methanogens are built up, where VFAs are used to produce methane and carbon dioxide. In a study by Park et al. 2008, the synthetic kitchen wastes were used at the single-stage and two-stage of thermophilic methane fermentation systems In both systems, the highest methane yield achieved was 90% was at OLR of 15 g COD/L/d. In another work, they compared single and two stages AD fermentation systems on wheat feed pellets obtained from a flour mill processing (Massanet-Nicolau et al. 2013, where the methane yield in the two-stage fermentation was improved by 37% when operated at shorter hydraulic retention time (HRTs) and higher loading rates. In the work of Lee and Chung 2010 also proved that two stages hydrogen/methane fermentation has significantly greater potential for producing methane. Several studies have concluded that the ability to operate two stages fermenters at much shorter HRTs and higher ORLs while still achieving an increase in methane yield which is still in excess compared to those predicted by using a single-stage BMP.

The accumulation of VFAs in the bioreactor that inhibits methanogenic activity could cause instability of AD systems feeding. (Pavi et al. 2017). One way for reducing VFA accumulation and improving the stability of the AD process, is mixing ash with organic waste. The alkali metals in the ash contribute to increasing buffer capacity and pH (Lo 2005). AD is a proper process for dealing with MSW or FW for the production of biogas and bio-fertilizer. However, the low methane production and low stability of AD process are related to the high moisture of most of the used biomass and due to low pH caused by the fast acidify of the carbohydrates (Markphan et al. 2020). The effects of the addition of incineration fly ash and organic loading in single-stage AD and two-stage AD processes on the production of methane were investigated. The results showed that the pH adjustment of two-stage AD process with the addition of fly ash could improve the biogas production and reduce the lag phase and enhanced biodegradability efficiency, the biogas production increased 87–92% compared to without adding ash (Markphan et al. 2020). The biogas production using two-stage AD has an 18.5% higher energy recovery than a single-stage AD. For high energy recovery, with a small area, and low greenhouse gas emissions, the suitable treatment with MSW is the tow-stage AD process (Markphan et al. 2020). For the methane production from olive mill waste the two-stage AD process improved the production by 10% compared to the single-stage. (Rincón et al. 2009).

The single-stage AD systems could have longer HRT and lower OLR than the separation of hydrolysis and methanogenesis stages using different types of reactors at the two-stage systems (Van et al. 2020). The advantages of these two-stage AD systems could be over the construction expenses, digestion efficiency, robust, methane content, operational flexibility, and could be used with low C/N (<10) (Van et al. 2020).

6.13 COMBINING BIOETHANOL AND BIOGAS PRODUCTION FROM THE FW BIOMASS

The use of a single substrate such as FW may create an imbalance in the AD process; therefore, a codigestion is the best option to increase biogas yields and to keep the C/N ratio, and pH balanced, and reduces the HRT. The codigestion approach decreases toxicity in the reactor and carries out the AD process effectively leading the process to become more profitable and easy to replace fossil fuel-based energy sources. Recent developments of substrates include aquaculture, FW, and agricultural waste, which are lignocellulose-based residuals, in addition to the utilization of the biomass for biogas production as well as bioethanol fermentation

Although bioethanol production from lignocellulosic materials has been widely studied, its production still has environmental, economic, and energetic limitations. From the environmental view, the leftover generated after the distillation step has a high pollutant potential and the alternative management still need to be further studied. Economically, the pretreatment stage still needs high energy costs, making bioethanol production less competitive compared to fossil fuels. Energetically, ethanol from biomass has a low-energy return on energy invested when compared to coal, oil, and gas. For that reason, there is need for solutions that can add value to the bioethanol production technology to make it more competitive; therefore, AD has been suggested as an option for handling waste recovery from bioethanol process (D'Adamo et al. 2019; Ferella et al. 2019; Rocha-Meneses et al. 2017).

Improving the total energy balance of biomass can be achieved with the combination of both, fermentation and AD processes (Figure 6.1). This combination in a biorefinery is considered a good strategy for fermentation plants' competitiveness. Such strategy goes in for the integration of the material flows of different bioprocess, so that the remnant from bioethanol production becomes the substrate for biogas production (Ferella et al. 2019; Rocha-Meneses et al. 2017). In addition, this technology of obtaining two biofuels enables the production process from waste to be more cost effective and environmentally more sustainable.

According to Kotarska et al. 2019, it was determined that the leftover residue after the production of ethanol can provide biodegradable components for methane microbe. This results in a rapid acid phase and efficient conduction of subsequent stages of methanogenesis, leading to the conception of high-power biogas production. Moshi et al. 2015 demonstrated that Cassava peels can be valorized into sequential bioethanol and biogas production. Sequential ethanol and methane production resulted in 1.2–1.3-fold fuel energy yield compared to only methane, and 3–4-fold compared to only ethanol production. Kotarska et al. 2019 studied the sequential combination of bioethanol and biogas production from corn straw. They obtained 17 g/L ethanol, while the ethanol yield was 31 L EtOH/100 kg of cellulose. The cellulose stillage used for biogas production was characterized by fast digestion, with 330 L/kg DM organic matter efficiency of biogas. Bioethanol production combined with AD looks a promising process for a more competitive solution for treating FW.

REFERENCES

Abbasi, T., Tauseef, S., Abbasi, S.A. 2011. *Biogas Energy*, vol. 2. Springer, New York.

Abu Amr, S.S., Aziz, H.A., Adlan, M.N. 2013. Optimization of stabilized leachate treatment using ozone/persulfate in the advanced oxidation process. *Waste Manage.* 33 (6), 1434–1441.

Abu Tayeh, H., Levy-Shalev, O., Azaizeh, H., Dosoretz, C.G. 2016. Subcritical hydrothermal pretreatment of olive mill solid waste for biofuel production. *Bioresour. Technol.* 199, 164–172.

Abu Tayeh, H., Najami, N., Dosoretz, C.G., Tafesh, A., Azaizeh, H. 2014. Potential of bioethanol production from olive mill solid wastes. *Bioresour. Technol.* 152, 24–30. 10.1016/j.biortech.2013.10.102

Ariunbaatar, J., Panico, A., Esposito, G., Pirozzi, F., Lens, P.N.L. 2014. Pretreatment methods to enhance anaerobic digestion of organic solid waste. *Appl. Energy* 123, 143–156. 10. 1016/j.apenergy.2014.02.035

Ariunbaatar, J., Panico, A., Yeh, D.H., Pirozzi, F., Lens, P.N.L., Esposito, G. 2015. Enhanced mesophilic anaerobic digestion of food waste by thermal pretreatment: Substrate versus digestate heating. *Waste Manageme.* 46, 176–181. 10.1016/j.wasman.2015.07.045

Aruwajoye, G.S., Faloye, F.D., Kana, E.G. 2020. Process optimisation of enzymatic saccharifcation of soaking assisted and thermal pretreated cassava peels waste for bioethanol production. *Waste Biomass Valorization.* 11, 2409–2420. 10.1007/s12649-018-00562-0

Ashoor, S., Comitini, F., Ciani, M. 2015. Cell-recycle batch process of *Scheffersomyces stipitis* and *Saccharomyces cerevisiae* co-culture for second generation bioethanol production. *Biotechnol. Lett.* 37, 2213–2218.

Atelge, M.R., Krisa, D., Kumar, G. 2018. Biogas production from organic waste: Recent progress and perspectives. *Waste Biomass Valor.* 11, 1019–1040. 10.1007/s12649-018-00546-0

Aylak, B.C., Sukan, F.V. 1998. Comparison of different production processes. *Turk. J. Chem.* 22, 351–359.

Azaizeh, H., Jadoun, J. 2010. Codigestion of olive mill wastewater and swine manure using UASB bioreactors for biogas production. *Water Resour. Prot.* 2(4), 314–321. doi: 10.4236/jwarp.2010.24036

Balat, M., Balat, H. 2009. Recent trends in global production and utilization of bio-ethanol fuel. *Appl. Energy.* 86, 2273–2282. 10.1016/j.apenergy.2009.03.015

Banu, J.R., Kavitha, S., Kannah, R.Y., Poornima Devi, T., Gunasekaran, M., Kim, S.-H. 2019. A review on biopolymer production via lignin valorization. *Bioresour. Technol.* 290, 121790.

Banu, J.R., Merrylin, J., Usman, T.M.M., Kannah, R.Y., Gunasekaran, M., Kim, S-H., Kumar, G. 2020. Impact of pretreatment on food waste for biohydrogen production: A review. *Int. J. Hydrog. Energy.* 45 (36): 18211–18225. 10.1016/j.ijhydene.2019.09.176

Barampouti, E.M.P., Mai, S.T., Vlyssides, A.G. 2005. Dynamic modeling of the ratio volatile fatty acids/bicarbonate alkalinity in a UASB reactor for potato processing wastewater treatment. *Environ. Monit. Assess.* 110, 121–128. 10.1007/s10661-005-6282-1

Bušić, A., Marđetko, N., Kundas, S., Morzak, G., Belskaya, H., Šantek, M.I., Komes, D., Novak, S., Šantek, B. 2018. Bioethanol production from renewable raw materials and its separation and purification: A review. *Food Technol. Biotechnol.* 56(3), 289–311. 10.17113/ftb.56.03.18.5546.

Cadete, R.M., Rosa, C.A. 2018. The yeasts of the genus *Spathaspora*: Potential candidates for second-generation biofuel production. *Yeast* 35, 191–199. 10.1002/yea.3279.

Cavaleiro, A.J., Pereira, M.A., Alves, M. 2008. Enhancement of methane production from long chain fatty acid based effluents. *Bioresour. Technol.* 99, 4086–4095. 10.1016/j.biortech.2007.09.005

Chen, Y., Cheng, J.J., Creamer, K.S. 2008. Inhibition of anaerobic digestion process: A review. *Bioresour. Technol.* 99(10), 4044–4064. 10.1016/j.biortech.2007.01.057

Cho, S.-H., Kim, J., Han, J., Lee, D., Kim, H.J., Kim, Y.T., Cheng, X., Xu, Y., Lee, J., Kwon, E.E. 2019. Bioalcohol production from acidogenic products via a twostep process: A case study of butyric acid to butanol. *Appl. Energy* 252, 113482.

Cho, S.-H., Kim, T., Baek, K., Lee, J., Kwon, E.E. 2018. The use of organic waste derived volatile fatty acids as raw materials of C4-C5 bioalcohols. *J. Clean. Prod.* 201, 14–21.

Chohan, N.A., Aruwajoye, G.S., Sewsynker-Sukai, Y., Kana, E.B.G. 2020. Valorisation of potato peel wastes for bioethanol production using simultaneous saccharification and fermentation: Process optimization and kinetic assessment. *Renew. Energy.* 146, 1031–1040.

Deheri, C., Acharya, S.K. 2021. Effect of calcium peroxide and sodium hydroxide on hydrogen and methane generation during the co-digestion of food waste and cow dung. *J.Clean. Prod.* 279, 123901. 10.1016/j.jclepro.2020.123901.

D'Adamo, I., Falcone, P.M., Ferella, F. 2019. A socio-economic analysis of biomethane in the transport sector: The case of Italy. *Waste Manag.* 95, 102–115.

Dahiya, S., Kumar, A.N., Sravan, J.S., Chatterjee, S., Sarkar, O., Mohan, S.V. 2018. Review Food waste biorefinery: Sustainable strategy for circular bioeconomy. *Bioresour. Technol.* 248, 2–12. 10.1016/j.biortech.2017.07.176.

Dahunsi, S.O., Oranusi, S., Owolabi, J.B., Efeovbokhan, V.E. 2016. Comparative biogas generation from fruit peels of fluted pumpkin (*Telfairia occidentalis*) and its optimization. *Bioresour. Technol.* 221, 517–525. 10.1016/j.biortech.2016.09.065

De Baere, L. 2006. Will anaerobic digestion of solid waste survive in the future? *Water Sci. Technol.* 53, 187–194. 10.2166/wst.2006.249

De Bari, I., de Canio, P., Cuna, D., Liuzzi, F., Capece, A., Romano, P. 2013. Bioethanol production from mixed sugars by *Scheffersomyces stipitis* free and immobilized cells, and co-cultures with *Saccharomyces cerevisiae*. *New Biotechnol.* 30 (6), 591–597. 10.1016/j.nbt.2013.02.003.

Dhar, H., Kumar, P., Kumar, S., Mukherjee, S., Vaidya, A.N. 2016. Effect of organic loading rate during anaerobic digestion of municipal solid waste. *Bioresour. Technol.* 217, 56–61. 10.1016/j.biortech.2015.12.004

Ebner, J.H., Labatut, R.A., Lodge, J.S., Williamson, A.A., Trabold, T.A. 2016. Anaerobic co-digestion of commercial food waste and dairy manure: Characterizing biochemical parameters and synergistic effects. *Waste Manag.* 52, 286–294. 10.1016/j.wasman.2016.03.046.

Edwiges, T., Frare, L., Mayer, B., Lins, L., Triolo, J.M., Flotats, X., de Mendonça, Costa, M.S.S. 2018. Influence of chemical composition on biochemical methane potential of fruit and vegetable waste. *Waste Manag.* 71, 618–625. 10.1016/j.wasman.2017.05.030

El Gnaoui, Y., Karouach, F., Bakraoui, M., Barz, M., El Bari, H. 2020. Mesophilic anaerobic digestion of food waste: Effect of thermal pretreatment on improvement of anaerobic digestion process. *Energy Rep.* 6, 417–422. 10.1016/j.egyr.2019.11.096.

El-Mashad, H.M., McGarvey, J.A., Zhang, R. 2008. Performance and microbial analysis of anaerobic digesters treating food waste and dairy manure. *Biol. Eng. Trans.* 1(3), 233–242.

Fan, Z. 2014. Chapter 7 - Consolidated bioprocessing for ethanol production. *Biorefin. Integr. Biochem. Process. Liquid Biofuels* 141–160 10.1016/B978-0-444-59498-3.00007-5

Farias, D., Filho, F.M. 2019. Co-culture strategy for improved 2G bioethanol production using a mixture of sugarcane molasses and bagasse hydrolysate as substrate. *Biochem. Eng. J.* 147, 29–38. 10.1016/j.bej.2019.03.020

Fei, X., Chen, T., Jia, W., Shan, Q., Hei, D., Ling, Y., Feng, J., Feng, H. 2020. Enhancement effect of ionizing radiation pretreatment on biogas production from anaerobic fermentation of food waste. *Radiat. Phys. Chem.* 168, 108534. 10.1016/j.radphyschem.2019.108534.

Ferella, F., Cucchiella, F., D'Adamo, I., Gallucci, K. 2019. A techno-economic assessment of biogas upgrading in a developed market. *J. Clean. Prod.* 210, 945–957. 10.1016/j.jclepro.2018.11.073.

Fisgativa, H., Tremier, A., Dabert, P. 2016. Characterizing the variability of food waste quality: A need for efficient valorisation through anaerobic digestion. *Waste Manag.* 50, 264–274. 10.1016/j.wasman.2016.01.041

Forster-Carneiro, T., Pérez, M., Romero, L.I. 2008. Influence of total solid and inoculum contents on performance of anaerobic reactors treating food waste. *Bioresour. Technol.* 99(15), 6994–7002.

Galbe, M., Wallberg, O. 2019. Pretreatment for biorefneries: A review of common methods for efficient utilisation of lignocellulosic materials. *Biotechnol. Biofuels* 12, 294 10.1186/s13068-019-1634-1.

Gil, A., Toledo, M., Siles, J.A., Martín, M.A. 2018. Multivariate analysis and biodegradability test to evaluate different organic wastes for biological treatments: Anaerobic co-digestion and co-composting. *Waste Manag.* 78, 819–828. 10.1016/j.wasman.2018.06.052

Gonzales, H.B., Takyu, K., Sakashita, H., Nakano, Y., Nishijima, W., Okada, M. 2005. Biological solubilization and mineralization as novel approach for the pretreatment of food waste. *Chemosphere* 58:57–63.

Gustavsson, J., Cederberg, C., Sonesson, U., van Otterdijk, R., Meybeck, A. 2011. *Global Food Losses and Food Waste - Extent, Causes and Prevention.* FAO, Rome.

Gutiérrez-Rivera, B., Beningo, O.M., Gómez-Rodrígues, J., Cárdenas-Cágal, A., González, J.M.D., Aguillar-Uscanga, A. 2015. Bioethanol production from hydrolyzed sugarcane bagasse supplemented with molasses in a mixed yeast culture. *Renew. Energy*. 74, 399–405. 10.1016/j.renene.2014.08.030

Hafid, H.S., Abdul Rahman, N.A., Mokhtar, M.N., Talib, A.T., Baharuddin, A.S., Umi Kalsom Md Shah, U.K.M. 2017. Over production of fermentable sugar for bioethanol production from carbohydrate-rich Malaysian food waste via sequential acid-enzymatic hydrolysis pretreatment. *Waste Manageme*. 67, 95–105. 10.1016/j.wasman.2017.05.017

He, M-X., Feng, H., Bai, F., Li, Y., Liu, X., Zhang, Y-Z. 2009. Direct production of ethanol from raw sweet potato starch using genetically engineered*Zymomonas mobilis*. *Afr. J. Biotechnol.* 3(11), 721–726.

Hegde, S., Trabold, T.A. 2019. Anaerobic digestion of food waste with unconventional co-substrates for stable biogas production at high organic loading rates. *Sustainability* 11, 3875. 10.3390/su11143875

Hegde, S., Lodge, J.S., Trabold, T.A. 2018. Characteristics of food processing wastes and their use in sustainable alcohol production. *Renew. Sust. Energ. Rev.* 81, 510–523. 10.1016/j.rser.2017.07.012

Hong, Y.S., Yoon, H.H. 2011. Ethanol production from food residues. *Biomass Bioenergy*. 35(7), 3271–3275. 10.1016/j.biombioe.2011.04.030

Hossain, T., Miah, A.B., Mahmud, S.A. 2018. Enhanced bioethanol production from potato peel waste via consolidated bioprocessing with statistically optimized medium. *Appl. Biochem. Biotechnol.* 186, 425–442. 10.1007/s12010-018-2747-x

Ilhak, M.I., Tngoz, S., Akansu, S.O., Kahraman, N. 2019. Alternative fuels for internal combustion engines: The future of internal combustion engines (book). 10.5772/intechopen.85446

Izumi, K., Okishio, Y.K., Nagao, N., Niwa, C., Yamamoto, S., Tod, T. 2010. Effects of particle size on anaerobic digestion of food waste. *Int. Biodeterior. Biodegradation.* 64, 601–608. doi:10.1016/j.ibiod.2010.06.013

Jayaraj, S., Deepanraj, B.V.S. 2014. Study on the effect of pH on biogas production from food waste by anaerobic digestion. In Proceedings of the 9th Annual Green Energy Conference, Tianjin, China, 25–28 May 2014.

Jeffries, T.W. 2006. Engineering yeasts for xylose metabolism. *Curr. Opin. Biotechnol.* 17, 320–326. 10.1016/j.copbio.2006.05.008

Jeffries, T.W., Van Vleet, J.R.H. 2009. *Pichia stipitis* genomics, transcriptomics, and gene clusters. *FEMS Yeast Res.* 9, 793–807. 10.1111/j.1567-1364.2009.00525.x

Jeong, S.M., Kim, Y.J., Lee, D.H. 2012. Ethanol production by co-fermentation of hexose and pentose from food wastes using *Saccharomyces coreanus* and *Pichia stipitis*. *Korean J. Chem. Eng.* 29, 1038–1043. 10.1007/s11814-011-0282-3

Kannah, R.Y., Merrylin, J., Devi, T.P., Kavitha, S., Sivashanmugam, P., Kumar, G., Banu, J.R. 2020. Food waste valorization: Biofuels and value added product recovery. *Bioresour. Technol. Rep.* 11, 100524. 10.1016/j.biteb.2020.100524

Karmee, S.K., Lin, C.S.K. 2014. Valorisation of food waste to biofuel: Current trends and technological challenges. *Sustain Chem Process* 2, 22. 10.1186/s40508-014-0022-1

Katami, T., Yasuhara, A., Shibamoto, T. 2004. Formation of dioxins from incineration of foods found in domestic garbage. *Environ. Sci. and Technol.* 38(4), 1062–1065.

Kavitha, S., Banu, J.R., Subitha, G., Ushani, U., Yeom, I.T. 2016. Impact of thermo-chemo-sonic pretreatment in solubilizing waste activated sludge for biogas production: Energetic analysis and economic assessment. *Bioresour. Technol.* 219, 479–486. 10.1016/j.biortech.2016.07.115

Kavitha, S., Kannah, R.Y., Gunasekaran, M., Banu, J.R., Kumar, G. 2019. Rhamnolipid induced deagglomeration of anaerobic granular biosolids for energetically feasible ultrasonic homogenization and profitable biohydrogen. *Int. J. Hydrog. Energy*. 45(10), 5890–5899. 10.1016/j.ijhydene.2019.04.063

Kennes, D., Abubackar, H.N., Diaz, M., Veiga, M.C., Kennes, C. 2016. Bioethanol production from biomass: Carbohydrate vs syngas fermentation. *J. Chem. Technol. Biotechnol.* 91, 304–317.

Khalid, A., Arshad, M., Anjum, M., Mahmood, T., Dawson, L., 2011. The anaerobic digestion of solid organic waste. *Waste Manage.* 31, 1737–1744.

Kim, J.S., Lee, Y.Y., Kim, T.H. 2016. A review on alkaline pretreatment technology for bioconversion of lignocellulosic biomass. *Bioresour. Technol.* 199, 42–48. 10.1016/j.biortech.2015.08.085

Kim, Y.S., Jang, J.Y., Park, S.J., Um, B.H. 2018. Dilute sulfuric acid fractionation of Korean food waste for ethanol and lactic acid production by yeast. *Waste Manage.* 74, 231–240. 10.1016/j.wasman.2018.01.012

Kiran, E.U., Trzcinski, A.P., Ng, W.J., Liu, Y. 2014. Bioconversion of food waste to energy: A review. *Fuel.* 134, 389–399. 10.1016/j.fuel.2014.05.074

Kondusamy, D., Kalamdhad, A.S., 2014. Pre-treatment and anaerobic digestion of FW for high rate methane production–A review. *J. Environ. Chem. Eng.* 2 (3), 1821–1830.

Kotarska, K., Dziemianowicz, W., Swierczy´nska, A. 2019. Study on the sequential combination of bioethanol and biogas production from corn straw. *Molecules* 24, 4558. 10.3390/molecules24244558.

Kumar, D., Singh, V. 2016. Dry-grind processing using amylase corn and superior yeast to reduce the exogenous enzyme requirements in bioethanol production. *Biotechnol. Biofuels.* 9, 228.

Kumar, P., Barrett, D.M., Delwiche, M.J., Stroeve, P. 2009. Methods for pretreatment of lignocellulosic biomass for efficient hydrolysis and biofuel production. *Ind. Eng. Chem. Res.* 48(8), 371337-29.

Lee, Y.W., Chung, J. 2010. Bioproduction of hydrogen from food waste by pilot-scale combined hydrogen/methane fermentation. *Int. J. Hydrog Energy.* 35(21), 11746–11755.

Li, L., Peng, X., Wang, X., Wu, D. 2018. Anaerobic digestion of food waste: A review focusing on process stability. *Bioresour. Technol.* 248, 20–28.

Liu, J., Zhao, M., Lv, C., Yue, P. 2020. The effect of microwave pretreatment on anaerobic co-digestion of sludge and food waste: Performance, kinetics and energy recovery. *Environ. Res.* 189, 109856. 10.1016/j.envres.2020.109856

Liu, Z-H., Chen, H-Z. 2016. Simultaneous saccharification and co-fermentation for improving the xylose utilization of steam exploded corn stover at high solid loading. *Bioresour. Technol.* 201, 15–26. 10.1016/j.biortech.2015.11.023

Lo, H-M. 2005. Metals behaviors of MSWI bottom ash co-digested anaerobically with MSW. *Resour. Conserv. Recycl.* 43, 263–280. 10.1016/j.resconrec.2004.06.004

Loow, Y.L., Wu, T.Y., Md. Jahim, J. 2016. Typical conversion of lignocellulosic biomass into reducing sugars using dilute acid hydrolysis and alkaline pretreatment. *Cellulose* 23, 1491–1520. 10.1007/s10570-016-0936-8

Ma, J., Duong, T.H., Smits, M., Verstraete, W., Carballa, M. 2011. Enhanced biomethanation of kitchen waste by different pre-treatments. *Bioresour. Technol.* 102, 592- 599. 10.1 016/j.biortech.2010.07.122

Markphan, W., Mamimin,C., Suksong, W., Prasertsan, P., Thong, S.O. 2020. Comparative assessment of single-stage and two-stage anaerobic digestion for biogas production from high moisture municipal solid waste. *PeerJ.* 8:e9693. 10.7717/peerj.9693

Massanet-Nicolau, J., Dinsdale, R., Guwy, A., Shipley, G. 2013. Use of real time gas production data for more accurate comparison of continuous single-stage and two-stage fermentation. *Bioresour. technol.* 129, 561–567.

Melikoglu, M., Lin, C.S.K., Webb, C. 2013. Analysing global food waste problem: pinpointing the facts and estimating the energy content. *Cent. Eur. J. Eng*. 3, 157–164. 10.2478/s13531-012-0058-5

Molino, A., Nanna, F., Ding Y., et al. 2013. Biomethane production by anaerobic digestion of organic waste. *Fuel* 103, 1003–1009.

Moon, H.C., Song, I.S., Kim, J.C., Shirai, Y., Lee, D.H., Kim, J.K., Chung, S.O., Kim, D.H., Oh, K.K., Cho, Y.S. 2009. Enzymatic hydrolysis of food waste and ethanol fermentation. *Int. J. Renew. Energy Res*. 33, 164–172. 10.1002/er.1432.

Morita, M., Sasaki, K. 2012. Factors influencing the degradation of garbage in methanogenic bioreactors and impacts on biogas formation. *Appl. Microbiol. Biotechnol.* 94, 575–582.

Moshi, A.P., Temu, S.G., Nges, I.A., Malmo, G., Hosea, K.M.M., Elisante, E., Mattiasson, B. 2015. Combined production of bioethanol and biogas from peels of wild cassava *Manihot glaziovii. Chem. Eng. J.* 279, 297–306. 10.1016/j.cej.2015.05.006.

Mosier, N., Hendrickson, R., Ho, N., Sedlak, M., Ladisch, M.R. 2005. Optimization of pH controlled liquid hot water pretreatment of corn stover. *Bioresour. Technol.* 96(18), 1986–1993.

Nagao, N., Tajima, N., Kawai, M., Niwa, C., Kurosawa, N., Matsuyama, T., Yusoff, F.M., Toda, T. 2012. Maximum organic loading rate for the single-stage wet anaerobic digestion of food waste. *Bioresour. Technol.*, 118, 210–218.

Nasir, I.M., Ghazi, T.I.M., Omar, R. 2012. Production of biogas from solid organic wastes through 36 anaerobic digestion: A review. *Appl. Microbiol. Biotechnol.* 95, 321–329.

Nielsen, F., Galbe, M., Zacchi, G. 2020. The effect of mixed agricultural feedstocks on steam pretreatment, enzymatic hydrolysis, and cofermentation in the lignocellulose-to-ethanol process. *Biomass Convers. Biorefin.* 10, 253–266. 10.1007/s13399-019-00454-w

Oh, J-I., Lee, J., Lin, K-Y., Kwon, E., Tsang, Y. 2018. Biogas production from food waste via anaerobic digestion with wood chips. *Energy Environ.* 29(8), 1365–1372. 10.1177/0958305X18777234

Ortigueira, J., Martins, L., Pacheco, M., Silva, C., Moura, P. 2019. Improving the non-sterile food waste bioconversion to hydrogen by microwave pretreatment and bioaugmentation with *Clostridium butyricum. Waste Manageme.* 88, 226–235. 10.1016/j.wasman.2019.03.021.

Ozdingis, A.G.B., Kocar, G. 2018. Current and future aspects of bioethanol production and utilization in Turkey. *Renew. Sust. Energ. Rev.* 81, 2196 - 2203. 10.1016/j.rser.2017.06.031

Paritosh, K., Kushwaha, S.K., Yadav, M., Pareek, N., Chawade, A., Vivekanand, V. 2017. Food waste to energy: An overview of sustainable approaches for food waste management and nutrient recycling. *Biomed. Res. Int.* 2, 1–19. 10.1155/2017/2370927.

Park, Y.J., Hong, F., Cheon, J.H., Hidaka, T., Tsuno, H. 2008. Comparison of thermophilic anaerobic digestion characteristics between single-phase and two-phase systems for kitchen garbage treatment. *J. Biosci. Bioeng.* 105(1), 48–54.

Pavi, S., Kramer, L.E., Gomes, L.P., Miranda, L.A.S. 2017. Biogas production from co-digestion of organic fraction of municipal solid waste and fruit and vegetable waste. *Bioresour. Technol.* 228, 362–367. 10.1016/j.biortech.2017.01.003

Pereira, G.S., Cipriani, M., Wisbeck, E., Souza, O., Strapazzon, J.O., Gern, R.M.M. 2017. Onion juice waste for production of *Pleurotus sajor*–caju and pectinases. *Food Bioprod. Process.* 106, 11–18. 10.1016/j.fbp.2017.08.006

Rincón, B., Borja, R., Martín, M.A., Martín, A. 2009. Evaluation of the methanogenic step of a two-stage anaerobic digestion process of acidified olive mill solid residue from a previous hydrolytic-acidogenic step. *Waste Manageme.* 29, 2566–2573. 10.1016/j.wasman.2009.04.009

Rocha-Meneses, L., Raud, M., Orupõld, K., Kikas, T. 2017. Second-generation bioethanol production: A review of strategies for waste valorisation. *Agron. Res.* 15, 830–847.

Sánchez, Ó.J., Cardona, C.A. 2008. Trends in biotechnological production of fuel ethanol from different feedstocks. *Bioresour. Technol.* 99, 5270–5295. 10.1016/j.biortech.2007.11.013

Serrano, A., Siles, J.A., Gutiérrez, M.C., Martín, M. 2015. Improvement of the biomethanization of sewage sludge by thermal pre-treatment and co-digestion with strawberry extrudate. *J. Clean. Prod.* 90, 25–33. 10.1016/j.jclepro.2014.11.039

Shah, A.A., Nawaz, A., Kanwal, L., Hasan, F., Khan, S., Badshah, M. 2015. Degradation of poly(ε-caprolactone) by a thermophilic bacterium *Ralstonia* sp. Strain MRL-TL isolated from hot spring. *Int. Biodeterior. Biodegrad.* 98, 35–42.

Shaohua, X., Bing, L., Yimei, W., Zhongfeng, F., Zehui, Z. 2014. Efficient conversion of cellulose into biofuel precursor 5-hydroxymethylfurfural in dimethyl sulfoxide–ionic liquid mixtures. *Bioresour. Technol.* 151, 361–366.

Solomon, S., Qin, D., Manning, M., Chen, Z., Marquis, M., Averyt, K.B. 2007. *Intergovernmental panel on climate change. Climate change 2007: the physical science basis; contribution of working group I to the fourth assessment report of the intergovernmental panel on climate change.* Cambridge University Press, New York.

Sreemahadevan, S., Singh, V., Roychoudhury, P.K., Ahammad, S.Z. 2018. Mathematical modeling, simulation and validation for co-fermentation of glucose and xylose by *Saccharomyces cerevisiae* and *Scheffersomyces stipites. Biomass Bioenergy.* 110, 17–24. 10.1016/j.biombioe.2018.01.008.

Takei, T., İkeda, K., İjima, H., Kawakami, K. 2011. Fabrication of poly(vinyl alcohol) hydrogel beads crosslinked using sodium sulfate for microorganism immobilization. *Process Biochem.* 46, 566–571.

Taniguchi, M., Tohma, T., Itaya, T., Fuji, M. 1997. Ethanol production from a mixture of glucose and xylose by co-culture of *Pichia stipitis* and a respiratory-deficient mutant of *Saccharomyces cerevisiae. J. Ferment. Bioeng.* 83 (4), 364–370. 10.1016/S0922-338X(97)80143-2

Tye, Y.Y., Lee, K.T., Abdullah, W.N.W., Leh, C.P. 2011. Second-generation bioethanol as a sustainable energy source in Malaysia transportation sector: Status, potential and future prospects. *Renew. Sust. Energy Rev.* 15, 4521– 4536. 10.1016/j.rser.2011.07.099

Van Vleet, J.H., Jeffries, T.W. 2009. Yeast metabolic engineering for hemicellulosic ethanol production. *Curr. Opin. Biotechnol.* 20, 300–306. 10.1016/j.copbio.2009.06.001

Van, D.P., Fujiwara, T., Tho, B.L., Pam P.P., Toan, P.S., Minh, G.H. 2020. A review of anaerobic digestion systems for biodegradable waste: Configurations, operating parameters, and current trends. *Environ. Eng. Res.* 25(1), 1–17. 10.4491/eer.2018.334

Vavouraki, A.I., Angelis, E.M., Kornaros, M. 2013. Optimization of thermo-chemical hydrolysis of kitchen wastes. *Waste Manage.* 33, 740–745.

Wang, F.S. Lin, H.T. 2010. Fuzzy optimization of continuous fermentations with cell recycling for ethanol production. *Ind. Eng. Chem. Res.* 49(5), 2306–2311. 10.1021/ie901066a

Wang, G-J., Wang, Z-S., Zhang, Y-W., Zhang, Y-Z. 2012. Cloning and expression of amyE gene from *Bacillus subtilis* in *Zymomonas mobilis* and direct production of ethanol from soluble starch. *Biotechnol. Bioprocess Eng.* 17, 780–786. 10.1007/s12257-011-0490-z

Wang, Q., Peng, L., Su, H. 2013. The effect of a buffer function on the semi-continuous anaerobic digestion. *Bioresour. Technol.* 139, 43–49. 10.1016/j.biortech.2013.04.006

WRAP. Operators Briefing Note. Defra Research Project, U.K. 2012. Available online: http://www.wrap.org.uk/sites/files/wrap/Operators%20Briefing%20Note.pdf (accessed on 12 November 2016).

Xu, F., Li, Y., Ge, X., Yang, L., Li, Y. 2018. Anaerobic digestion of food waste–Challenges and opportunities. *Bioresour. Technol.* 247, 1047–1058.

Yan, S., Chen, X., Wu, J. 2013. Pilot-scale production of fuel ethanol from concentrated food waste hydrolysates using Saccharomyces cerevisiae H058. *Bioprocess Biosyst Eng.* 36, 937–946. 10.1007/s00449-012-0827-9

Yan, S., Li, J., Chen, X., Wu, J., Wang, P., Ye, J., Yao, J. 2011. Enzymatical hydrolysis of food waste and ethanol production from the hydrolysate. *Renew. Energy.* 36, 1259–1265. 10.1016/j.renene.2010.08.020

Yin, Y., Liu, Y-J., Meng, S-J., Kiran, E.U., Liu, Y. 2016. Enzymatic pretreatment of activated sludge, food waste and their mixture for enhanced bioenergy recovery and waste volume reduction via anaerobic digestion. *Appl. Energy* 179, 1131–1137. 10.1016/j.apenergy.2016.07.083

Yu Y., Lou X., Wu H. 2008. Some recent advances in hydrolysis of biomass in hot-compressed water and its comparisons with other hydrolysis methods. *Energy fuels.* 22, 46–60.

Zhang, C., Su, H., Baeyens, J., Tan, T. 2014. Reviewing the anaerobic digestion of food waste for biogas production. *Renew. Sustain. Energy Rev.* 38, 383–392. 10.1016/j.rser.2014.05.038

Zhang, J., Wang, Q., Zheng, P., Wang, Y. 2014. Anaerobic digestion of food waste stabilized by lime mud from papermaking process. *Bioresour. Technol.* 170, 270–277.

Zhang, Q.Q., Tian, B.H., Zhang, X., Ghulan, A., Fang, C.R., He, R. 2013. Investigation on characteristics of leachate and concentrated leachate in three landfill leachate treatment plants. *Waste Manage.* 33(11), 2277–2286.

Zhang, Z., Zhang, G., Li, W., Li, C., Xu, G. 2016. Enhanced biogas production from sorghum stem by co-digestion with cow manure. *Int. J. Hydrog. Energy.* 41, 9153–9158. 10.1016/j.ijhydene.2016.02.042

7 Sugarcane Waste: Current Scenario in Microbial Valorization for Biomolecules

Juliana da Conceição Infante
Universidade de São Paulo (USP) campus Ribeirão Preto, Faculdade de Medicina de Ribeirão Preto, Departamento de Bioquímica e Imunologia, Ribeirão Preto, SP, Brazil

Rosymar Coutinho de Lucas
Universidade Federal de Alfenas, Departamento de Bioquímica, Alfenas, MG, Brazil

Tássio Brito de Oliveira
Universidade Federal da Paraíba, Centro de Ciências Exatas e da Natureza, Departamento de Ecologia e Sistemática, João Pessoa, PB, Brazil

Thiago Machado Pasin
Universidade de São Paulo (USP) campus Ribeirão Preto, Faculdade de Medicina de Ribeirão Preto, Departamento de Bioquímica e Imunologia, Ribeirão Preto, SP, Brazil

Vivian Machado Benassi
Universidade Federal dos Vales do Jequitinhonha e Mucuri (UFVJM) campus JK, Instituto de Ciência e Tecnologia Diamantina, MG, Brazil

David Lee Nelson
Universidade Federal dos Vales do Jequitinhonha e Mucuri (UFVJM) Campus JK Diamantina, MG, Brazil

CONTENTS

DOI: 10.1201/9781003128977-7

7.1 INTRODUCTION

Many global actions are being taken to minimize the rampant consumption of fossil sources (FAO 2010). However, these attitudes alone are not sufficient to reduce this impact. Thus, an alternative to solve these problems is the use of biomass to produce energy and other bioproducts (de Lucas et al. 2021).

Biomass is defined as a renewable resource from all organic material (plant or animal) that can be used to produce energy or biomolecules of greater value (Goldemberg 2011). The most abundant biomass in nature is the lignocellulosic raw material, such as agro-industrial and municipal waste and woods from angiosperms and gymnosperms (Segato et al. 2014). Biomass can be used for the generation of energy and bioproducts, and its composition can vary according to the species, variety, and season (Kowalczyk, Benoit and de Vries 2014; Mäkelä, Donofrio and de Vries 2014). The main sources of lignocellulosic waste produced by global agribusiness are sugarcane, corn, rice, and wheat (Saini et al. 2015).

The use of sugarcane or other residues for the production of high-value biomolecules is called waste valorization (Arancon et al. 2013). This practice has been gaining special attention in local communities and, mainly, in the industrial environment because of the indiscriminate use of natural resources and their possible depletion for future generations. It is a sustainable way of consuming and producing eco-friendly products that are focused on the preservation of the environment, and the planet on which we live (Mateo and Maicas 2015).

7.1.1 Sugarcane

Grasses comprise a group of monocotyledonous plants of well-known economic importance. They are distributed across all continents, and their evolutionary success is mainly due to their genetic diversity. The use of these plants as a nutritional source is closely related to their photosynthetic capacity, which results in the accumulation of carbohydrates. Corn, oats, wheat, and sugarcane are among the most representatives used for cultivation and subsistence (Hodkinson 2018).

Tropical grasses have a special characteristic, which is the use of an alternative photosynthetic pathway, the C4 pathway. The use of this pathway leads to an increase in the efficiency of sugar synthesis in these plants, which results in greater production than those plants that depend on the C3 pathway. The consequence of this event is an increase in the production of biomass (Furbank 2011).

In Brazil, the representatives of the genus *Saccharum* L, popularly known as sugarcane, are crops currently dominated by hybrids of the species *Saccharum officinarum* L and should be underscored. Originating in southern and southeastern Asia, it was introduced in this country in the 16th century in the sugar mill period (Cheavegatti-Gianotto et al. 2011). Because of its high capacity to store sucrose, which is transformed into alcohol through fermentation processes, this plant has taken over the economic scenario, making Brazil the world's largest producer. Approximately 590.36 thousand tons were produced in the south-central region of the country in 2019/2020 (Annual Sustainability Report 2019/2020).

The sugarcane cell wall is formed by arabinoxylans, xyloglucans, β-glucan, pectins (homogalacturonan and rhamnogalacturonan I), phenolic compounds, and cellulose microfibrils (Carpita and Gibeaut 1993; Carpita 1996; Vogel 2008; Buckeridge et al. 2019). Additionally, lignin, which is formed from coniferyl and sinapyl alcohols and *p*-coumaryl monomers, is present in the secondary cell wall. It connects the entire structure (all the polymers), including pectin (Carpita and Gibeaut 1993; de Lucas et al. 2020). Buckeridge et al. (2019) proposed the existence of a glycomic code, which consists of a complex network of covalent, ionic, and specific and nonspecific noncovalent bonds that connect the entire structure of the sugarcane cell wall, making it very recalcitrant.

To overcome this barrier, the pretreatment of lignocellulosic biomass is essential to ensure the conversion of polysaccharides into monomers that can be used by different industries. This type of strategy, which can be chemical, physical, or biological, has the function of exposing cellulose fibers, as well as the hemicellulose structure of sugarcane, to facilitate the enzymatic attack and, consequently, the formation of the monomers (Pasin et al. 2020). This biomass, whose polymeric constitution is so rich, contains about 38–43% glucan, 25–32% hemicellulose, 17–24% lignin, and 1.6–7.5% of extractives, and it can be explored in its "*in natura*" or waste form for the production of biomolecules of greater value (de Souza et al. 2012).

7.1.2 Sugarcane Industry Wastes

The increase in production and the improvement of the final product quality, combined with the increase in areas cultivated for sugarcane, result in the co-production of large amounts of waste. The main by-products of the sugarcane industry are sugarcane bagasse, molasses, vinasse, and press mud (or filter cake). These residues are produced at different processing stages of sugar/ethanol production, and they are potential pollutants unless they are recycled.

The ethanol production process begins with the arrival of raw materials at the mill. The cane is then washed to eliminate soil and impurities and conveyed to the basic preparation where its density and grinding capacity increase so that the maximum rupture of the cells to release the juice occurs (BNDE 2008). After harvesting and washing, the cane is milled to extract the juice, which is used to produce first-generation (1G) ethanol.

Sugarcane bagasse (SCB) is the fibrous residue that remains after the extraction of the juice, and it is usually burned in a boiler to produce heat and electricity (Anukam et al. 2016). The annual global production was recently estimated at approximately 1.6 billion tons (Khattab and Watanabe 2019). Considering that about 0.28 tons of SCB are generated for each ton of sugarcane produced annually worldwide, this fact means that 448 million tons of SCB were produced. The cell walls of sugarcane straw and bagasse are composed of a recalcitrant polymeric structure containing 31–48.6% cellulose, 25–32% hemicellulose, 17–24% lignin, and 1.2–7% ash or extractives as the main components (de Souza et al. 2012).

Press mud is produced during the clarification of the sugarcane juice by the carbonation or sulphitation process, in which the clear juice is separated at the top and the mud at the bottom. Approximately 26–40 kg of this residue is produced after crushing one ton of sugarcane (Bhosale et al. 2012). It is a residue rich in phosphorus, nitrogen, organic matter (with a C/N ratio of approximately 29), and high moisture content, and it has been largely used as fertilizer for crops (Oliveira et al. 2016).

Molasses is a dark, viscous, and sugar-rich syrup remaining after the crystallization of sugar from cane juice. It is separated from the sugar crystals by repeated centrifugation, and it is produced at the rate of 40 kg/ton of cane (Santos et al. 2020). Its composition varies with the variety of cane, climate, and processes employed for the extraction of sugar. Molasses is composed of readily biodegradable sugars including approximately 34% of sucrose and 11% reducing sugars (glucose and fructose) and several minerals (Sindhu et al. 2016).

Another byproduct, vinasse, is generated in the distillation stage at an average rate of 13 L per liter of ethanol produced (BNDES 2008). The fermented wort (wine) is composed of 7–10% by volume of alcohol, in addition to other components of a liquid, solid, and gaseous nature. The alcohol present in this wine is recovered by distillation, a process of separating components of a mixture based on its volatility at a given temperature and pressure. In distillation, the mixture is heated to boiling, and the vapors are cooled for condensation. Thus, the final effect is to increase the concentrations of the most volatile component in the vapor and the least volatile component in the liquid. Because of the difference in density between the liquid and vapor phases, the liquid descends the distillation column, whereas the vapor rises, and it is collected separately. The nonvolatile fraction of this process is vinasse, which will be retained at the base of the column in the boiler. Vinasse contains organic carbon components, such as residual fermentable sugars, non-fermentable sugars, and organic acids, which result in high chemical oxygen demand (COD). It also contains nitrogen, phosphorus, potassium, and sulfur, which result in a high-fertilization potential for the crop. Therefore, it is sent back to the cultivation area in the fertigation process (da Silva Neto et al. 2020).

7.2 MICROBIAL VALORIZATION OF BIOMOLECULES

7.2.1 Microbial Enzymes in the Biodegradation of Sugarcane Wastes

The use of sugarcane waste as a carbon source for the production of microbial enzymes has awakened a new change in the bioeconomy in recent years. The decrease in the world's dependence on the use of raw materials from fossil sources and its substitution by sustainable processes has resulted in the development of products of high environmental, economic, and social value. This change has contributed to the achievement of the sustainable development goals set by the United Nations (UN) for 2030. The sugarcane wastes and byproducts are almost totally utilized, and they can be used in human and animal nutrition, in soil fertilization, in the bioconversion to 2G ethanol, and in the co-generation of energy (Cintra et al. 2020). Microbial enzymes are easily produced and economically viable. They are secreted by a wide variety of bacteria and fungi, they are widely used in industry, and they can act in free or immobilized form, purified, or in enzyme pools. Because of their practicality, they are used in a wide variety of applications in various sectors of the economy.

The use of microbial enzymes in the biodegradation of sugarcane wastes has a range of advantages, including low energy consumption through the use of mild conditions, lower cost of the substrate, and the nonproduction of byproducts. Also, other factors are of great importance for the choice of a specific enzyme or an enzyme consortium in an industrial process, including illumination, temperature, pH, stirring rate, inhibitors, specificity, purity, activators, analysis methods, costs, and substrates (Siqueira and Filho 2010; Farinas et al. 2010; Barchuk et al. 2016).

The degradation of lignocellulosic biomass by microorganisms occurs extracellularly, and it is attributed to the action of a series of enzymes that act synergistically. The microbial strategies in the production of enzymes for the depolymerization of lignocellulosic biomass are very complex because of the high complexity of the structures and their recalcitrance. The combination of hemicellulose and lignin forms a protective barrier around the cellulose, which must be removed. The removal of lignin is a key challenge for the access of enzymes to hemicellulose and cellulose. Also, the crystalline structure of cellulose makes it insoluble and resistant to enzymatic catalysis (Jaramillo et al. 2015; Andlar et al. 2018) (Table 7.1).

The degradation of polysaccharides to sugar monomers requires the synergistic action of several classes of enzymes (CAZymes). Among the CAZymes, the Carbohydrate-Active Enzymes Database (CAZy) classification, glycosyl hydrolases

TABLE 7.1
Main enzymes for the Biodegradation of Lignocellulosic Biomass from Sugarcane Wastes

Cellulose	Hemicellulose		Pectin	Lignin
Exo-β-glucanase (EC 3.2.1.91)	Endo-β-xylanase (EC 3.2.1.8)	Endo-β-mannanase (EC 3.2.1.78)	Endo-polygalacturonase (EC.3.2.1.15)	Manganese peroxidase (EC1.11.1.13)
Endo-β-glucanase (EC 3.2.1.4)	Exo-β-xylanase, β-xylosidase (EC 3.2. 1.37)	Exo-β-mannanase (EC 3.2.1.25)	Exo-polygalacturonase (EC.3.2.1.67; (EC.3.2.1.82)	Lignin peroxidase (EC1.11.1.14)
β-glucosidase (EC 3.2.1.21)	Acetyl xylan esterase (EC 3.1.1.72)	Acetyl esterase (EC 3.1.1.6)	Pectinesterase (EC 3.1.1.11)	Laccase (EC 1.10.3.2)
	Feruloyl esterase (EC 3.1.1.73)	α-galactosidase (EC 3.2.1.22)	Pectin Lyase (EC 4.2.2.10)	
	α-L-arabinofuranosidase (EC 3.2.1.55)	α-glucuronidase (EC 3.2.1.131)	Pectate lyase (EC 4.2.2.2)	
			Proto-pectinases (3.2.1.99)	

(GHs) are the main families of enzymes involved in the degradation of polysaccharides such as cellulose and hemicellulose (Siqueira and Filho 2010; Berlemont and Martiny 2016; Nguyen et al. 2018). Thus, the performance in the enzymatic biodegradation of lignocellulosic biomass depends on the structure of the substrate and the interaction of enzymes with the substrate, which varies with the nature and source of the enzyme complex (Brienzo et al. 2016).

Many cellulases and hemicellulases are linked to structures called carbohydrate-binding modules (CBM). The main function of these domains is to concentrate the enzyme molecules on the surface of the substrate, in addition to allowing the catalyst to bind specifically to its substrate, but they have no catalytic activity. In this context, the removal of CBM from the enzyme can result in a decrease in enzyme activity and enzyme stability (Morril et al. 2015; Andlar et al. 2018; Teo et al. 2019; Sidar et al. 2020). The CBM domains can be single or multiple (duplicates) and connected to the C and/or N-terminus of the catalytic domain, and they consist of approximately 30–200 amino acids. The organization of the CBM domain of glycoside hydrolases (GHs) varies within organisms (Lombard et al. 2014; Sidar et al. 2020).

7.2.2 Cellulose – Degrading Enzyme System

Cellulases are glycosyl hydrolases produced by bacteria and fungi, and few species are capable of producing sufficient cellulases to dissolve crystalline cellulose (Bhat and Bhat 1997). The complete hydrolysis of cellulose requires the action of three groups of enzymes acting synergistically. These groups of enzymes correspond (i) to endo-1,4-β-glucanases (EC 3.2.1.4, endocellulase), which cleave the cellulose molecule to create new reducing and non-reducing ends and release smaller fragments that will be used as substrates for (ii) exo-1,4-β-glucanases (EC 3.2.1.91, exo-glucosidase or cellobiohydrolase) that cleave cellobiose molecules produced from cellulose or cellulose-derived oligosaccharides. These enzymes are specific for reducing and non-reducing ends. Finally, (iii) 1,4-β-glucosidases (EC 3.2.1.21, cellobiase) release D-glucose from the nonreducing ends of oligosaccharides obtained from cellulose, especially cellobiose (Jovanovic et al. 2009).

Some cellulases are organized in supramolecular structures (multienzyme complexes) called cellulosomes, which can efficiently degrade cellulosic substrates and improve the efficiency of hydrolysis (Bayer et al. 2004; Hu and Zhu 2019; Pasin et al. 2020). These structures are produced by anaerobic bacteria (Smith et al. 2017; Hu and Zhu 2019) and fungi (Haitjema et al. 2017; Gilmore et al. 2020). They are composed of a complex framework such as a structural subunit and several enzymatic subunits (Pasin et al. 2020). Scientific studies have shown that various substrates such as lignocellulosic biomass, crystalline cellulose, xylan, pectin, etc., can be efficiently degraded by cellulosomes (Hu and Zhu 2019).

Endo-β-1,4-glucanases (EC 3.2.1.4) catalyze the hydrolysis of the β-(1,4) glycosidic bonds of the cellulose chains. They act randomly in the amorphous regions of the cellulose to create new terminals of long-chain oligomers. They exhibit a rapid rate of dissociation and cause a reduction in the viscosity of the medium (Lynd et al. 1999). Endoglucanases (EG) can also be called carboxymethylcellulases (CMCase) because the artificial substrate carboxymethylcellulose (CMC) is used to measure enzyme activity (Annamalai et al. 2016a). They belong to CAZy families GH5, GH6, GH7, GH9, GH12, GH45, and GH74 (Andlar et al. 2018). EGs from GH5, 7, 12, and 45 families are the most common naturally occurring fungal cellulases (Annamalai et al. 2016b). Zhao et al. (2019) reported the discovery of an endo-β-1,4-glucanase belonging to the GH10 carboxymethylcellulase CAZy family. It was isolated from *Arcticibacterium luteifluviistationis* (*Al*CMCase), an arctic marine bacterium that exhibits tolerance to extreme cold and salt.

Exoglucanases (EC 3.2.1.91), also known as cellobiohydrolases (CBH), are produced by a wide diversity of bacteria and fungi with catalytic modules that belong to the GH5, GH6, GH7, GH9, GH48, and GH74 CAZy families (Annamalai et al. 2016b). GH6 and GH7 are found mainly in filamentous fungi (Gusakov et al. 2017). They catalyze the cleavage of disaccharides (cellobiose) and a small number of trisaccharides (cellotrioses) from the end of the cellulose chain. Some CBHs can catalyze the hydrolysis only at the reducing ends (EC 32.1.176), whereas others only act at the nonreducing terminals (EC 3.2.1.91) (Annamalai et al. 2016b; Andlar et al. 2018; Pasin et al. 2020). Enzymes that exhibit high activities with microcrystalline cellulose (Avicel) and low activities with carboxymethylcellulose (CMC) are usually identified as exoglucanases (CBHs) (Figure 7.1).

β-Glucosidases (β-D-glucoside, glucohydrolase, EC 3.2.1.21) catalyze the cleavage of short-chain oligosaccharides and cellobiose to liberate glucose. They belong to a heterogeneous family of enzymes that can hydrolyze glycosidic bonds

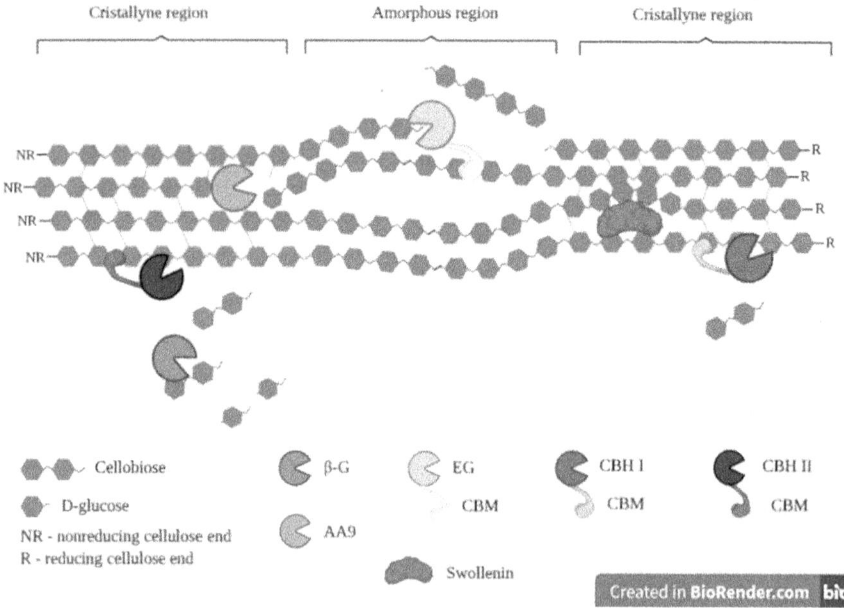

FIGURE 7.1 Illustration of enzymatic degradation of crystalline cellulose and crystal structures with synergistic interaction of cellulases (endoglucanase, exoglucanase, and β-glucosidase), auxiliary enzyme AA9, and the protein swollenin (Created with BioRender.com).

of alkyl-, amino-, or aryl-β-D-glycosides, cyanogenic glycosides, disaccharides, and short oligosaccharides. They are also able to catalyze the synthesis of glycosyl bonds between various molecules to produce different oligosaccharides through transglycosylation. (Salgado et al. 2018; Pasin et al. 2020). They are components of cellulosomes (Singh et al. 2016). These enzymes are present in the Archaea, Bacteria, and Eukarya domains, and they are involved in several cellular functions, such as the hydrolysis of the plant cell wall by microorganisms, breakdown of glycolipids and lignification processes involving defense against pests, activation of phytohormones, etc. (Ketudat Cairns and Esen 2010; Singh et al. 2016; Salgado et al. 2018). They are found mainly in the GH1 or GH3 CAZy families, although these enzymes are also found in GH5, GH9, and GH30 families (Singhania et al. 2013). Many β-glucosidases are inhibited by the reaction product, glucose (Salgado et al. 2018). The complete conversion of cellulose to glucose by β-glucosidases is limited because they are susceptible to inhibition by oligosaccharides. Thus, the production of microbial β-glucosidases with a high glucose tolerance has been extensively researched, and they have been added in large quantities to enzymatic cocktails to weaken the inhibition exhibited by cellobiose on exoglucanases (CBH) and endoglucanases (Zhenming et al. 2009; Andríc et al. 2010; Ketudat Cairns and Esen 2010; Singhania et al. 2013).

The fungal polysaccharide lytic monooxygenases (LPMO), belonging to the AA9 (formerly GH61) family, are copper-dependent oxidative enzymes. They perform a fundamental role in the degradation of the recalcitrant lignocellulosic biomass because they enhance the action of cellulases during hydrolysis. They promote the oxidative cleavage of cellulose chains by oxidation of carbons C1 or C4 and C6, or all three, by acting in the crystalline regions, causing a rupture in the cellulose chain, permitting access to the canonical cellulases, and improving the overall hydrolysis. They can connect to a carbohydrate-binding module (CBM). The oxidative action promoted by the LPMO requires an external electron donor, such as ascorbic acid, cellobiose dehydrogenase, lignin, and other plant biomass-derived phenols. The oxidation of LPMOs only occurs in the presence of O_2 or H_2O_2 as a co-substrate, and it even exhibits a preference for H_2O_2 over O_2 (Monclaro et al. 2017; Bertini et al. 2018; Várnai et al. 2018; Chalak et al. 2019; Pasin et al. 2020).

Swollenins (SWOs) are a class of accessory (non-enzymatic) proteins homologous to plant expansins. They are known to act in the separation of the crystalline cellulose structure by loosening the macrofibrils. They increase the accessibility and efficiency of the other enzymes involved in deconstructing the cellulose structure (Santos et al. 2017; Pasin et al. 2020).

7.2.3 Hemicellulose – Degrading Enzyme System

Hemicellulases are produced by several microorganisms, including bacteria, filamentous fungi, and yeasts, that have already been investigated and reported in the literature. They are key components in the degradation of biomass that requires a combined action of several enzymes, acting synergistically, for its complete degradation (Shallom and Shoham 2003; Polizeli et al. 2005). Hydrolysis occurs like the hydrolysis of cellulose, where the enzymes involved in this process are specific hydrolases that cleave certain types of bonds that exist in the polymer. Thus, hemicellulases are divided into two groups: endo-hemicellulases and exo-hemicellulases. Xylans and mannans generally have different substituent groups attached to the main chain, and auxiliary enzymes are required to remove these groups and provide access to the enzymes responsible for the degradation of the main chain (Shallom and Shoham 2003; Polizeli et al. 2005; Meyer et al. 2009).

Xylan, a β-1,4-polymer of xylose, is a principal component of hemicellulose (Pasin et al. 2020). A complex system of several enzymes acting synergistically is necessary for the complete hydrolysis of xylan. Xylans generally have different substituent groups attached to the main chain, and auxiliary enzymes are needed to remove these groups and provide access to enzymes that are responsible for the degradation of the main chain. The auxiliary enzymes α-L-arabinofuranosidases (EC 3.2.1.55), α-glucuronidase (EC 3.2.1.131), acetylxylan esterase (EC 3.1.1.72), and feruloyl esterase (EC 3.1.1.73) specifically cleave the links between hemicellulose and lignin. The α-L-arabinofuranosidases have different specificities; some cleave α-1,2-, α-1,3-, or α-1,5- bonds, whereas others can cleave arabinoxylan and arabinose residues. When the β-1,4-glycosidic bonds of the xylan main chain are cleaved, the products formed are xylooligosaccharides, xylobiose, and xylose (Polizeli et al. 2005; Meyer et al. 2009; Alokika and Singh 2019; Cintra et al. 2020). The acetyl groups next to the glucuronosyl substituents can partially obstruct α-glucuronidase activity (Polizeli et al. 2005). Endo-1,4-β-xylanases (EC 3.2.1.8) cleave the internal β-1,4-glycosidic bond of xylan to release branched xylooligosaccharides and arabinoxylooligosaccharides of different lengths. Exo-β-xylanases or β-xylosidases (EC 3.2.1.37) cleave the oligosaccharides at the nonreducing end, releasing xylose, whereas α-L-arabinofuranosidases remove the residues of α-L-arabinofuranosyl from the xylan chain (Polizeli et al. 2005; Heinen et al. 2018; Cintra et al. 2020).

Mannanases hydrolyze the mannan chains present in the hemicellulosic fraction. Microbial mannanases occur in extracellular form, and they can be induced by several substrates containing β-mannans in their structure, such as locust bean gum, guar gum, and coconut flour (Kote et al. 2009). Some mannanases have a carbohydrate-binding module (CBM). The mannanases are classified in the GH5, GH26, and GH113 CAZy families (Malgas et al. 2015; Pasin et al. 2020). The complete mannan hydrolysis involves β-mannanase, β-mannosidase, and β-glucosidase, in addition to accessory enzymes such as acetylmannan esterase and α-galactosidase, which will remove groups from the side chains (Moreira and Filho 2008; de Marco et al. 2015; Malgas et al. 2015). The main chain of mannan is hydrolyzed by endo-β-D-mannanase (EC 3.2.1.78), whereas α-galactosidase (EC 3.2.1.22) and acetylmannan esterase (EC 3.1.1.6) release galactose and acetyl groups, respectively. The products generated by endo-β-D-mannanase – mannose dimers and glycomannoses – are then hydrolyzed by β-mannosidase (EC 3.2.1.25) and β-glucosidase (EC 3.2.1.21) to liberate mannose, glucose, and galactose (Van Zyl et al. 2010, de Marco et al. 2015). The addition of mannanases to enzyme pools facilitates the contact of cellulase with cellulose fibrils by removing glucomannans

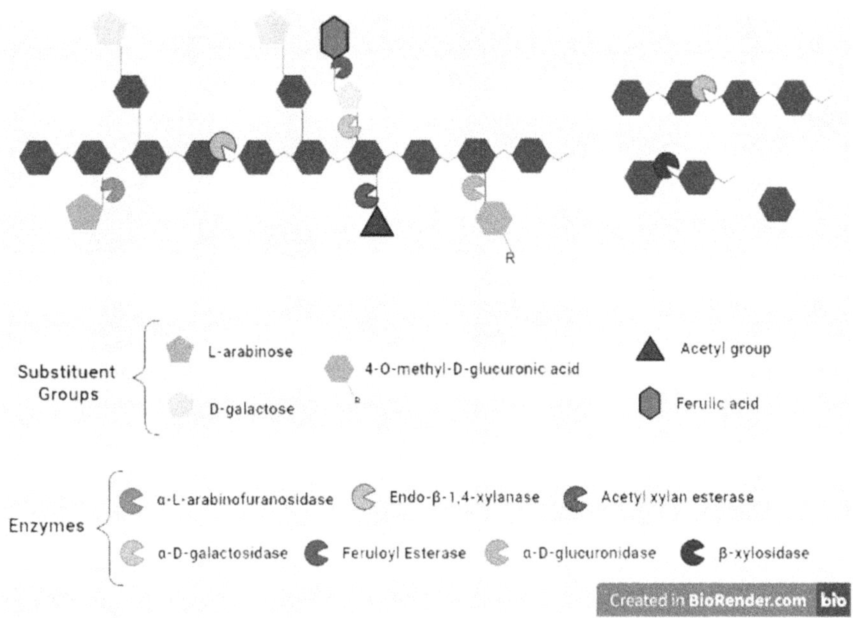

FIGURE 7.2 Illustration of enzymatic degradation of xylan structure with synergistic interaction of xylanases (endo-β-1,4-xylanase and β-xylosidase) and auxiliary enzymes (α-L-arabinofuranosidase, α-galactosidase, α-D-glucuronidase, acetylxylan esterase, and feruloyl esterase) (Created with BioRender.com).

present on the cell surface and increasing the release of reducing sugars (Pham et al. 2011) (Figure 7.2).

7.2.4 Pectin – Degrading Enzyme System

Pectinases (EC.3.2.1.15) is a group of complex hydrolases that degrade pectin. Because of the wide variety of pectins present in vegetables, pectinases are endowed with a complex system that acts synergistically to cause the complete depolymerization of pectin. Hydrolysis reactions occur through depolymerizing enzymes (hydrolases and lyases), de-esterifying enzymes (pectin esterases), and protopectinases. The enzymes include polygalacturonase (PGs), pectinesterases (PEs), pectin lyases (PLs), and pectate lyases (PGLs) with different substrate specificities (Tarik and Latif 2012; Polizeli 2013; Amin et al. 2019). Depending on the site of action, they can also be classified as endopectinases or exopectinases (John et al. 2020).

The esterases involved in the degradation of pectin are pectinesterases (EC 3.1.1.11) or pectin methylesterases, which cleave the methoxyl group of pectin to form pectic acid. Pectin acetyl esterase (EC 3.1.1.6) or acetyl esterase cleaves acetyl esters in the pectin HG regions. Polygalacturonases (PGs) are hydrolases that cleave the α-1,4-glycosidic bond in the structure of HG, and they belong to the GH28 CAZy family. Polygalacturonases are classified as endo and exoenzymes.

Endopolygalacturonases (EC 3.2.1.15) hydrolyze polygalacturonic acid randomly by cleaving the (1–4)-α-D-galacturonoside bonds in pectate and other galacturonans. Exopolygalacturonase type I (EC 3.2.1.67) acts on the nonreducing end by successively hydrolyzing D-galacturonic acid. Exopoligalacturonase type II (EC 3.2.1.82) hydrolyzes the polygalacturonic acid chain at the nonreducing ends to release di-galacturonate (Khan et al. 2013; Amin et al. 2019; Nighojkar et al. 2019). The lyases are classified into pectin lyases (EC 4.2.2.10) or endopectin lyases that randomly cleave pectin by trans-elimination reactions, forming double bonds between C4 and C5 of galacturonic acid residues at the nonreducing end. Endopectate lyases (EC 4.2.2.2) and exopectate lyases (EC 4.2.2.9) randomly cleave the non-reducing end of α-1,4-polygalacturonic acid through a trans-elimination process (Yang 2018; Nighojkar et al. 2019; John et al. 2020). Protopectinases (3.2.1.99) cleave the structures of protopectins that are insoluble in polymerized soluble pectin (John et al. 2020).

7.2.5 Lignin – Degrading Enzyme System

Lignin cannot be cleaved by hydrolytic enzymes like most other natural polymers (cellulose, starch, proteins, etc.). The biodegradation of lignin is an oxidative process that requires delignifying enzymes that act on lignin to relax the recalcitrant structure of the lignocellulosic biomass. They can also remove some inhibitors (mainly phenols) that interfere with the fermentation process. The degradation of lignin is mainly accomplished by laccases and heme-peroxidases such as manganese peroxidases and lignin peroxidases (Andlar et al. 2018; Aragão et al. 2020).

Laccases (EC 1.10.3.2; CAZy AA1) are glycoproteins that belong to the group of oxy-reductases. They are multi-copper enzymes capable of degrading the lignin present in the lignocellulosic residues and oxidizing a variety of phenolic and non-phenolic compounds. They can be monomeric, dimeric, or tetrameric, the last of which usually contains four copper atoms per monomer, distributed in three redox sites called T1, T2, and T3. It can oxidize a wide range of substrates, and it is referred as laccase-mediator system (LMS) (Yao and Ji 2014; Andlar et al. 2018; Brenelli et al. 2018; Pinheiro et al. 2020). Manganese peroxidases (EC1.11.1.13) and lignin peroxidases (EC1.11.1.14) are produced by white-rot basidiomycetes and catalyze hydrogen peroxide-dependent oxidative degradation of lignin from phenolic and nonphenolic substrates, respectively (Plácido and Capareda 2015).

7.3 MICROBIAL BIOPRODUCTS OBTAINED FROM DIFFERENT SUGARCANE WASTE

As was previously discussed, cellulose, hemicellulose, and lignin are the principal chemical components of all sugarcane solid waste materials. Therefore, they can be recycled and utilized for a variety of products of commercial interest. Some applications of the lignocellulosic by-products of the sugarcane industry are in the production of biofuels, organic acids, alcohols, enzymes, biosorbents, biofertilizers, food, prebiotics, and active compounds, as previously mentioned.

7.3.1 Biofuels

The term biofuel refers to liquid or gaseous fuels for the transport sector that is predominantly produced from biomass. Biomass appears to be an attractive feedstock for three main reasons. First, it is a renewable resource that could be sustainably developed in the future. Second, it appears to have formidably positive environmental properties resulting in no net release of carbon dioxide and very low sulfur content. Third, it appears to have significant economic potential, provided that fossil fuel prices continue to increase in the future (Demirbas 2008). Also, two major crises resulting from the growing global consumption of fossil fuels (80% of the required energy) have occurred in recent years: environmental pollution and a growing acceleration in the decrease in energy resources (Parsaee et al. 2019). Therefore, the search for clean and renewable energy sources is at the top of the world's agenda. Biomass represents the principal renewable resource in the world that can replace fossil fuels (Parsaee et al. 2019). One of the important advantages of energy production from biomass, that is, sugarcane waste, is a smaller requirement for capital investment compared to other renewable energies such as hydro, solar, and wind (Parsaee et al. 2019).

7.3.2 Bioethanol

Second-generation bioethanol comprises all processes for obtaining ethanol whose raw material is cellulosic biomass, such as sugarcane, corn, and cellulose. It is a type of biofuel. The bioethanol production from lignocellulosic biomass is composed of the following main steps: hydrolysis of cellulose and hemicellulose, sugar fermentation, separation of lignin residues, and finally recovery and purification of ethanol to meet the fuel specifications. This compound can be produced from several resources, including sugar crops (e.g., sugar beets, sugarcane, molasses, sweet sorghum, and grapes), starch crops (e.g., corn, wheat, rice, cassava, and barley), tuber crops (e.g., potato and cassava), dairy products and cellulose materials (e.g., rice straw, bagasse, wood, and municipal solid waste), and the agave plant, in a mesophilic fermentation using mainly *Saccharomyces cerevisiae* (Siles et al. 2011).

The two main types of waste that result from the production of sugar and first-generation bioethanol from sugarcane are agricultural residues and industrial by-products. The former consists of burned or crushed sugarcane in the harvesting process. The latter consists of five types of waste in addition to the main product: ash, bagasse, press mud, molasses, and vinasse (Nogueira et al. 2015). All of these wastes can be used in the second-generation production of bioethanol, which makes it a viable and eco-friendly way of turning residues into valuable components. Brazil currently produces 50% of the total sugarcane-based bioethanol in the world (Parsaee et al. 2019).

The leading sugarcane producers in the world are Brazil, India, China, Thailand, Pakistan, Mexico, Colombia, Indonesia, the Philippines, and the United States, with the contribution of 739.000, 341.200, 124.500, 100.100, 63.800, 61.200, 34.900, 33.700, 31.900, and 27.900 Mt per year, respectively. The corresponding shares are

38.1, 17.3, 6.1, 4.3, 3.2, 2.8, 1.8, 1.3, 1.1, and 1.4% (Pasin et al. 2020). However, an even larger market can be reached by improving the use of waste, mainly from sugarcane processing. Therefore, scientific research is essential to make the process more profitable in economic terms, less aggressive in environmental terms, and more accessible in social terms.

7.3.3 Biobutanol

The growing demand for liquid fuels has motivated the exploitation of renewable resources in the substitution of fossil fuels. It has been demonstrated that the introduction of butanol from lignocellulosic materials can result in higher revenues for biorefineries, compared to base scenarios, wherein ethanol is produced exclusively (Pratto et al. 2020). Thus, the opportunities for developing efficient refining of biomass that envisions the butanol market can contribute to the future implementation of biorefineries and result in progress for society and the economy. In addition to the economic aspects, butanol stands out from other biofuels because its fuel properties are similar to those of gasoline. Biobutanol is superior to bioethanol in many ways, such as a higher calorific value, lower volatility, and lower corrosiveness (Guan et al. 2016). Additionally, butanol can be blended with gasoline up to 40% (v/v) without any adverse effects on the performance of the spark-ignition engine (Pratto et al. 2020). It can also be used as biojet fuel with further processing (Silva-Braz and Pinto-Mariano 2018).

Therefore, biobutanol is considered to be an advanced combustible and an excellent "green" substitute for fossil fuel. It is naturally produced by *Clostridium* bacteria via a conventional acetone–ethanol–butanol (ABE) fermentation process, with a ratio of 3:6:1. ABE fermentation occurs in two stages: an acidogenic phase during the exponential growth; followed by the solventogenic phase at the end of the exponential growth. In the acidogenic phase, the pathways for acid formation are activated to produce acetic and butyric acids, carbon dioxide, and hydrogen as the main products. At this stage, the pH of the medium drops from 6.5–7.0 to 4.5–5.0. In response to the lower pH, the clostridial metabolic pathway is directed toward solvent production. Finally, in the solventogenic phase, acids are re-assimilated to produce ABE and gases (Pratto et al. 2020). Therefore, as in the case of bioethanol, sugarcane wastes represent promising sources of second-generation sugars for the production of fuels and chemicals such as butanol in the principal sugarcane-producing countries around the world: Brazil, India, Colombia, China, and the United States (Chacón et al. 2020).

7.3.4 Biomethanol

Methanol (CH_3OH), also known as methyl alcohol, has been used as fuel blended with gasoline in internal combustion engines (85% methanol and 15% gasoline) or as pure methanol (100% methanol), but the latter is still in the research and development stage. Methanol also serves as a raw material for chemical products, such as formaldehyde, acetic acid, and a wide variety of other chemical products, including polymers, paints, adhesives, construction materials, synthetic chemicals, and others (Renó et al. 2011).

The biomethanol produced from renewable biomass sources is the most cost-effective alternative, and its cost is lower than that of light oil, which is used in power stations (Kasmuri et al. 2019). The comparison of biomethanol with gasoline (petroleum fuel) reveals that biomethanol has a higher average octane number at 98.65, rather than 85 to 96, respectively. Thus, the use of biomethanol can help to increase fuel energy efficiency, and it can become an attractive alternative in the biofuel sector. There are several methods in new and conventional processes for producing biomethanol, including pyrolysis, gasification, fermentation, electrolysis, and photoelectrochemical processes (Kasmuri et al. 2019). Therefore, it is imperative to gradually increase the use of biomass as feedstock in the production of biomethanol through the different processes to maintain renewable sources because it is environmentally friendly and its use is economically favorable, especially in developing countries.

7.3.5 Biohydrogen

Hydrogen is an alternative energy source to replace conventional fossil fuels. It is a clean and environmentally friendly fuel that produces water instead of greenhouse gases upon combustion. It has a high energy yield of 122 kJ/g, which is 2.75 times greater than that of hydrocarbon fuels. Hydrogen can be directly used to produce electricity through fuel cells (Pattra et al. 2008). Sugars such as glucose, fructose, galactose, arabinose, lactose, and sucrose have been widely studied as substrates for biohydrogen production by several bacterial strains, especially those of *Clostridium* genus (Rai et al. 2014). Because the cost of the substrate is of primary concern for the economics of biohydrogen production, there is a need to develop cheaper and abundant feedstocks to make the process cost-effective.

The use of organic residues from industries and agriculture not only supports green energy generation but also aids in bioremediation by reducing waste disposal. A large fraction of wastes from the municipal, industrial, and agricultural sectors composed of lignocellulosic biomass is renewable, inexpensive, and well-suited for biohydrogen production (Rai et al. 2014). One of the principal lignocellulosic materials found in large quantities, especially in tropical countries such as Brazil and India, is sugarcane waste produced during cane milling. There are well-established methods for the saccharification of sugarcane residue so it could represent an excellent low-cost alternative for biohydrogen production.

7.3.6 Biogas

Biogas is the common name given to a mixture of gases that are produced by the biological decomposition of organic matter in the absence of oxygen. It usually consists of a gas mixture composed mainly of methane (CH_4) and carbon dioxide (CO_2), with small amounts of hydrogen sulfide (H_2S) and moisture. Biogas production occurs naturally in any submerged location where atmospheric oxygen cannot penetrate, such as in swamps, the bottom of water glasses, animal intestines, or in an anthropogenic manner such as in landfills and biogas plants. It can be classified as a biofuel because it represents a renewable energy source and a way to obtain energy that can aid human beings in emancipating themselves from dependence on fossil fuels.

Biogas can be obtained from several different biomasses, such as food scraps, wood residues, rice straws, sugarcane bagasse, vinasse, animal manure, and other forms. Its production from organic waste has several advantages over other alternatives such as incineration, bio-oil, and other biological products (bioethanol, biohydrogen, and biodiesel) and electricity (Parsaee et al. 2019). Higher energy-yielding, lower environmental impact, and lower capital investment requirements are among the advantages (Parsaee et al. 2019). Biogas is obtained through a process involving four stages such as hydrolysis, acidogenesis, acetogenesis, and methanogenesis, using a microbial consortium containing different types of bacteria and archea.

The hydrolysis of organic matter is performed by different bacteria, including *Clostridium*, *Cellulomonas*, *Bacillus*, *Thermomonospora*, *Ruminococcus*, *Baceriodes*, *Acetovibrio*, and *Microbispora* genera. *Lactobacillus*, *Streptococcus*, *Bacillus*, and *Escherichia* are mainly responsible for acidogenesis (Christy et al. 2014). In acetogenesis, different genera, including *Acetobacterium*, *Syntrophomonas*, *Clostridium*, *Sporomusa*, *Syntrophospora*, *Thermosyntropha*, and *Eubacterium*, are involved. Finally, different archeas, including *Methanococcus*, *Methanosarcina*, and *Methanolobus*, are responsible for methanogenesis (Parsaee et al. 2019). In addition to the use of different organisms to produce biogas from wastes that damage the environment, the fact that, during the production of bioethanol from sugarcane wastes, a large volume of vinasse wastewater is generated. Vinasse is already known as the main source of biogas production and represents an excellent alternative, being a simple method for controlling environmental pollution, providing biofertilizers, and producing clean and cheap fuel. Vinasse has a low pH, a high chemical oxygen demand, and is rich in potassium.

7.3.7 Organic Acids

Organic acids result from the synthetic activities of microorganisms, plants, and animals. They are generally weak acids, soluble in water and organic solvents, and they can be produced by the metabolic activity of living beings. The most common organic acids are carboxylic acids, whose acidity is associated with the carboxyl group (COOH). Large-scale industrial production of bulk and specialty organic acids by microbial fermentation has undergone continuing growth, progressively expanding its market niche and portfolio within the chemical industry (Alonso et al. 2015).

The range of commercial applications of organic acids has increased. They have emerged as innovative building blocks and platform chemicals for the synthesis of novel bio-based materials that can replace unsustainable petroleum-based polymers (Lee et al. 2011). Whereas bulk commodity carboxylic acids such as lactic and citric acid have consolidated their place in the market (Alonso et al. 2015), the chemical industry has recently focused on emerging high-value organic acids with broader industrial applications. Succinic, fumaric, propionic, itaconic, and lactobionic acids have accordingly burst onto the industrial scene as relevant starting materials for the synthesis of several fine compounds, including pharmaceuticals, chemicals, and food additives (Alonso et al. 2015).

The potential of next-generation carboxylic acids has even been underscored by the USA Department of Energy (DOE), which has included up to nine organic acids among the future top 12 value-added platform chemicals (Alonso et al. 2015). Some of the most common organic acids obtained from the utilization of sugarcane wastes are on that list: lactic acid, citric acid, and succinic acid.

7.3.8 Lactic Acid

Lactic acid is a valuable chemical platform that has extensive applications in the food, cosmetics, textile, pharmaceutical, and chemical industries. Demand for lactic acid has grown substantially in recent years because of its great potential as a building block for the production of poly-lactic acid (PLA) materials, biodegradable agro-based products that are more environmentally friendly, and alternatives to petroleum-based plastics (Wischral et al. 2019). In 2013, global demands for lactic acids and PLA were 714.2 and 360.8 kilotons, respectively. Considering an expected annual growth of 15.5% and 18.8%, demands could reach 1,960 and 1,205 kilotons, respectively, by 2021.

Lactic acid has optical isomers, L-lactic acid, and D-lactic acid, which can be produced by chemical synthesis (DL-lactic acid) or microbial fermentation (L-lactic, D-lactic, or DL-lactic). Of these two methods, microbial fermentation processes have greater advantages because they make use of renewable substrates for lactic acid bacteria and require mild production conditions and low energy consumption (Wischral et al. 2019). Lactic acid bacteria are capable of producing lactic acid by the homo- or heterofermentative degradation of sugars. During the anaerobic growth of obligatory homofermentative lactic acid bacteria under conditions of an excess substrate, energy sources such as glucose are converted into pyruvate via the Embden-Meyerhoff-Parnas pathway (EMP-P), and pyruvate is further metabolized to lactate.

Throughout the heterofermentative pathway, sugars are catabolized via the phosphoketolase pathway (PK-P), resulting in equimolar amounts of CO_2, lactate, and acetate or ethanol. On the other hand, heterofermentative lactic acid bacteria can be divided into obligatory heterofermentative species, which ferment both hexoses and pentoses via PK-P, and facultative heterofermentative organisms, which degrade hexoses via the EMP-P and pentoses via the PK-P.

Final products and proportions may vary, depending on the presence of other proton and electron acceptors (Wischral et al. 2019). In general terms, several countries around the world have a thriving economy based on agriculture, and they have a high potential for the use of lignocellulosic materials from sugarcane, mainly Brazil, as it is the largest producer of this material. In 2015–2016, the sugarcane crop from Brazil reached 666 million tons of sugarcane, and for each ton, approximately 250–270 kg of waste was generated (Wischral et al. 2019). This abundant residue is an important feedstock for the production of second-generation bioproducts such as lactic acid.

7.3.9 Citric Acid

Citric acid is an important commercial product. Its global production has reached 1.7 million tons per year, and its annual growth rate is 5% (Farooq et al. 2013). The largest amount of citric acid is consumed in the food industry, which uses almost 70% of the total production, followed by about 12% in the pharmaceutical industry and 18% for other applications (Rodrigues et al. 2013).

The production of citric acid by *Aspergillus niger* is one of the most commercially utilized examples of fungal overflow metabolism. Many microorganisms such as fungi and bacteria can produce citric acid. The various fungi that have been found to accumulate citric acid in their culture media include strains of *Aspergillus niger, Aspergillus awamori, Penicillium restrictum, Trichoderma viride, Mucor piriformis*, and *Yarrowia lipolytica* (Farooq et al. 2013).

Researchers are engaged in exploring the suitability of various waste materials for bioconversion. These materials might be cheaper sources for the production of various fermented and high-value products. Among various industrial wastes, sugarcane wastes and their hydrolysates might be desirable raw materials for citric acid fermentation because of their availability at relatively low prices, and they can be the basic substrates for citric acid fermentation (Farooq et al. 2013). The importance of citric acid and the utilization of sugarcane wastes to prevent environmental pollution and generate products with great industrial importance such as citric acid should be kept in mind.

7.3.10 Succinic Acid

Succinic acid, also known as amber acid or butanedioic acid, is a dicarboxylic acid with the molecular formula $C_4H_6O_4$. It can be used as a precursor for the production of many chemicals for use in the agricultural, food processing, and pharmaceutical industries such as surfactants, detergents, adipic acid, 1,4-butanediol, tetrahydrofuran, *N*-methylpyrrolidinone, 2-pyrrolidinone, succinate salts, gamma-butyrolactone, several green solvents, biodegradable polymers, such as polybutyrate succinate (PBS), and ingredients for stimulating animal and plant growth (Borges and Pereira 2011).

Several research teams have been working on the development of industrial-level fermentation processes for succinic acid production using strains of *Anaerobiospirillum succiniciproducens*, *Actinobacillus succinogenes*, *Mannheimia succiniciproducens*, *Corynebacterium glutamicum*, and recombinant *Escherichia coli* bacteria (Borges and Pereira 2011). All these strains can produce a relatively large amount of succinic acid from a broad range of carbon sources such as arabinose, cellobiose, fructose, galactose, glucose, lactose, maltose, mannitol, mannose, sorbitol, sucrose, and xylose under anaerobic conditions (presence of CO_2). Thus, it is estimated that the combination of the uses of sugarcane waste and greenhouse gas consumption by the different strains can create a green and viable pathway for balancing the initial large capital of a biorefinery.

7.3.11 Biosorbents

Modern agricultural industries yearly produce a million tons of waste and by-products that have the potential as useful resources. These agro-industrial residues (organic materials produced as byproducts from the harvesting and industrial processing of crops) are a promising alternative to traditional adsorbents because they are cheap, readily available, highly sorptive, easily modifiable, insensitive to toxic substances, and easy to manipulate during treatment (Tran et al. 2015).

Recently, a variety of pollutants have been removed from aqueous solution using different agro-industry-derived adsorbents, including sugarcane bagasse (do Carmo-Ramos et al. 2016), olive seeds (Moubarik and Grimi 2015), orange peels (Romero-Cano et al. 2016), and many other residues. Sugarcane bagasse is one of the most cost-effective and available agro-industrial residues, especially in tropical regions (Sarker et al. 2017). It has been proven by different authors that sugarcane wastes can efficiently remove a wide range of target adsorbates from aqueous solutions, including toxic heavy metals (do Carmo-Ramos et al. 2016), dyes (Ferreira et al. 2015), petroleum (Boni et al. 2016), phenolic compounds (Deokar et al. 2016), and organic nutrients (Sarker et al. 2017).

Sugarcane wastes are fibrous residues containing unique functional groups that can bind polluting ions (Sarker et al. 2017). The sorption efficiency of sugarcane wastes depends on the biosorbent dosage, initial pollutant concentration, solution pH and temperature, contact time, and sorbent particle size. There is still a great need for studies in this area to compare different biosorbents from different residues and test the applicability of new residues in materials for which no ecologically acceptable solutions have been found. Sugarcane wastes can represent eco-friendly and low-cost biosorbents that could be widely used after careful studies regarding their applications.

7.3.12 Biofertilizers

Industries in all countries are consuming agricultural products as their raw material and generating various types of residues. Sugarcane industries are one of them, generating huge amounts of byproducts, such as bagasse and press mud, which are creating a storage problem across the country.

These compounds are considered rejected waste materials from sugarcane industries, and they can cause serious problems of storage and pollution of the area surrounding sugar mills when it accumulates. It furnishes a large amount of organic manure, and it can represent an alternative source of plant nutrients, acting to ameliorate the soil (Dotaniya et al. 2016). It contains significant amounts of iron, manganese, calcium, magnesium, silicon, and phosphorus, which enhances the suitability of sugarcane press mud as a source of nutrients for different crops, and it can also be obtained through the distillation of alcohol during the fermentation of sugarcane molasses. It contains a huge volume of water and plant nutrients.

Therefore, it is necessary to treat the sugarcane press mud as a valuable biofertilizer with the capacity for being applied in a wide range of crops. This residue has the potential for use as one of the substrates for biocomposting, a step needed to remove potential pathogens and opportunistic species, and produce a highly valuable fertilizer (Oliveira et al. 2016). Thus, recycling sugarcane waste by applying it to agricultural land seems to be a good option to diminish the waste storage problem and decrease the necessity for plant nutrients. In general, all the residues that are produced in the different processes in which sugarcane is used can be applied in this biotransformation for biofertilizers to reduce the ecological impact of residues, add value to previously discarded compounds, and improve the health of essential crops.

7.3.13 Food and Prebiotics

In response to the growing awareness of consumers regarding the relationship between health and diet, there has been a large amount of interest in the use of nondigestible oligosaccharides with prebiotic properties as functional food ingredients. Indeed, prebiotics can be selectively fermented by beneficial bacteria in the human colon and promote intestinal health (Khodaei and Karboune 2018). Health-promoting properties of prebiotics are dependent on their structural properties, including monosaccharide composition, degree of polymerization, and level of branching (Holck et al. 2011).

Because of the complexity and heterogeneity of the polysaccharides from the cell wall, they can be good candidates for the production of new prebiotics with unique properties. These polysaccharides are present in abundant and accessible by-products from the sugarcane industry, such as bagasse, molasses, vinasse, filter cake, leaves, etc. There is great potential for the enzymatic treatment of these residues to obtain many food and prebiotic products.

Vinasse has also been demonstrated to be a valuable source of carbon for protein production through its incubation with the yeast torula (*Candida utilis*). The protein produced can be used for animal or human consumption. This process also has the additional advantage that the COD of the vinasse residue decreases and the pH increases so that it can safely be discarded or used for fertigation (Lezcano 2005; Ortiz et al. 2013; Pires et al. 2018).

7.3.14 Xylooligosaccharides (XOS)

XOS are sugar oligomers formed by xylose units, and they are naturally present in fruits, vegetables, milk, and honey. Industrial production is mainly obtained from the hydrolysis of lignocellulosic materials (such as sugarcane wastes) by the action of hemicellulases. These biomolecules are stable over a wide pH range, and they possess high resistance to heat, characteristics that favor their commercial viability.

XOS can be used as antioxidants in the prevention of oxidative stress, anemia, atherosclerosis, diabetes, osteoporosis, and certain types of cancer (Heinen et al. 2018). Besides, these compounds have been shown to possess important immunomodulatory, antimicrobial, antiallergic, and anti-inflammatory activities (Gupta et al. 2016). XOS have also been used as prebiotic components in the development of new functional foods because they favor mineral absorption, induce peristalsis, and act as a regulatory substrate for intestinal flora (Heinen et al. 2017).

Additionally, xylose residues can be converted to xylitol, a sweetener widely used to replace common sugar (sucrose) by diabetics (Sena et al. 2016). Finally, xylobiose has recently gained great attention because of its use as a valuable food supplement (prebiotic). It promotes the proliferation of Bifidobacteria, a microorganism beneficial to the human intestine that reduces the risk of colon cancer (Chen et al. 2019). Therefore, the production of new products from residues such as sugarcane waste is essential. XOS can represent an important new form of reusing these residues to produce valuable compounds.

7.3.15 Cello-oligosaccharides (COS)

COS are biologically important molecules, and they have been associated with food, animal feed, and bioenergy industrial sectors (Barbosa et al. 2020a). COS are defined as oligomers composed of two to six β-1,4-linked glucose units (Zhao et al. 2009), and they have been proposed as novel substrates for ethanol fermentation with potential advantages over glucose, including a reduced risk of process contamination, shorter total fermentation times and limited process inhibition by high concentrations of glucose (Barbosa et al. 2020a).

Moreover, COS are considered to be important functional oligosaccharides (Song et al. 2013), and they are significant for the food and feed industrial sectors as a probiotic compound (Karnaouri et al. 2019) because it has been confirmed that COS can support the growth of different probiotic strains of the *Lactobacilli* and *Bifidobacteria* species. However, limited information regarding the large-scale production of COS is available (Barbosa et al. 2020a). In general, these compounds can be derived from the cellulose fraction of biomass through different pathways, such as acid hydrolysis, hydrolysis over carbon catalysts, mild thermal conversion, and controlled enzymatic hydrolysis.

The latter pathway is considered to be more selective and greener than other approaches, and it can be designed according to the degree of polymerization desired with minimal production of monomers (Barbosa et al. 2020b). The choice of the right pretreatment for sugarcane residue is essential for reducing the recalcitrance of the biomass, thereby increasing the accessibility of cellulose for COS production (Leu and Zhu 2013) and accessing the great potential for industrial application.

7.3.16 Oligogalacturonides (OGAs)

OGAs are molecules that contain from 2 to 20 residues of α-1,4-D-galacturonic acid derived from the cell wall of plants during hydrolysis (Hahn et al. 1981). OGAs are released by the action of pectinases, which act on the pectic substances present in the middle lamella and on the primary cell wall of the plants (Ridley et al. 2001).

There is greater interest in the production and commercialization of oligogalacturonides (OGAs) by physical, chemical, or enzymatic methods because of their bioactivity (Gullon et al. 2013). OGAs exert a series of effects on human health such as prebiotic properties, decrease in glycemic and cholesterol levels, relief of constipation, promotion of mineral absorption, anticancer, anti-inflammatory, antioxidant, and antiobesity effects (Yang et al. 2020). Moreover, OGAs have become a research hot spot of plant defenses recently because they are well-known defense elicitors in plants. They play an important role in plant disease defense and regulation of plant growth and development, such as damage-associated molecular patterns that stimulate plant immunity (Benedetti et al. 2018).

OGAs have been produced commercially from citrus peel residues (pectin: 15–25%), apple pomace (pectin: 10–15%), sunflower (pectin: 15–25%), and sugar beets (pectin: 10–20%) (Yang et al. 2020). However, there is also great potential for using sugarcane residues for the production of OGAs because sugarcane wastes can contain high concentrations of pectin in the cell wall. Therefore, the exploitation of efficient and economical methods for the production of oligogalacturonides from sugarcane residues needs to receive special attention because of their biological function and great potential.

7.3.17 Arabino-oligosaccharides (AOS)

The arabino-oligosaccharides are chains of L-arabinose moieties linked by α-(1,5) bonds and branched via α-(1,2) or α-(1,3) bonds with other L-arabinose moieties (Fernández et al. 2015). Recently, arabino-oligosaccharides and L-arabinose produced by arabinan-degrading enzymes have attracted much attention because of their potential applications in diverse fields, including food technology, nutrition research, bioenergy production, and organic synthesis (Hong et al. 2009). AOS fermentation can diminish the inflammatory conditions in ulcerative colitis (UC) patients. An *in vitro* approach to this effect has been achieved by fermenting AOS with fecal samples from UC patients and healthy control people. These experiments showed that AOS stimulated bacteria genera such as *Bifidobacterium* and *Lactobacillus* and lead to an increase in short-chain fatty acids such as acetic acid, which are known to elicit anti-inflammatory responses.

AOS might, therefore, represent a new prebiotic candidate for reducing the risk of flare-ups in UC patients in the future (Vigsnæs et al. 2011); however, many applications and safety studies are needed. Also, the search for other sources of these compounds is very important. In this regard, sugarcane wastes are excellent candidates for use in enzymatic hydrolysis. Large amounts of AOS with potential for medical and commercial use can be generated because of the high concentrations of arabinan in sugarcane residues.

7.3.18 Aroma Constituents

Aroma compounds play a major role in a great variety of applications, such as in the food, cosmetic, chemical, and pharmaceutical industries, by improving the organoleptic properties of the products. Currently, the annual fragrance and flavor market is estimated at US$ 26.5 billion, and it increases by almost 4% each year (Martínez et al. 2017). This quantity represents over a quarter of the world market for food additives, of which a 13% share corresponds to aroma compounds (Martínez et al. 2017).

Furthermore, the rising consumers demand for natural products demonstrates the need for alternative processes. A promising substitute route for aroma production is based on microbial biosynthesis, bioconversion, or enzymatic hydrolysis (Akacha and Gargouri 2015). These bioprocesses that utilize microorganisms and enzymes involve the synthesis or hydrolysis of flavors as secondary metabolites during or after the fermentation of nutrients such as sugars, amino acids, or agro-residues. These processes are also encouraged by the current legislation based on their qualification as Generally Recognized As Safe (GRAS) products by entities like the Food and Drug Administration (FDA) (Martínez et al. 2017). Biotechnology-derived products based on microorganisms, plant cell cultures, and enzymes are considered to be natural aromas.

Additionally, bioconversion for producing aroma compounds is following the aim of exploiting renewable sources and agro-industrial wastes to promote clean production, fewer energy-intensive systems, and more economical processes (Laufenberg et al. 2003). Agro-industrial residues represent a particularly good choice as substrates for microbial biosynthesis because they are rich in carbohydrates and other nutrients (Sarma et al. 2014).

Sugarcane residues are commonly used for energy production in the same sugar industry; but more recently, it has been used as a source for several biotechnological applications (Martínez et al. 2017). Some applications of sugarcane waste in the production of aromatic compounds such as in the production of coconut-like aroma using *Trichoderma* strains (Fadel et al. 2015) and for producing fruity aromas by *Ceratocystis fimbriata* (Martínez et al. 2017) have been found. Therefore, there is still a long way to go for the discovery of other applications in the production of compounds from sugarcane residues. For that reason, it is essential that scientific research increasingly maintain this horizon as an objective.

7.4 CONCLUDING REMARKS

Brazil is the largest sugarcane producer in the world and a reference in the international community in the use of renewable energy sources, as it has been carrying out scientific research for years with the aim of reusing the waste generated in a more sustainable and bioeconomic way. The use of microorganisms to deconstruct the plant cell wall and ferment these residues has been the key to the production of molecules, enzymes, and bioproducts that are increasingly replacing products based on fossil residues and pollutants. Efforts have been continually made for the entire world to not only fit into the UN's 17 Sustainable Development Goals (SDGs) but also to change the exploratory to compensatory mentality. We are walking in steps of living in harmony and balance with the Planet.

REFERENCES

Akacha, N. B., Gargouri, M. 2015. Microbial and enzymatic technologies used for the production of natural aroma compounds: synthesis, recovery modelling, and bioprocesses. *Food. Bioprod. Process.*, 94:675–706.

Alokika Singh, B. 2019. Production, characteristics, and biotechnological applications of microbial xylanases. *Appl. Microbiol. Biotechnol.*, 103(21-22):8763–8784.

Alonso, S., Rendueles, M., Díaz, M. 2015. Microbial production of specialty organic acids from renewable and waste materials. *Crit. Rev. Biotechnol.*, 35(4):497–513.

Amin, F., Bhatti, H. N., Bilal, M. 2019. Recent advances in the production strategies of microbial pectinases – A review. *Int. J. Biol. Macromol.*, 122:1017–1026.

Andlar, M., Rezi, T., Marđetko, N., Kracher, D., Ludwig, R., Santek, B. 2018. Lignocellulose degradation: An overview of fungi and fungal enzymes involved in lignocellulose degradation. *Eng. Life Sci.*, 18:768–778.

Andríc, P., Meyer, A. S., Jensen, P. A., Dam-Johansen, K. 2010. Reactor design for minimizing product inhibition during enzymatic lignocellulose hydrolysis: I. Significance and mechanism of cellobiose and glucose inhibition on cellulolytic enzymes. *Biotechnol. Adv.*, 28:308–324.

Annamalai, N., Rajeswari, M. V., Balasubramanian, T. 2016a. Chapter 3: Endo-1,4-β-glucanases: role, applications and recent developments. In: *Microbial Enzymes in Bioconversions of Biomass, Biofuel and Biorefinery Technologies 3*, ed. V. K. Gupta, 37–45. Springer International Publishing, Switzerland.

Annamalai, N., Rajeswari, M. V., Sivakumar, N. 2016b. Chapter 2: Cellobiohydrolases: role, mechanism, and recent developments. In: *Microbial Enzymes in Bioconversions of Biomass, Biofuel and Biorefinery Technologies 3*, ed. V. K. Gupta, 29–35. Springer International Publishing, Switzerland.

Annual Sustainability Report 2019/2020. 2020. https://fsbioenergia.com.br/wp-content/uploads/2020/09/FS-RAS19-EN-V2.pdf (accessed December 12, 2020).

Anukam, A., Mamphweli, S., Reddy, P., Meyer, E., Okoh, O. 2016. Pre-processing of sugarcane bagasse for gasification in a downdraft biomass gasifier system: a comprehensive review. *Renew. Sust. Energ. Rev.*, 66:775–801.

Aragão, M. S., Menezes, D. B., Ramos, L. C., et al. 2020. Mycoremediation of vinasse by surface response methodology and preliminary studies in air-lift bioreactors. *Chemosphere*, 244:1–9.

Arancon, R. A. D., Lin, C. S. K., Chan, K. M., Kwan, T. H., Luque, R. 2013. Advances on waste valorization: new horizons for a more sustainable society. *Energ. Sci. Eng.*, 1:53–71.

Barbosa, F. C., Kendrick, E., Brenelli, L. B., et al. 2020a. Optimization of cello-oligosaccharides production by enzymatic hydrolysis of hydrothermally pretreated sugarcane straw using cellulolytic and oxidative enzymes. *Biomass Bioenerg.*, 141:105697.

Barbosa, F. C., Martins, M., Brenelli, L. B., et al. 2020b. Screening of potential endoglucanases, hydrolysis conditions and different sugarcane straws pretreatments for cello-oligosaccharides production. *Bioresour. Technol.*, 316:123918.

Barchuk, M. L., Díaz, G. V., Coll, P. A. F., et al. 2016. Selection of Trichoderma strain to enhanced cellulase-poor xylanase production using sugarcane bagasse as sole carbon source under light. *Int. J. Rec. Biotechnol.*, 4:25–34.

Bayer, E. A., Belaich, J. P., Shoham, Y., Lamed, R. 2004. The cellulosomes: multienzyme machines for degradation of plant cell wall polysaccharides. *Annu. Rev. Microbiol.*, 58:521–554.

Benedetti, M., Verrascina, I., Pontiggia, D., et al. 2018. Four arabidopsis berberine bridge enzyme-like proteins are specific oxidases that inactivate the elicitor-active oligogalacturonides. *Plant J.*, 94(2):260–273.

Berlemont, R., Martiny, A. C. 2016. Glycoside hydrolases across environmental microbial communities. *PLoS. Comput. Biol.*, 12(12):16

Bertini, L. Lambrughi, M. Fantucci, P. et al. 2018. Catalytic mechanism of fungal lytic polysaccharide monooxygenases investigated by first-principles calculations. *Inorg. Chem.*, 57:86–97.

Bhat, M. K., Bhat, S. 1997. Cellulose degrading enzymes and their potential industrial applications. *Biotechnol. Adv.*, 15:583–620.

Bhosale, P. R., Chonde, S. G., Nakade, D. B., Raut, P. D. 2012. Studies on physico-chemical characteristics of waxed and dewaxed pressmud and its effect on water holding capacity of soil. *ISCA J. Biological Sci.*, 1:35–41.

Boni, H. T., de Oliveira, D., de Souza, A. A. U., de Souza, S. U. 2016. Bioadsorption by sugarcane bagasse for the reduction in oil and grease content in aqueous effluent. *Int. J. Environ. Sci. Technol.*, 13:1169–1176.

Borges, E. R., Pereira, N. 2011. Succinic acid production from sugarcane bagasse hemicellulose hydrolysate by *Actinobacillus succinogenes*. *J. Ind. Microbiol. Biotechnol.*, 38(8):1001–1011.

Brenelli, L., Squina, F. M., Felby, C., Cannella, D. 2018. Laccase-derived lignin compounds boost cellulose oxidative enzymes AA9. *Biotechnol. Biofuels.*, 11:10.

Brienzo, M., Carvalho, A. F. A., Figueiredo, F. C., Neto, P. O. 2016. Chapter 8: Sugarcane bagasse hemicellulose properties, extraction technologies and xylooligosaccharides production. In: *Food Waste*, ed. G. L. Riley, 155–188. Nova Science Publishers, Inc.

Buckeridge, M. S., Grandis, A., Tavares, E. Q. P. 2019. Disassembling the Glycomic Code Of Sugarcane Cell Walls To Improve Second-generation Bioethanol Production. In: *Bioethanol Production from Food Crops*, ed. R. Ray and R. Ramachandran, 31–43. Elsevier.

Cano, L. A. R., Gutierrez, L. V. G., Perez, L. A. B. 2016. Biosorbents prepared from orange peels using instant controlled pressure drop for Cu(II) and phenol removal. *Ind. Crop. Prod.*, 84:344–349.

Carpita, N. C. 1996. Structure and biogenesis of the cell walls of grasses. *Annu. Rev. Plant Physiol. Plant Mol. Biol.*, 47:445–476.

Carpita, N. C., Gibeaut, D. M. 1993. Structural models of primary cell walls in flowering plants: consistency of molecular structure with the physical properties of the walls during growth. *Plant J.*, 3:1–30.

Chacón, S. J., Matias, G., Vieira, C. F. S., Ezeji, T. C., Maciel Filho, R., Mariano, A. P. 2020. Enabling butanol production from crude sugarcane bagasse hemicellulose hydrolysate by batch-feeding it into molasses fermentation. *Ind. Crops Prod.*, 155:112837.

Chalak, A., Villares, A., Moreau, C., et al. 2019. Influence of the carbohydrate-binding module on the activity of a fungal AA9 lytic polysaccharide monooxygenase on cellulosic substrates. *Biotechnol. Biofuels.*, 12:206.

Cheavegatti-Gianotto, A., et al. 2011. Sugarcane (Saccharum X officinarum): a reference study for the regulation of genetically modified cultivars in Brazil. *Trop. Plant Biol.*, 4:62–89.

Chen, Z., Liu, Y., Zaky, A. A., Liu, L., Chen, Y., Li, S., Jia, Y. 2019. Characterization of a novel xylanase from *Aspergillus flavus* with the unique properties in production of xylooligosaccharides. *J. Basic Microbiol.*, 59:351–358.

Christy, P. M., Gopinath, L., Divya, D. 2014. A review on anaerobic decomposition and enhancement of biogas production through enzymes and microorganisms. *Renew. Sustain. Energy Rev.*, 34:167.

Cintra, L. C., Costa, I. C., Oliveira, I. C. M., et al. 2020. The boosting effect of recombinant hemicellulases on the enzymatic hydrolysis of steam-treated sugarcane bagasse. *Enzyme Microb. Technol.*, 133:1–8.

da Silva-Neto, J. V., Gallo, W. L., Nour, E. A. A. 2020. Production and use of biogas from vinasse: implications for the energy balance and GHG emissions of sugar cane ethanol in the brazilian context. *Environ. Prog. Sustain. Energy*, 39:1–11.

de Lucas, R. C., Oliveira, T. B., Lima, M. S., et al. 2020. Effect of enzymatic pretreatment of sugarcane bagasse with recombinant hemicellulases and esterase prior to the application of the cellobiohydrolase CBH I Megazyme®. *Biomass Conv. Biorefinery*, (May):1–9. 10.1007/s13399-020-00719-9.

de Lucas, R. C., Oliveira, T. B., Lima, M. S., et al. 2021. The profile secretion of *Aspergillus clavatus*: different pre-treatments of sugarcane bagasse distinctly induces holocellulases for the lignocellulosic biomass conversion into sugar. *Renew. Energy*, 165(1):748–757.

de Marco, J. C. I. De Souza-Neto, G. P. Castro, C. F. S. Michelin, M. Polizeli, M. L. T. M. Filho, E. X. F. 2015. Partial purification and characterization of a thermostable β-mannanase from *Aspergillus foetidus*. *Appl. Sci.*, 5:881–893.

de Oliveira, T. B., Lopes, V. C. P., Barbosa, F. N., et al. 2016. Fungal communities in pressmud composting harbour beneficial and detrimental fungi for human welfare. *Microbiology*, 162(7):1147–1156.

de Souza, A. P., Leite, D. C. C., Pattathil, S., Hahn, M. G., Buckeridge, M. S. 2012. Composition and structure of sugarcane cell wall polysaccharides: implications for second-generation bioethanol production. *Bioenerg. Res.*, 6(2):1–16.

Demirbas, A. 2008. The Importance of bioethanol and biodiesel from biomass. *Energy Sources*, 3(2):177–185.

Deokar, S. K., Mandavgane, S. A., Kulkarni, B. D. 2016. Adsorptive removal of 2,4-dichlorophenoxyacetic acid from aqueous solution using bagasse fly ash as adsorbent in batch and packed-bed techniques. *Clean. Technol. Environ. Policy.*, 18:1971–1983.

Dotaniya, M. L., Datta, S. C., Biswas, D. R., et al. 2016. Use of sugarcane industrial by-products for improving sugarcane productivity and soil health. *Int. J. Recycl. Org. Waste. Agricult.*, 5:185–194.

Fadel, H. H. M., Mahmoud, M. G., Asker, M. M. S., Lotfy, S. N. 2015. Characterization and evaluation of coconut aroma produced by *Trichoderma viride* EMCC-107 in solid state fermentation on sugarcane bagasse. *Electron. J. Biotechnol.*, 18:5–9.

FAO (Food and Agriculture Organization). 2010. *Global forest resources assessment 2010*. FAO forestry paper 163. Food and Agriculture Organization, Rome, Italy.

Farinas, C. S., Loyo, M. M., Baraldo, A. J., Tardioli, P. W., Neto, V. B., Couri, S. 2010. Finding stable cellulase and xylanase: evaluation of the synergistic effect of pH and temperature. *New Biotechnol.*, 27:810–815.

Farooq, U., Anjum, F. M., Zahoor, T., Rahman, S., Hayat, Z., Akram, K., Ashraf, E. 2013. Citric acid production from sugarcane molasses by *Aspergillus niger* under different fermentation conditions and substrate levels. *Int. J. Agric. Appl. Sci.*, 5(1):8–16.

Fernández, J., Redondo-Blanco, S., Miguélez, E. M., Villar, C. J., Clemente, A., Lombó, F. 2015. Healthy effects of prebiotics and their metabolites against intestinal diseases and colorectal cancer. *AIMS Microbiol.*, 1:48–71.

Ferreira, B. C. S., Teodoro, F. S., Mageste, A. B., Gil, L. F., de Freitas, R. P., Gurgel, L. V. A. 2015. Application of a new carboxylate-functionalized sugarcane bagasse for adsorptive removal of crystal violet from aqueous solution: kinetic, equilibrium and thermodynamic studies. *Ind. Crop. Prod.*, 65:521–534.

Furbank, R. T. 2011. Evolution of the C4 photosynthetic mechanism: are there really three C4 acid decarboxylation types? *J. Experiment. Bot.*, 62:3103–3108.

Gilmore, S. P., Lillington, S. P., Haitjema, C. H., Groot, R., O'Malle, M. A. 2020. Designing chimeric enzymes inspired by fungal cellulosomes. *Synth. Syst. Biotechnol.*, 5(1):23–32.

Goldemberg, J. 2011. Chapter 1: The Role of Biomass in the World's Energy System. In: *Routes to Cellulosic Ethanol*, eds. M. S. Buckeridge, and G. H. Goldman. 3–14. New York: Springer.

Guan, W., Shi, S., Tu, M., Lee, Y. Y. 2016. Acetone-butanol-ethanol production from Kraft paper mill sludge by simultaneous saccharification and fermentation. *Bioresour. Technol.*, 200:713–721.

Gullon, B., Gomez, B., Martinez-Sabajanes, M., Yanez, R., Parajo, J. C., Alonso J. L. 2013. Pectic oligosaccharides: manufacture and functional properties *Trends Food Sci. Technol.*, 30(2):153–161.

Gupta, P. K., Agrawal, P., Hegde, P., Shankarnarayan, N., Vidyashree, S., Singh, S. A., Ahuja, S. 2016. Xylooligosaccharide – A valuable material from waste to taste: A review. *J. Environ. Res. Develop.*, 10:555–563.

Gusakov, A. V., Dotsenko, A. S., Rozhkova, A. M., Sinitsyn, A. P. 2017. N-Linked glycans are an important component of the processive machinery of cellobiohydrolases. *Biochimie.,* 132:102–108.

Hahn, M. G., Darvill, A. G., Albersheim, P. 1981. Host-pathogen interactions: XIX. The endogenous elicitor, a fragment of a plant cell wall polysaccharide that elicits phytoalexin accumulation in soybeans. *Plant. Physiol.*, 68(5):1161–1169.

Haitjema, C. H., Gilmore, S. P., Henske, J. K., et al. 2017. A parts list for fungal cellulosomes revealed by comparative genomics. *Nat. Microbiol.*, 2:1–8.

Heinen, P. R., Bauermeister, A., Ribeiro, L. F., et al. 2018. GH11 xylanase from *Aspergillus tamarii Kita*: Purification by one-step chromatography and xylooligosaccharides hydrolysis monitored in real-time by mass spectrometry. *Int. J. Biol. Macromol.*, 108:291–299.

Heinen, P. R., Pereira, M. G., Rechia, C. G. V. 2017. Immobilized endo-xylanase of *Aspergillus tamarii* Kita: an interesting biological tool for production of xylooligosaccharides at high temperatures. *Process Biochem.*, 53:145–152.

Hodkinson, T. R. 2018. Evolution and taxonomy of the grasses (Poaceae): a model family for the study of species-rich groups. *Annual Plant Rev.* 1:1–39.

Holck, J., Hjernø, K., Lorentzen, A., et al. 2011. Tailored enzymatic production of oligosaccharides from sugar beet pectin and evidence of differential effects of a single DP chain length difference on human faecal microbiota composition after in vitro fermentation. *Process Biochem.*, 46(5):1039–1049.

Hong, M., Park, C., Oh, D. 2009. Characterization of a thermostable endo-1,5-α-L-arabinanase from *Caldicellulorsiruptor saccharolyticus. Biotechnol. Lett.*, 31:1439.

Hu, B. B., Zhu, M. J. 2019. Reconstitution of cellulosome: Research progress and its application in biorefinery. *Biotechnol. Appl. Biochem.*, 66(5):720–730.

Jaramillo, P. M. D., Gomes, H. A. R., Monclaro, A. V., Silva, C. O. G., Filho, E. X. F. 2015. Lignocellulose degrading enzymes: an overview of the global market In: *Fungal Biomolecules: Sources, Applications and Recent Developments*, eds. V. K. Gupta, R. L. Mach, and S. Sreenivasaprasad, 73–85. John Wiley and Sons.

John, J., Kaimal, K. K. S., Smith, M. L., Rahman, P. K. S. M., Chellam, P. V. 2020. Advances in upstream and downstream strategies of pectinase bioprocessing: a review. *Int. J. Biol. Macromol.*, 162:1086–1099.

Jovanovic, I., Magnuson, J. K., Collart, F., et al. 2009. Fungal glycoside hydrolases for saccharification of lignocellulose: outlook for new discoveries fueled by genomics and functional studies. *Cellulose*, 16(4):687–697.

Karnaouri, A., Matsakas, L., Krikigianni, E., Rova, U., Christakopoulos P. 2019. Valorization of waste forest biomass toward the production of cello-oligosaccharides with potential prebiotic activity by utilizing customized enzyme cocktails. *Biotechnol. Biofuels.*, 12:1–19.

Kasmuri, N. H., Kamarudin, S. K., Abdullah, S. R. S., Hasan, H. A., Som A. M. 2019. Integrated advanced nonlinear neural network-simulink control system for production of bio-methanol from sugar cane bagasse via pyrolysis. *Energy*, 168:261–272.

Ketudat, C. J. R., Esen, A. 2010. β-Glucosidases. *Cell. Mol. Life. Sci.* 67:3389–3405.

Khan, M., Nakkeeran, E., Umesh-Kumar, S. 2013. Potential application of pectinase in developing functional foods. *Annu Rev. Food. Sci. Technol.* 4:21–34.

Khattab, S. M. R., Watanabe, T. 2019. Chapter 10 - Bioethanol From Sugarcane Bagasse: Status and Perspectives. In *Bioethanol Production from Food Crops*, eds. R. C. Ray, S. Ramachandran, 187–212. Elsevier.

Khodaei, N., Karboune, S. 2018. Optimization of enzymatic production of prebiotic galacto/galacto(arabino)-oligosaccharides and oligomers from potato rhamnogalacturonan I *Carbohydr. Polym.* 181:1153–1159.

Kote, N. V., Patil, A. G. G., Mulimani, V. H. 2009. Optimization of the production of thermostable endo-β-1,4 mannanases from a newly isolated *Aspergillus niger* gr and *Aspergillus flavus* gr. *Appl. Biochem. Biotechnol.*, 152:213–223.

Kowalczyk, J. E., Benoit, I., de Vries, R. P. 2014. Regulation of plant biomass utilization in Aspergillus. *Adv. Appl. Microbiol.*, 88:31–56.

Laufenberg, G., Kunz, B., Nystroem M. 2003. Transformation of vegetable waste into value added products: (A) the upgrading concept; (B) practical implementations. *Bioresour. Technol.*, 87:167–198.

Lee, J. W., Kim, H. U., Choi, S., Yi, J., Lee, S. Y. 2011. Microbial production of building block chemicals and polymers. *Curr. Opin. Biotechnol.*, 22(6):758–767.

Leu, S. Y., Zhu, J. Y. 2013. Substrate-related factors affecting enzymatic saccharification of lignocelluloses: our recent understanding. *Bioenerg. Res.*, 6:405–415.

Lezcano, P. 2005. Desarrollo de una fuente de proteína en Cuba. Levadura torula (*Candida utilis*). *Rev. Cuba. Cien. Agríc.*, 39:459–463.

Lombard, V., Golaconda, R. H., Drula, E., Coutinho, P. M., Henrissat, B. 2014. The carbohydrate-active enzymes database (CAZy) in 2013. *Nucleic Acids Res.* 42:D490–D495.

Lynd, L. R., Wyman, C. E., Gerngross, T. U. 1999. Biocommodity engineering. *Biotechnol. Progr.*, 15:777.

Mäkelä, M. R., Donofrio, N., de Vries, R. P. 2014. Plant biomass degradation by fungi. *Fungal Genet Biol.*, 72:2–9.

Malgas, S., van Dyk, J. S., Pletschke, B. I. 2015. A review of the enzymatic hydrolysis of mannans and synergistic interactions between β-mannanase, β-mannosidase and α-galactosidase. *World. J. Microbiol. Biotechnol.*, 31:1167–1175.

Martínez, O., Sánchez, A., Font, X., Barrena, R. 2017. Valorization of sugarcane bagasse and sugar beet molasses using *Kluyveromyces marxianus* for producing value-added aroma compounds via solid-state fermentation. *J. Clean. Prod.*, 158:8–17.

Mateo, J. J., Maicas, S. 2015. Valorization of winery and oil mill wastes by microbial technologies. *Food Res. Int.*,73:13–25.

Meyer, A. S., Rosgaard, L., Sørensen, H. R. 2009. The minimal enzyme cocktail concept for biomass processing. *J. Cereal Sci.*, 50(3):337–344.

Monclaro, A. V., Filho, E. X. F. 2017. Fungal lytic polysaccharide monooxygenases from family AA9: Recent developments and application in lignocellulose breackdown. *Int. J. Bio. Macromol.*, 102:771–778.

Moreira, L. R. S., Filho, E. X. F. 2008. An overview of mannan structure and mannan-degrading enzyme systems. *Appl. Microbiol. Biotechnol.*, 79(2):165–178.

Morril, J., Kulcinkaja, E., Sulewska, A. M., Lahtinen, S., Stalbrand, B. S., Hachem, M. A. 2015. The GH5 1,4-β-mannanase from *Bifidobacterium animalis* subsp. lactis Bl-04 possesses a low-affinity mannan-binding module and highlights the diversity of mannanolytic enzymes. *BMC Biochem.*, 16–26.

Moubarik, A., Grimi, N. 2015. Valorization of olive stone and sugar cane bagasse by-products as biosorbents for the removal of cadmium from aqueous solution. *Food. Res. Int.*, 73:169–175.

National Bank for Economic and Social Development (BNDES-Brazil). Management and Strategic Studies Center. 2008. *Sugarcane bioethanol: energy for sustainable development*. 1st ed. Rio de Janeiro: National Bank for Economic and Social Development, 314 p.

Nguyen, S. T. C., Freund, H. L., Kasanjian, J., Berlemont, R. 2018. Function, distribution, and annotation of characterized cellulases, xylanases, and chitinases from CAZy. *Appl. Microbiol. Biotechnol.*, 102:1629–1637.

Nighojkar, A., Patidar, M. K., Nighojkar, S. 2019. Pectinases: Production and Applications for fruit juice beverages. Processing and Sustainability of Beverages. In: *The Science of Beverages*, eds. A. M. Grumezescu and A. M. Holban, 2:235–273. Elsevier.

Nogueira, C. E. C., Souza, S. N. M., Micuanski, V. C., Azevedo, R. L. 2015. Exploring possibilities of energy insertion from vinasse biogasin the energy matrix of Paraná State, Brazil. *Renew. Sustain. Energy Rev.*, 48:300–305.

Ortiz, A., Ferreira, W. M., Ramirez, M. A., Hosken, F. M., Lezcano, P. 2013. Soybean meal substitution by torula yeast (*Candida utilis*) grown on vinasses in pelleted diets for fattening rabbits. *Cuba. J. Agric. Sci.*, 47:389–393.

Parsaee, M., Kiani, M. K. D., Karimi, K. 2019. A review of biogas production from sugarcane vinasse. *Biomass Bioenergy.*, 122:117–125.

Pasin, T. M., Almeida, P. Z., Scarcella, A. S. A., Infante, J. C., Polizeli, M. L. T. M. 2020. Bioconversion of Agro-industrial Residues to Second-Generation. In *Biorefinery of Alternative Resources: Targeting Green Fuels and Platform Chemicals*, eds. S. Nanda, D. V. N. Vo, and P. K. Sarangi, 23–47. Springer Nature, Singapore.

Pattra, S., Sangyoka, S., Boonmee, M., Reungsang, A. 2008. Bio-hydrogen production from the fermentation of sugarcane bagasse hydrolysate by *Clostridium butyricum*. *Int. J. Hydrogen. Energy.*, 33:5256–5265.

Pham, T. A., Berrin, J. G., Record, E., To, K. A., Sigoillot, J. 2011. Hydrolysis of softwood by *Aspergillus* mannanase: role of a carbohydrate-binding module. *J. Biotechnol.*, 70:148–163.

Pinheiro, V. E., Michelin, M., Vici, A. C., Almeida, P. Z., Polizeli, M. L. T. M. 2020. *Trametes versicolor* laccase production using agricultural wastes: a comparative study in Erlenmeyer flasks, bioreactor and tray. *Bioprocess. Biosyst. Eng.*, 43:507–514.

Pires, J. R. M., Grazziotti, P. H., Pinto, N. V. D., et al. 2018. Optimization of protein production by C*andida utilis* in industrial vinasse with applicability in food. *Aust. J. Basic Appl. Sci.*, 12:13–26.

Plácido, J., Capareda, S. 2015. Ligninolytic enzymes: a biotechnological alternative for bioethanol production. *Bioresour. Bioprocess.*, 2:23.

Polizeli, M. L. T. M., Damasio, A., Maller, A., Cabral, H., Polizeli, A. 2013. Pectinases produced by microorganisms. In: *Fungal Enzymes*, eds. M. L. T. M. Polizeli and M. Rai. 1:316–340. Boca Raton, USA: CRC Press.

Polizeli, M. L. T. M., Rizzati, A. C., Monti, R., Terenzi, H. F., Jorge, J. A., Amorim, D. S. 2005. Xylanases from fungi: properties and industrial applications. *Appl. Microbiol. Biotechnol.*, 67(5):577–591.

Pratto, B., Chandgude, V., de Sousa, R., Cruz, A. J. G., Bankar, S. 2020. Biobutanol production from sugarcane straw: defining optimal biomass loading for improved ABE fermentation. *Ind. Crops Prod.*, 148:112265.

Rai, P. K., Singh, S. P., Asthana, R. K., Singh, S. 2014. Biohydrogen production from sugarcane bagasse by integrating dark- and photo-fermentation. *Bioresour. Technol.*, 152:140–146.

Ramos, S. N. D. C., Xavier, A. L. P., Teodoro, F. S. E., Gil, L. F., Gurgel, L. V. A. 2016. Removal of cobalt (II), copper (II), and nickel (II) ions from aqueous solutions using phthalate-functionalized sugarcane bagasse: mono-and multicomponent adsorption in batch mode. *Ind. Crop. Prod.*, 79:116–130.

Renó, M. L. G., Lora, E. E. S., Palacio, J. C. E., Venturini, O. J., Buchgeister, J., Almazan, O. 2011. A LCA (life cycle assessment) of the methanol production from sugarcane bagasse *Energy*, 36(6):3716–3726.

Ridley, B. L., O'Neill, M. A., Mohnen, D. 2001. Pectins: structure, biosynthesis, and oligogalacturonide-relate signaling. *Phytochemisty*, 57(6):929–967.

Rodrigues, C., Vandenberghe, L. P., Sturm, W., et al. 2013. Effect of forced aeration on citric acid production by *Aspergillus sp.* mutants in SSF. *World J. Microbiol. Biotechnol.*, 29(12):2317–2324.

Saini, J. K., Saini, R., Tewari, L. 2015. Lignocellulosic agriculture wastes as biomass feedstocks for second-generation bioethanol production: concepts and recent developments. *Biotech.* 5:337–353.

Salgado, J. C. S., Meleiro, L. P., Carli, S., Ward, R. J. 2018. Glucose tolerant and glucose stimulated β-glucosidases – A review. *Bioresour. Technol.*, 267:704–713.

Santos, C. A., Ferreira-Filho, J. A., O'Donovan, A., Gupta, V. K., Tuohy, M. G., Souza, A. P. 2017. Production of a recombinant swollenin from *Trichoderma harzianum* in *Escherichia coli* and its potential synergistic role in biomass degradation. *Microb. Cell. Fact.*, 16:83.

Santos, F., Eichler, P., Machado, G., Mattia, J., de Souza, G. 2020. By-Products of the Sugarcane Industry. In: *Sugarcane Biorefinery, Technology and Perspectives*, eds. F. Santos, S. R. M. De Matos, and P. Eichler, 21–48. Elsevier: Academic Press.

Sarker, T. C., Azam, S. M. G. G., El-Gawad, A. M. A., Gaglione, S. A., Bonanomi, G. 2017. Sugarcane bagasse: a potential low-cost biosorbent for the removal of hazardous materials. *Clean. Techn. Environ. Policy.*, 19:2343–2362.

Sarma, S. J., Dhillon, G. S., Hedge, K., Brar, S. K., Verma, M. 2014. Utilization of agro-industrial wastes for the production of aroma compounds and fragrances. In: *Biotransformation of Waste Biomass into High Value Biochemicals*, eds. S. K. Brar, G. S. Dhillon, and M. Fernandes, 99–115. Springer Books, New York.

Segato, F., Damásio, A. R. L., de Lucas, R. C., Squina, F. M., Prade, R. A. 2014. Genomics Review of holocellulose deconstruction by *Aspergilli. Microbiol. Mol. Biol. Rev.*, 78(4):588–513.

Sena, L. M. F., Morais, C. G., Lopes, M. R., et al. 2016. D-Xylose fermentation, xylitol production and xylanase activities by seven new species of *Sugiyamaella. Anton. v. Lee.*, 110:53–67.

Shallom, D., Shoham, Y. 2003. Microbial hemicellulases. *Curr. Opin. Microbiol.*, 6(3):219–228.

Sidar, A., Albuquerque, E. D., Voshol, G. P., Ram, A. F. J., Vijgenboom, E., Punt, P. J. 2020. Carbohydrate binding modules: diversity of domain architecture in amylases and cellulases from filamentous microorganisms. *Front. Bioeng. Biotechnol.*, 8:1–15.

Siles, J. A., García-García, I., Martín, A., Martín, M. A. 2011. Integrated ozonation and biomethanization treatments of vinasse derived from ethanol manufacturing. *J. Hazard Mater.*, 188:247.

Silva-Braz, D., Pinto-Mariano A. 2018. Jet fuel production in eucalyptus pulp mills: economics and carbon footprint of ethanol vs. Butanol pathway. *Bioresour. Technol.*, 268:9–19.

Sindhu, R., Gnansounou, E., Binod, P., Pandey, A. 2016. Bioconversion of sugarcane crop residue for value added products – An overview. *Renew. Energy*, 98:203–215.

Singh, G., Verma, A. K., Kumar, V. 2016. Catalytic properties, functional attributes and industrial applications of b-glucosidases. *3 Biotech.*, 6:3.

Singhania, R. R., Patel, A. K., Sukumaran, R. K., Larroche, C., Pandey, A. 2013. Role and significance of beta-glucosidases in the hydrolysis of cellulose for bioethanol production. *Bioresour. Technol.*, 127:500–507.

Siqueira, F. G., Filho, E. X. F. 2010. Plant cell wall as a substrate for the production of enzymes with industrial applications. *Bent. Sci. Publ.*, 7(1):55–60.

Smith, S. P., Bayer, E. A., Czjzek, M. 2017. Continually emerging mechanistic complexity of themulti-enzyme cellulosome complex. *Curr. Opin. Struc. Biol.*, 44:151–160.

Song, J., Jiao, L. F., Xiao, K., Luan, Z. S., Hu, C. H., Shi, B., Zhan, X. A. 2013. Cello-oligosaccharide ameliorates heat stress-induced impairment of intestinal microflora, morphology and barrier integrity in broilers. *Anim. Feed Sci. Technol.*, 185:175–181.

Tarik, A., Latif, Z. 2012. Isolation and biochemical characterization of bacterial isolates producing different levels of polygalacturonases from various sources. *African J. Microbiol. Res.*, 6(45):7259–7264.

Teo, S. C., Liew, K. J., Shamsir, M. S., Chong, C. S., Bruce, N. C., Chan, K. G., Goh, K. M. 2019. Characterizing a halo-tolerant GH10 Xylanase from *Roseithermus sacchariphilus* strain RA and its CBM-truncated variant. *Int. J. Mol. Sci.*, 20:2284.

Tran, V. S., Ngo, H. H., Guo, W., Zhang, J., Liang, S., Ton-That, C., Zhang, X. 2015. Typical low-cost biosorbents for adsorptive removal of specific organic pollutants from water. *Bioresour. Technol.*, 182:353–363.

Van Zyl, W. H., Rosea, S. H., Trollopeb, K., Görgensb, J. F. 2010. Fungal β-mannanases: Mannan hydrolysis, heterologous production and biotechnological applications. *Process. Biochem.*, 45(8):1203–1213.

Várnai, A., Umezawa, K., Yoshida, M., Eijsink, V. G. H. 2018. The pyrroloquinolinequinone-dependent pyranose dehydrogenase from *Coprinopsis cinerea* drives lytic polysaccharide monooxygenase action. *Appl. Environ. Microbiol.*, 84:156–158.

Vigsnæs, L. K., Holck, J., Meyer, A. S., Licht, T. R. 2011. In vitro fermentation of sugar beet arabino-oligosaccharides by fecal microbiota obtained from patients with ulcerative colitis to selectively stimulate the growth of *Bifidobacterium* spp. and *Lactobacillus* spp. *Appl. Environ. Microbiol.*, 77(23):8336–8344.

Vogel, J. 2008. Unique aspects of the grass cell wall. *Curr. Opin. Plant. Biol.*, 11:301–307.

Wischral, D., Arias, J. M., Modesto, L. F., de França Passos, D., Pereira, N. 2019. Lactic acid production from sugarcane bagasse hydrolysates by *Lactobacillus pentosus*: integrating xylose and glucose fermentation. *Biotechnol. Prog.*, 35(1):e2718.

Yang, G., Tan, H., Li, S., et al. 2020. Application of engineered yeast strain fermentation for oligogalacturonides production from pectin-rich waste biomass. *Bioresour. Technol.*, 300:122645.

Yang, Y., Zhang, Y., Li, B., Yang, X., Dong, Y., Qiu, D. 2018. A *Verticillium dahliae* pectate lyase induces plant immune responses and contributes to virulence. *Front. Plant. Sci.*, 9:1–15.

Yao, B., Ji, Y. 2014. Lignin biodegradation with laccase-mediator systems. *Front. Energy. Res.*, 2:12.

Zhao, F., Cao, H.-Y., Zhao, L. S., et al. 2019. A novel subfamily of endo-1,4-glucanases in glycoside hydrolase family 10. *Appl. Environ. Microbiol.*, 85(18):e01029–e01019

Zhao, Y., Lu, W. J., Wang, H. T. 2009. Supercritical hydrolysis of cellulose for oligosaccharide production in combined technology. *Chem. Eng. J.*, 150:411–417.

Zhenming, C., Zhe, C., Guanglei, L., Fang, W., Liang, J., Tong, Z. 2009. *Saccharomycopsis fibuligera* and its applications in biotechnology. *Biotechnol. Adv.*, 27:423–431.

8 Advances on the Valorization of Wastes Generated by Different Coffee Processing

Cintia Lacerda Ramos
Department of Basic Science, Federal University of the Jequitinhonha and Mucuri Valeys, Diamantina, MG, Brazil

Disney Ribeiro Dias
Department of Food Science, Federal University of Lavras, Lavras, MG, Brazil

Rosane Freitas Schwan
Department of Biology, Federal University of Lavras, Lavras, MG, Brazil

CONTENTS

DOI: 10.1201/9781003128977-8

8.1 INTRODUCTION

Discovered approximately 1000 years ago, coffee became one of the most consumed products in the world. Coffee, in addition to petroleum, has represented the most important commodities. *Coffea arabica* (called Arabica coffee) and *Coffea canephora* (called Robusta coffee) are mainly employed for economic and commercial purposes. The consumption of moderate coffee beverages has been associated with several health benefits due to its antioxidant components, especially chlorogenic acids, melanoidins formed during the Maillard reaction, and caffeine. The high consumption and production of coffee have generated large quantities of wastes, which has become an increasing problem to the environment (Chanakya and Alwis, 2004; Clay, 2004; Dias et al., 2015).

The coffee cherry is the ripe fruit of the coffee tree, which is rapidly processed to dry coffee beans (moisture content around 10–12%) to preserve its properties. During the coffee processing, the ripe coffee fruits are transformed into dried beans. There are different methods for coffee processing, including the dry (also known as natural), wet, and semi-dry (also known as semi-wet) methods (Schwan et al., 2012; Schwan et al., 2014). Dry processing is the traditional and most common method used in Indonesia, Ethiopia, Brazil, and Haiti. In this processing, coffee berries are distributed in a delimited area on the ground, fermented, and sun-dried (Schwan et al., 2014).

On the other hand, wet processing comprises sequential steps where selected coffee berries are mechanically peeled, pulped, separated, and placed into vats with water to ferment. During the fermentation, the micro-organisms consume the remaining mucilage, and then the coffee beans are spread on the ground to dry. This coffee processing is most common in India, Ethiopia, Colombia, and Central American countries. Finally, semi-dry processing combines the two methods, where the coffee berries are mechanically pulped and transferred to the ground for fermentation and dry steps (Schwan et al., 2014; Poltronieri and Rossi, 2016).

Coffee processing generates enormous amounts of solid and liquid wastes. These wastes are usually employed in composting and animal feed; however, a large part is disposed of in the environment (Chanakya and Alwis, 2004; Janissen and Huynh, 2018). Therefore, adequate waste management is crucial. The main byproducts generated during coffee processing are spent coffee ground, defective and premature coffee beans, silverskin, coffee pulp, coffee husk, and wastewater, according to the coffee processing method used (Dias et al., 2015; Hoseini et al., 2021). This chapter describes the most recent advances regarding the valorization of wastes generated by different coffee processing.

8.2 COFFEE PROCESSING

Coffee must be rapidly processed after the harvesting to avoid undesired fermentation and deterioration and keep the seeds' quality. Furthermore, postharvest

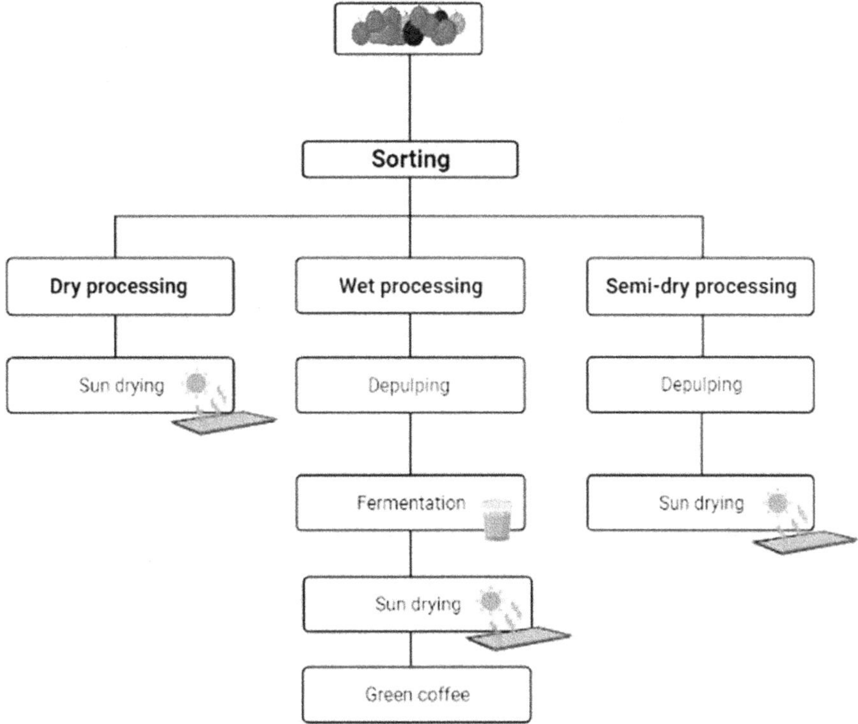

FIGURE 8.1 Flowchart of dry, wet, and semi-dry coffee processing.

practices are crucial to enhance the intrinsic quality of the coffee. As already mentioned in this chapter, there are three coffee processing methods: dry, wet, and semi-dry (Figure 8.1). The different coffee processing generates variable liquid and substantial quantities of wastes.

8.2.1 Dry Method

The dry method, also called natural, is the most common and traditional process. This coffee processing has few steps and produces less solid and liquid wastes when compared to wet and semi-dry methods. In dry processing, the fruits are harvested and spread onto the ground to allow fermentation and drying. The coffee cherries distributed on the ground are sun-dried with frequent revolving to reach homogenous and lower moisture. Some producers use variable drying supports, including patio pavements, raised beds, or combining these methods with mechanical dryers. When the cherries reach around 10–12% of moisture, they are inserted into hulling machines for outer layers remotion. The obtained coffee beans, also called green coffee, are sorted by size, roasted, and packed.

Dry processing is a simple and cheap way to transform the cherries into green coffee, employed mainly for Robusta coffee and approximately 80% of Yemen Arabica coffee and 60% of Brazilian and Ethiopian Arabica coffees (Poltronieri and

Rossi, 2016). In general, the green coffee obtained by the dry method has less acidity, sweetness, softness, and a more complex flavor than coffees processed by the other methods. Furthermore, more care is needed to obtain high-quality coffee (Brando, 2004). Although fewer steps are involved in this processing, it is more time-consuming than the sun-drying step. The longer time of drying makes the coffee more exposed to external agents and susceptible to damage.

8.2.2 Wet Method

The wet method removes the peel, pulp, and mucilage from selected coffee berries using a pulping machine. The obtained coffee berries are then transferred to water vats, where micro-organisms consume the remaining mucilage during the fermentation process (Hamdouche et al., 2016). This process contributes to preserving the intrinsic parameters of the coffee bean, resulting in a homogeneous green coffee with few undesired beans.

After fermentation, the coffee beans are sun-dried or dried using mechanical methods. The drying step may be shortened in the wet process due to the outer layers and part of the mucilage remotion; thus, avoiding excessive fermentation and undesired microbial colonization. It is known that the several steps associated with this coffee processing may become more expensive. However, when properly executed, the wet method may provide very high-quality coffee, increasing the product value (Brando, 2004).

A disadvantage of the wet method is water use. This wet coffee processing generates more wastewater than the dry method. Besides, for each ton of mature coffee berries, it produces 0.5 tons of coffee pulp as a byproduct (Alves et al., 2017).

8.2.3 Semi-dry Method

In the semi-dry method, the coffee berries are depulped to remove skin and pulp, and variable mucilage amounts (20–80%) are left. Then, the berries follow to dry step performed in a similar way to the dry method. The duration of the drying step is variable according to the different mucilage content, producing coffee products with variable quality. The main generated waste in this processing is the coffee pulp (Bonilla-Hermosa et al., 2014).

8.3 GENERATED WASTES

Coffee production has increased each year, and consequently, the coffee wastes resulting from coffee processing have become a major environmental problem. According to Cameron (2016), the coffee industry produces approximately 6 Mt of waste per year. In this sense, strategies to utilize coffee waste are crucial to decrease their impact on the environment. Researchers have demonstrated the use of coffee waste in several ways, such as biofuel and energy generation, animal feed, soil amendment, composting, mushroom cultivation, recovery of biocompounds, and others. The generated wastes depend on the coffee processing employed on the farm. The main byproducts associated with coffee production include defective and premature coffee beans, silverskin, coffee pulp, coffee husk, and wastewater (Dias et al., 2015; Hoseini et al., 2021). Furthermore, to obtain a cup of coffee by extraction of coffee solution from the coffee beans, a significant amount of spent coffee ground is produced as waste.

8.3.1 Defective and Premature Coffee Beans

Immature and defective beans are generally separated from the valuable mass of coffee during the harvesting and roasting process. Due to these beans not reaching the desired maturity level, they may have higher free amino acids and phenol and lower carbohydrate contents, affecting the quality of the final coffee (Franca and Oliveira. 2009; Mazzafera, 1999). However, these undesired coffee beans are byproducts containing important biocompounds, such as caffeine, chlorogenic acid, and pectin, which have pharmaceutical and food applications (Dias et al., 2015; Alves et al., 2017; Murthy and Naidu, 2012a). These compounds are present in skin and pulp and can also be obtained from other wastes generated during coffee processing.

8.3.2 Coffee Silverskin

Silverskin is a fragile layer surrounding de coffee seed (Figure 8.2) eliminated after roasting during coffee processing (Mussatto et al., 2011). This layer contains dietary fiber and antioxidant activity due to phenolic compounds and melanoidins formed by Maillard reaction during roasting (Ballesteros et al., 2014; Napolitano et al., 2007; Mussatto et al., 2011). Moreover, silverskin has shown prebiotic properties by favoring the probiotic bifidobacteria growth (Borrelli et al., 2004). Based on these exciting properties, this byproduct has been suggested to be employed in food and pharmaceutical products, such as developing functional food and cosmetics (Martinez-Saez et al., 2014; Mussatto et al., 2011; Rodrigues et al., 2015). Studies have demonstrated that coffee silverskin contains around 60–70% total dietary fiber and considerable antioxidant activity (approximately 2 mmol of Trolox/100 g), making this byproduct an interesting food ingredient (Borrelli et al., 2004; Napolitano et al., 2007).

Silverskin has also been evaluated as a substrate for biotechnological processes. It has been used for the β-fructofuranosidase and fructo-oligosaccharides by the fungus *Aspergillus japonicus* under solid-state fermentation (Mussatto and Teixeira, 2010).

8.3.4 Coffee Pulp

The pulp is located between the husk (outer layer) and the pectin layer of the coffee, as demonstrated in Figure 8.2. The coffee pulp contains higher amounts of phenolic compounds (around 1.5%) than the coffee husk (around 1.2%), represented mainly by flavan-3-ols, hydroxycinnamic acids, flavanols, and anthocyanidins (Ramirez-Coronel et al., 2004). Fresh cherries contain around 350 g of pulp per kilogram

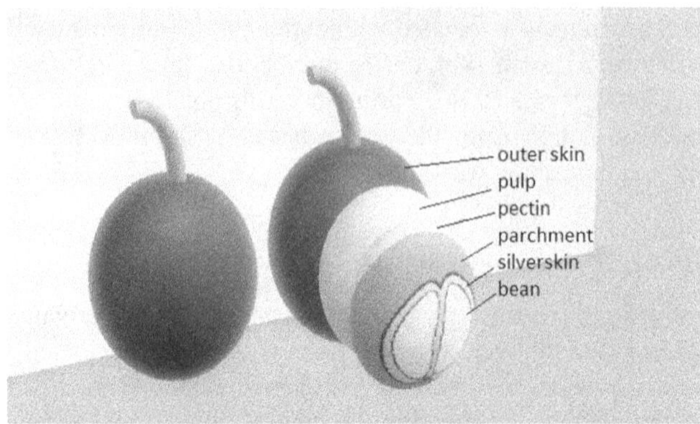

FIGURE 8.2 Scheme representing the main parts that make up the coffee bean.

which corresponds to approximately 30–35% of the dry weight of the berry (Braham and Bressani, 1979; Roussos et al., 1995).

Generally, the coffee pulp is a waste mainly obtained during wet and semi-dry coffee processing (Bonilla-Hermosa et al., 2014). This byproduct has been used to recover essential biocompounds such as caffeine and phenolic compounds, including chlorogenic and ferulic acids, and tannins which may show antimicrobial activity (Duangjai et al., 2016). Further, coffee pulp has been used as substrate for the mushrooms (e.g., genera *Pleurotus, Lentinula.,* and *Auricularia*) cultivation (Martínez-Carrera et al., 1996; 2000; Rodríguez and Zuluaga, 1994). After mushrooms harvesting, the substrate shows a decrease (around 20%) of caffeine concentration, and it is a source (% of dry weight) of crude fiber (31.5%), carbohydrates (30%), protein (21.5%), ash (15.2%), and fat (1.8%) (Martínez-Carrera et al., 2000).

8.3.5 Coffee Husk

Generally, the coffee husk is composed of all coffee layers (outer skin, pulp, parchment, and silverskin) (Figure 8.2) and is obtained after the hulling step during the dry coffee processing (Pandey 2000; Saenger et al., 2001). Coffee husk corresponds to around 12–18% of dried coffee beans, and its composition varies according to coffee species, geographical origin of the coffee, and processing method (Bekalo and Reinhardt, 2010; Gouvea et al., 2009. Navya and Pushpa, 2013). In general, coffee husk contains carbohydrates, lipids, proteins, phenolic compounds, caffeine, and inorganic compounds, which allows this byproduct to be applied in animal feed and as substrates for various biotechnological processes. This waste is a cheap substrate to produce organic acids (e.g., gibberellin and citric acid) and enzymes by microbial fermentation (Battestin and Macedo, 2007; Machado et al., 2000; Murthy and Naidu, 2010; Shankaranand and Lonsane, 1994). Furthermore, husk has a moisture content of around 12% and a heat capacity ranging from around 16 to 18 MJ/kg, making this byproduct favorable for energy generation in different applications such as mechanical coffee dryers, bakery furnaces, and high-temperature steam gasification (Saenger et al., 2001; Suarez and Luengo 2003; Vélez et al., 2009; Gangaputra, 2013).

8.3.6 Spent Coffee

Spent coffee is considered the principal waste of the coffee industry. Spent coffee is an acceptable particulate result of coffee brewing obtained from different coffee preparations, such as homemade coffee, coffee machines, instant coffee, and coffee-based beverages. This byproduct is a rich source of vitamin E, especially those obtained from classic espresso coffee or coffee machine methods since only about 1–5% of vitamins are extracted. The spent coffee contains lipids (about 2% on fresh weight) composed mainly of palmitic and linoleic acids (around 35% of total extractable lipids) (Couto et al., 2009).

It has been estimated that 0.65 kg of dried spent coffee (or 2.0 kg of wet spent coffee) is produced for each kilogram of brewed coffee. Thus, due to the high availability and chemical composition of this waste, it has been studied and employed in several applications, including biodiesel and bioethanol production, heat generation, animal feed, substrates for microbial enzyme production, edible mushroom cultivation, soil amendment, composting, phenolic and dietary fiber recovery, adsorbents, and activated carbon.

8.3.7 Wastewater

Wastewater is considered the water used during coffee processing. Generally, dry coffee processing produces a lower amount of wastewater than wet processing. However, the residual water is dumped into the environment, and its usage constitutes a challenge.

In dry processing, water is used to wash and separate (by flotation) the coffee berries in the proper degree of maturation from the other berries and stones (Brando, 2004). The water may be recycled sometimes until it is considered inappropriate for new washes. The produced wastewater is traditionally employed for irrigation or soil amendment in the farm (Matos, 2003).

Large amounts of wastewater are produced when the wet method is used (around 20 L/kg of the coffee bean). This water is used in mechanical pulping, mucilage removal, and fermentation steps. Unlike dry processing, this byproduct is rich in organic matter, including cellulose, hemicellulose, sucrose, monosaccharides, pectin, proteins, lipids, polyphenols, and vitamins. These chemical compounds are provided by coffee pulp and mucilage remotion and products of the fermentation step. The wastewater has high levels of chemical oxygen demand (COD) and biochemical oxygen demand (BOD), which may cause severe problems if disposed of in the environment. Thus, it is crucial to treat this waste to decrease COD and BOD before returning to the environment.

The main challenge for the wastewater treatment from coffee processing has been to develop an anaerobic digestion system that does not require previous acidity neutralization of this waste. It is known that parameters such as temperature, acidity, inoculants, and reactor type may affect the process (Chanakya and Alwis, 2004;

Haddis and Devi, 2008). Various technologies and small-scale reactors for anaerobic treatment have been evaluated to minimize the influence of these parameters and obtain a waste with decreased levels of COD and BOD. The use of upflow anaerobic filter (UAF), upflow anaerobic sludge blanket (UASB), and a combination of both UASB/UAF have demonstrated to be a promising system to treat wastewater from coffee processing (Bello-Mendoza and Castillo-Rivera, 1998). Moreover, in addition to anaerobic treatment, a combination of chemical coagulation/flocculation and advanced oxidation processes employing UV/H_2O_2, UV/O_3, and UV/H_2O_2/O_3 reduced 87% of wastewater COD (Zayas et al., 2007). And a reduction of 99% wastewater COD and BOD was observed using commercially activated carbon (particle size < 0.25 mm) at an adsorbent dose of 4 g/100 mL during 70 minutes of treatment at pH 7 and agitation speed of 600 rpm (Devi et al., 2008).

An exciting alternative developed in Colombia to reduce the wastewater generated by wet coffee processing was to pulp the coffee beans without water, remove mucilage by a mechanical process, followed by washing and drying steps. This process uses from 1 to 5 L of water per kilogram of coffee berries and is called "Becolsub" which means "Beneficio Ecologico de los Subproductos", in Spanish (Zambrano 1994; Zambrano and Cárdenas, 2000; Chanakya and Alwis, 2004).

8.4 COFFEE WASTE VALORIZATION

The coffee processing and industry have continuously generated liquid and solid wastes. The most common practices for treating these wastes are animal feed, composting, or disposal in landfills. However, the large amounts of generated wastes, in addition to environmental and economic concerns, have awakened the interest in methods and technologies for wastes recycling and their exploitation.

8.4.1 Coffee Husk

Coffee husk generally comprises all coffee layers and is one of the main coffee wastes. Its management has been the subject of various studies to decrease waste pollutants and obtain high-value products. Coffee husk has been evaluated for several purposes, as described below.

8.4.1.1 Industrial Purpose

It is common to use coffee husk as solid fuel in coffee-producing regions. The coffee husk combustion results in a large ash amount composed of alkaline and alkaline-earth metals. These compounds can replace the expensive and scarce feldspars traditionally employed in clay-based ceramic production as a fluxing component. The coffee waste used to produce clay-based ceramic has shown improved quality and reduced the cost of the final product (Manni et al., 2019). According to the studies performed by Acchar et al. (2013), the addition of 25–40% of ashes in clay-based ceramic formulation produced high-quality ceramic. Further, coffee husk has been demonstrated to be an interesting alternative for replacing more than 50% of the wood in particleboard production (Bekalo and Reinhardt, 2010; Nuamsrinuan et al., 2019). The coffee husk-based wood board's obtained flexural and internal bound properties showed potential for structural and non-structural panel products.

Another potential use of coffee husk is to obtain food flavoring compounds. These compounds are produced by chemical synthesis, and, alternatively, they can be produced by microbial fermentation or extracted from vegetal material. Coffee husks are a source of flavor compounds and essential oils; however, the presence of antinutrients such as tannin and caffeine does not allow their direct use. According to the study performed by Soares et al. (2000), steam-treated coffee husk did not show caffeine and chlorogenic acids and was used as substrates for flavor production by the fungus *Ceratocystis fimbriata*. The authors found that variable glucose supplementation to the substrate produced different volatile compounds' profile conferring banana and pineapple flavor notes. Furthermore, by microbial fermentation, organic acids such as gibberellin and citric acid, and enzymes such as pectinase, tannase, caffeine, amylase, protease, xylanase, and others, can be produced using coffee husk as substrate (Battestin and Macedo, 2007; Machado et al., 2000; Murthy and Naidu, 2010; Shankaranand and Lonsane, 1994).

8.4.1.2 Healthy and Nutraceutical Compounds

Agricultural wastes, including the coffee husk, are important sources of dietary fiber such as lignin, cellulose, hemicelluloses, pectin, and other polysaccharides, which have been associated with other health benefits reducing the risks of cardiovascular and gastrointestinal diseases and obesity. In addition to dietary fiber, carotenoids are natural dyes and have also been reported to display interesting biological properties such as antioxidant and anticarcinogenic activity and the prevention of cardiovascular diseases and macular degeneration. Coffee pulp and coffee husk extract have been used as the substrate for carotenoid production by the yeast *Rhodotorula mucilaginosa* (Moreira et al., 2018). These carotenoids show antioxidant and antimicrobial activities against pathogens such as Escherichia coli, Listeria monocytogenes, Salmonella colorless, Staphylococcus aureus, and some toxigenic fungi as *Aspergillus flavus, A. carbonarius, A. ochraceus,* and *A. parasiticus*. Further, the cyanidin 3-rutinoside has been reported as the dominant anthocyanin (the compound responsible for red/blue color in flowers and fruits) in coffee husks, which could be used as a natural colorant in foods (Prata and Oliveira, 2007). These anthocyanins have also been identified by Esquivel et al. (2010) and Murthy et al. (2012) in coffee skin and pulp and demonstrated an *in vitro* inhibitory effect on human α-glucosidase and α-amylase, indicating a potential nutraceutical use.

8.4.1.3 Biofuel

The use of coffee husk, including coffee pulp, for biofuel production, has been extensively evaluated by several studies (Calzada et al., 1981; 1984a; 1984b; 1986; Gouvea et al., 2009; Menezes et al., 2013; Protásio et al., 2013; Shenoy et al., 2011; Suarez and Luengo, 2003). These wastes may be employed in different ways and techniques to produce solid fuel, biogas, and ethanol. The use of coffee husk as solid fuel is generally performed by carbonization in a furnace, followed by grounding, coagulation, and molding in briquettes form. Then, the briquettes are packed and commercialized. According to Nyangito (2000), around 70% of the coffee husks generated during coffee processing in Kenya are solid fuel for heat generation. The coffee husk briquettes have been a potential candidate to replace firewood as solid fuel in bakery furnaces as it has demonstrated similar combustion characteristics (Suarez and Luengo, 2003).

Furthermore, the combustion of coffee husk briquettes has produced lower amounts of ash and reduced environmental impact. An interesting comparison between the energetic density (GJ/m^3) of coffee wastes and fossil fuels has revealed that coffee husk showed the highest equivalent volume in fossil fuels. It means that 1 m^3 of coffee husk produces similar quantities of energy as 100 L of fuel oil, or 119 L of petroleum, or 1221 L of diesel oil, or 135 L of gasoline (Protásio et al., 2013).

Another possibility to employ coffee waste for energy generation is gasification, which consists of obtaining a mixture of gas (CO_2, CH_4, H_2, and N_2) by applying elevated temperatures and controlled heating rates for partial incineration of the coffee husk (Miito and Banadda, 2017). For optimized gasification from coffee wastes, it is crucial to develop technologies to increase biomass-to-energy conversion, such as the high-temperature air/steam gasification of biomass (Wilson et al., 2010). Moreover, biogas can be produced by using the coffee pulp in an anaerobic digestion system. Studies have been performed using fresh coffee pulp and aerobically composted coffee in one-phase and two-phase digestion systems for up to 60 days. The higher methane production was obtained using fresh coffee pulp and a two-phase anaerobic digestion system (Calzada et al., 1981; 1984a; 1984b). Further to the benefits of biogas generation, the anaerobic digestion of coffee wastes also contributes to reducing emissions of CH_4 and CO_2 to the atmosphere, reducing the environmental pollution.

Coffee husk has also been evaluated as a substrate for yeast fermentation in bioethanol production. Traditionally, bioethanol is produced from vegetable sources rich in carbohydrates such as sugarcane, sugarbeets, and maize. However, many studies and efforts have been made to use lignocellulosic materials to produce carbohydrates for bioethanol production. In this sense, several lignocellulosic, which include coffee husk, have been extensively studied. The cellulose, hemicellulose, and lignin content in the coffee husk is similar to other agricultural wastes such as rice husk, barley, wheat straws, sugarcane bagasse, etc. In addition to the high availability of coffee wastes, this property makes it a promising alternative for bioethanol production (Somashekar and Appaiah, 2013). However, studies regarding economic and technical viability for ethanol production by using coffee wastes have been done (Gouvea et al., 2009; Menezes et al., 2013; Shenoy et al., 2011).

Fermentation parameters, such as temperature and yeast concentration, and substrate treatment are crucial to improving the fermentation process and producing high ethanol yield. Gouvea et al. (2009) employed coffee husks without previous hydrolytic treatment as a substrate for ethanol production by *Saccharomyces cerevisiae* fermentation. These authors found an ethanol production of 8.5 g/100 g of the substrate (dry basis) by using 0.3 g yeast/100 g substrate at 30°C. A bioethanol yield of 0.46 g/g of total carbohydrates was obtained from the fermented pretreated coffee pulp substrate (Shenoy et al., 2011). The substrate was previously treated with 2% sulfuric acid and cooked at 120°C for 10 minutes at 15 psi and then cooked at 90°C for 90 minutes at the same pressure for solubilization. The pre-treated coffee pulp substrate was fermented by *S. cerevisiae* in a 5.0 g yeast/L substrate concentration at 30°C for 48 hours under 120 rpm (Shenoy et al., 2011). Four different coffee pulp treatments, which included manual pressing, mechanical process (milling and filtering), the thermo-mechanical process in the water bath at 60°C for 15 minutes (followed by milling and filtering), and thermo-mechanical processes in the autoclave at 121°C for 15 minutes (followed by manual pressing and filtering), were also evaluated for carbohydrate extraction (Menezes et al., 2013). These authors found that the mechanical process was the most suitable extraction process. Furthermore, five different substrates' composition: coffee pulp extract, sugarcane juice, coffee pulp extract added with sugarcane juice, diluted molasses, and coffee pulp extract added with molasses, were assayed for ethanol production by *S. cerevisiae* inoculated at 8 log CFU/mL and fermented at 30°C for up to 24 hours. According to the authors, the addition of coffee pulp extract did not affect the fermentation and yeast viability, and they suggest mixing it with sugarcane juice or molasses to produce bioethanol at a yield of about 70 g/L (Menezes et al., 2013).

8.4.1.4 Agricultural Use

Coffee husk use in agriculture may be performed differently, such as animal feeding, composting, mushroom cultivation, and others. It is important to highlight that coffee residues may be rich in some antinutritional compounds, such as caffeine and phenolic acids, limiting their direct use in the same agricultural application. Thus, before using the coffee husk in agriculture, pretreatments are often needed to reduce the antinutrient content and increase the nutritional value (e.g., protein profile). Microbial activity has been the main strategy employed in this sense, as described by several works (Mazzafera 2002; Mohapatra et al., 2006; Nayak et al., 2012; Orozco et al., 2008; Roussos et al., 1995).

The use of coffee husk as animal feed was stimulated some decades ago due to its excellent availability and nutrients profile. However, cattle can tolerate only a small portion of it due to the phenolic acid content. Coffee husk addition to animal feed does not show good palatability in animals, thereby reducing food intake. Furthermore, phenolic compounds can inhibit some rumen cellulolytic bacteria species and reduce cellulose hydrolysis and glucose intake. In this sense, coffee wastes can be added to animal feeds in suitable proportions, according to animal species, to provide a significant acceptability feed ratio and decrease the antinutrients negative impact. Coffee husks do not affect feed intake or digestibility when replacing up to 25% of corn in sheep concentrate feed (Souza et al., 2004). The replacement of up to 60% of ground corn with coffee husk for feeding Holstein-Brahman steers in a two-year feedlot trial showed a benefit/cost ratio of 7%. Although the weight gain of animals was reduced, the consumption was not affected (Barcelos et al., 1997).

In soil, coffee husk without any previous treatment can inhibit the plants, particularly the growth of the roots due to the presence of toxic compounds (e.g., caffeine, chlorogenic acid). Further, it may decrease the buffering capacity for soil pH maintenance. An efficient alternative to reduce the toxicity of coffee husk is the composting process which may transform an environmental problem into a valuable agricultural product. Coffee husk is composed of lignocellulosic material and has a C/N ratio of about 30, making this waste a suitable substrate for microbial activity

(Murthy and Naidu, 2012a; Shemekite et al., 2014). Composts using variable concentrations of coffee husk and different types of supplementation, including cow manure, lime, green wastes, and effective micro-organisms such as N_2 fixing bacteria (*Azobacter* sp., *Bacillus megaterium*), have been elaborated by several authors and further evaluated as fertilizer in comparison with fertilizer available in the agriculture market (Bidappa, 1998; Kassa and Workayehu, 2014; Sekhar et al., 2014; Shemekite et al., 2014; Tuan, 2005). These studies have demonstrated a great potential of coffee husk compost as soil fertilizer as observed by soil fertility and plant growth improvement.

The coffee pulp has also been used in composting process. Generally, coffee pulp mass is spread on coffee plantations, overturned every 15 days, and the composting finishes in approximately four months. However, this process can be induced by supplementation with micro-organisms, cow manure, and others, which reduces the composting time by approximately 40% (Pandey et al., 2000a).

Another agricultural application of coffee husk is its use as a bed for mushroom growth. Coffee husk and coffee pulp are suitable substrates for mushroom growth, including *Lentinula edodes* (shiitake), *Flammulina velutipes, Pleurotus ostreatus*, and *Pleurotus pulomonarius* (Catarina et al., 2015; Martínez-Carrera et al., 1996; 2000; Rodríguez and Zuluaga, 1994). Because of the fragmented nature of the coffee husk, no previous grinding is needed; however, a disinfection process is required. It has been shown that fermentation of coffee husk and pulp by *P. ostreatus* can increase protein and cellulose contents in the spent substrate (obtained after mushroom harvesting) further to decrease lignin, tannins, and caffeine contents (Catarina et al., 2015). According to Martínez-Carrera et al. (1996; 2000), the caffeine content in the fresh pulp decreases by at least 20% after mushroom cultivation, obtaining a spent substrate containing (% of dry basis) crude fiber (31.5%), carbohydrates (30%), protein (21.5%), ash (15.2%), and fat (1.8%).

8.4.2 Coffee Silverskin

The most common use of coffee silverskin reported in the literature is its antioxidant activity and potential substrate to produce fungal enzymes and other metabolites. Due to the antioxidant property of coffee silverskin, some authors have suggested its use as a food and cosmetic ingredient (Borrelli et al., 2004; Martinez-Saez et al., 2014; Mussatto et al., 2011; Rodrigues et al., 2015). According to these studies, coffee silverskin comprises 60% of total dietary fiber, a significant amount (around 14%) of soluble dietary fiber. Furthermore, fat and reducing sugars are present in low quantities, while high anti-oxidative activity is probably due to high melanoidins content. For this reason, coffee silverskin may be an exciting source for the recovery of phenolic antioxidants and chlorogenic acids (Murthy and Naidu, 2012b).

Coffee silverskin has demonstrated potential prebiotic effects as observed by its stimulation of bifidobacterial growth over other bacteria, including coliforms, lactobacilli, clostridia, and Bacteroides spp. (Borrelli et al., 2004). Coffee silverskin also increased *Aspergillus japonicus* growth under solid-state fermentation to produce fructooligosaccharides (FOS), showing an advantage to other wastes such as corn cobs and cork oak (Mussatto and Teixeira, 2010).

8.4.3 Spent Coffee Grounds

Spent coffee grounds are a huge coffee waste, and their degradation in the environment requires enormous amounts of oxygen. Furthermore, toxic material may be generated due to the presence of some organic constituents. It is estimated that 1 kg of green coffee can generate 650 g of spent coffee grounds, and from 1 g of soluble coffee grounds, 2 g of spent coffee grounds are produced (Pflunger, 1975). Due to their excellent availability and toxic potential to the environment, the valorization of this waste is of great interest. Biofuel production, especially biodiesel, is the most described use of spent coffee ground. However, several other applications have been proposed in the literature (Caetano et al., 2014; Kondamudi et al., 2008; Tuntiwiwattanapun et al., 2017). Antioxidant compounds can be recovered from spent coffee grounds using aqueous alcohol such as ethanol, methanol, and isopropanol throughout the solid–liquid extraction process (Acevedo et al., 2013; Mata et al., 2018). Furthermore, due to the cellulose and hemicellulose content of spent coffee grounds, this waste can be employed for bioethanol and biocomposite materials (Kwon et al., 2013; García-García et al., 2015; Wu et al., 2016). Spent coffee grounds can also be used to produce biochar and activated carbon; further, they can be directly burned to produce heat energy (Safarik et al., 2012; Vardon et al., 2013).

Biodiesel production using spent coffee grounds has shown some advantages, such as stability due to the natural antioxidants and phenolic contents that avoid oxidation during transportation and storage (Todaka et al., 2018; Yanagimoto et al., 2004). However, the large-scale biodiesel production using spent coffee grounds present some limitations, which includes low lipid content (7–13%), the yield of the high content of unsaponifiable materials (40%), high moisture content (1.18–65.70%), and high heating rate (19.0–26.9 MJ/kg) (Saratale et al. 2020). Optimizations of the process are crucial for the improvement of biodiesel production. *In-situ* transesterification process in two batches was evaluated to produce biodiesel from spent coffee grounds on a large scale, reaching a yield of 80.7 and 83.0% (based on oil of spent coffee grounds weight) (Tuntiwiwattanapun et al., 2017). Transesterification is a standard process employed for biodiesel generation from biomass-derived oils and may convert the oil contained in spent coffee grounds into biodiesel fuel. This process can also be used to extract other products of pharmaceutical and food interest such as tannins, antioxidants, and bioactive compounds (Acevedo et al., 2013; Bravo et al., 2013; Murthy and Naidu, 2012b). However, using coffee wastes to produce value-added products requires a previous and comprehensive evaluation that includes various criteria, such as environmental, energetic, technical, and economical criteria. Further, the obtained products from spent coffee grounds should exhibit characteristics comparable to those available in the market. Table 8.1 shows some references describing various applications of spent coffee grounds.

8.5 CONCLUDING REMARKS

Coffee consumption represents one of the most important economic activities in the world. There are at least three coffee processing methods: dry, wet, and semi-dry processing, generating liquid and solid wastes in different amounts. The main wastes

TABLE 8.1
Examples of Literature Describing Some Applications of Spent Coffee Grounds

Application of Spent Coffee Grounds	References
Heat generation	Silva et al. (1998); Zuorro and Lavecchia (2012)
Biodiesel and bioethanol production	Abdullah and Koc (2013); Al-Hamamre et al. (2012); Atabani et al. (2019); Burton et al. (2010); Kondamudi et al. (2008); Kwon et al. (2013)
Bio-composite materials production	García-García et al. (2015); Wu et al. (2016)
Phenolic and dietary fiber recovery	Acevedo et al. (2013); Andrade et al. (2012); Bravo et al. (2013); Machado et al. (2012); Murthy and Naidu (2012b); Mussatto et al. (2011); Passos and Coimbra (2013); Zuorro and Lavecchia (2012)
Adsorbents	Franca et al. (2009); Kyzas (2012a,b); Kyzas et al. (2012); Plaza et al. (2012); Roh et al. (2012); Tokimoto et al. (2005)
Activated carbon	Djilani et al. (2012); Nakamura et al. (2009); Namane et al. (2005); Safarik et al. (2012)
Substrate for biotechnological process	Jooste et al. (2013); Murthy et al. (2009); Murthy and Naidu (2012c); Mussatto and Teixeira (2010)
Animal feed	Givens and Barber (1986); Sikka et al. (1985); Sikka and Chawla (1986)
Mushroom cultivation	Leifa et al. (2001)
Biocomposting	Hachicha et al. (2012); Liu and Price (2011)

produced during these coffee processes include defective and premature coffee beans, silverskin, coffee pulp, coffee husk, and wastewater that varies according to the coffee processing. Moreover, tremendous amounts of spent coffee ground are produced from coffee brewing. Therefore, suitable management practices of coffee wastes recovery are crucial to avoid environmental impacts. In this sense, several studies have been performed using coffee residues in several ways, such as biofuel and energy generation, animal feed, soil amendment, composting, mushrooms cultivation, recovery of biocompounds, and others. However, considering the large quantities of coffee wastes generated every year, it has been needed to improve the utilization and evaluate novel processes to obtain high-value products. Although studies on coffee wastes valorization have substantially increased, it is essential to consider that the obtained products from coffee wastes should exhibit lower costs and similar or improved characteristics of those products available in the market.

REFERENCES

Abdullah, M., Koc, A.B. (2013). Oil removal from waste coffee grounds using two-phase solvent extraction enhanced with ultrasonication. *Renew. Energy*, 50:965–970.

Acchar, W., Dultra, E.J.V., Segadães, A.M. (2013). Untreated coffee husk ashes used as flux in ceramic tiles. *Appl. Clay Sci.*, 75–76:141–147, 10.1016/j.clay.2013.03.009.

Acevedo, F., Rubilar, M., Scheuermann, E., Cancino, B., Uquiche, E., Garces, M., Inostroza, K., Shene, C. (2013). Spent coffee grounds as a renewable source of bioactive compounds. *J. Biobased Mater. Bioenergy*, 7 (1):420–428.

Alves, R.C., Rodrigues, F., Antónia Nunes, M., Vinha, A.F., Oliveira, M.B.P.P. (2017). *State of the Art in Coffee Processing By-Products*, Elsevier Inc., 2017, 10.1016/B978-0-12-811290-8.00001-3

Al-Hamamre, Z., Foerster, S., Hartmann, F., Kröger, M., Kaltschmitt, M. (2012). Oil extracted from spent coffee grounds as a renewable source for fatty acid methyl ester manufacturing. *Fuel*, 96:70–76.

Andrade, K.S., Gonçalvez, R.T., Maraschin, M., Ribeiro-do-Valle, R.M., Martínez, J., Ferreira, S.R.S. (2012). Supercritical fluid extraction from spent coffee grounds and coffee husks: antioxidant activity and effect of operational variables on extract composition. *Talanta*, 88:544–552.

Atabani, A.E., Shobana, S., Mohammed, M.N., Uğuz, G., Kumar, G., Arvindnarayan, S., Aslam, M., Al-Muhtaseb, A.H. (2019). Integrated valorization of waste cooking oil and spent coffee grounds for biodiesel production: blending with higher alcohols, FT–IR, TGA, DSC and NMR characterizations. *Fuel*, 244: 419–430.

Ballesteros, L.F., Teixeira, J.A., Mussatto, S.I. (2014). Chemical, functional, and structural properties of spent coffee grounds and coffee silverskin. *Food Bioprocess Technol.*, 7:3493–3503, 10.1007/s11947-014-1349-z

Barcelos, A.F., Andrade, I.F., Tiesenhausen, I.M.E.V.V., Sette, R.S., Bueno, C.F.H., Ferreira, J.J., Amaral, R., Paiva, P.C.A. (1997). The utilization of coffee hulls to feed steers in a feedlot – results of second year. *Brazilian J. Animal Sci.*, 26:1215–1221.

Battestin, V., Macedo, G.A. (2007). Effects of temperature, pH and additives on the activity of tannase produced by *Paecilomyces variotii*. *Electron. J. Biotechnol.*, 10:191–199, 10.2225/vol10-issue2-fulltext-9

Bekalo, S.A., Reinhardt, H.-W. (2010). Fibers of coffee husk and hulls for the production of particleboard. *Mater. Struct.*, 43:1049–1060, 10.1617/s11527-009-9565-0

Bello-Mendoza, Castillo-Rivera (1998). Start-up of an anaerobic hybrid (UASB/Filter) reactor treating wastewater from a coffee processing plant. *Anaerobe*, 4:219–225.

Bidappa, C.C. (1998). Organic manure from coffee husk: Comparison of technologies for organic manure from plantation wastes. *J. Plant. Crops*, 26:120–126.

Bonilla-Hermosa, V.A., Duarte, W.F., Schwan, R.F. (2014). Utilization of coffee by-products obtained from semi-washed process for production of value-added compounds. *Bioresour. Technol.*, 166:142–150, 10.1016/j.biortech.2014.05.031

Borrelli, R.C., Esposito, F., Napolitano, A., Ritieni, A. Fogliano, V. (2004). Characterization of a new potential functional ingredient: Coffee silverskin. *J. Agricul. Food Chem.*, 52:1338–1343.

Braham, J.E., Bressani, R. (1979). Coffee pulp: Composition, technology, and utilization: IDRC. (Ottawa, ON, CA).

Brando, C.H.J. (2004). Harvesting and Green Coffee Processing. In: J.N. Wintgens (Ed.), *Coffee: Growing, Processing, Sustainable Production*. Wenheim, Germany, Wiley-VCH.

Bravo, J., Monente, C., Juániz, I., Peña, M.P., Cid, C. (2013). Influence of extraction process on antioxidant capacity of spent coffee. *Food Res. Intern.*, 50:610–616.

Burton, R., Fana, X., Austic, G. (2010). Evaluation of two-step reaction and enzyme catalysis approaches for biodiesel production from spent coffee grounds. *Intern. J. Green Energy*, 7:530–536.

Caetano, N.S., Silva, V.F.M., Melo, A.C., Martins, A.A., Mata, T.M. (2014). Spent coffee grounds for biodiesel production and other applications. *Clean Techn. Environ. Policy*, 16:1423–1430.

Calzada, J.F., Arriola, M.C., Castañeda, H.O., Godoy, J.E., Rolz, C. (1984a). Methane from coffee pulp juice: Experiments using polyurethane foam reactors. *Biotechnol. Lett.*, 6:385–388.

Calzada, J.F., de León, O.R., Arriola, M.C., Micheo, F., Rolz, C., de León, R., Menchú, J.F. (1981). Biogas from coffee pulp. *Biotechnol. Lett.*, 3:713–716.

Calzada, J.F., Porres, E., Yurrita, A., Arrioloa, M.C., Micheo, F., Rolz, C., Menchú, J.F. (1984b). Biogas production from coffee pulp juice: One and two-phase systems. *Agricult. Wastes*, 9:217–230.

Cameron, A., O'Malley, S. Coffee Ground Recovery Program: Summary Report. (2016). [accessed 1 April 2021]; Available from: https://www.gordianbusiness.com.au/s/Coffe-Ground-Recovery-Program-Report-2019_Planet-Ark.pdf

Catarina M., Kasuya M., Maria J., Moura C., Braga C., Bento P. (2015). Production of selenium-enriched mushrooms in coffee husks and use of this colonized residue. *Coffee Heal. Dis. Prev.*, 301–309, 10.1016/B978-0-12-409517-5.00033-4

Chanakya, H.N., Alwis, A.A.P. (2004). Environmental issues and management in primary coffee processing. *Process Saf. Environ. Protect.*, 82:291–300.

Clay, J. (2004). *World agriculture and the environment. A commodity-by-commodity guide to impacts and practices*. World Wildlife Fund-Island Press, Washington, DC, USA. 570 pp.

Couto, R.M., Fernandes, J., da Silva, M.D.R.G., Simões, P.C. (2009). Supercritical fluid extraction of lipids from spent coffee grounds. *J. Supercrit. Fluids*, 51:159–166, 10.1016/j.supflu.2009.09.009

Devi, R., Singh, V., Kumar, A. (2008). COD and BOD reduction from coffee processing wastewater using Avacado peel carbon. *Bioresour. Technol.*, 99:1853–1860.

Dias, D.R., Rodríguez, N.V., Zambrano, D.A.F., López-Núñez, J.C. (2015). Management and utilization of wastes from coffee processing. In: R.F. Schwan, G.H. Fleet. (Eds.), *Cocoa and Coffee Fermentations*. CRC Taylor & Francis, Boca Raton, Ch. 15, pp. 545–588.

Djilani, C., Zaghdoudi, R., Modarressi, A., Rogalski, M., Djazi, F., Lallam, A. (2012). Elimination of organic micropollutants by adsorption on activated carbon prepared from agricultural waste. *Chem. Engineer. J.*, 189-190:203–212.

Duangjai, A., Suphrom, N., Wungrath, J., Ontawong, A., Nuengchamnong, N., Yosboonruang, A. (2016). Comparison of antioxidant, antimicrobial activities and chemical profiles of three coffee (*Coffea arabica* L.) pulp aqueous extracts. *Integr. Med. Res.*, 5(4): 324–331.

Esquivel, P., Kramer, M., Carle, R., Jiménez, V.M. (2010). Anthocyanin profiles and caffeine contents of wet-processed coffee (*Coffea arabica*) husks by HPLCDAD-MS/MS. 28th International Horticultural Congress. *Book of Abstracts*, 2:129–130.

Franca A.S., Oliveira, L.S. (2009). Coffee Processing Solid Wastes: Current Uses and Future Perspectives, 2009.

Franca, A.S., Oliveira, L., Ferreira, M.E. (2009). Kinetics and equilibrium studies of methylene blue adsorption by spent coffee grounds. *Desalination*, 249:267–272.

Gangaputra, R. (2013). Production of electricity using coffee husk. Proceedings of the International Conference on Technology and Business Management, 912–917.

García-García, D., Carbonell, A., Samper, M.D., García-Sanoguera, D., Balart, R. (2015). Green composites based on polypropylene matrix and hydrophobized spend coffee ground (SCG) powder. *Compos. Part B Eng.*, 78:256–265.

Givens, D.I., Barber, W.P. (1986). *In vivo* evaluation of spent coffee grounds as a ruminant feed. *Agricult. Wastes*, 18:69–72.

Gouvea, B.M., Torres, C., Franca, A.S., Oliveira, L.S., Oliveira, E.S. (2009). Feasibility of ethanol production from coffee husks. *Biotechnol. Lett.*, 31:1315–1319, 10.1007/s10529-009-0023-4

Hachicha, R., Rekik, O., Hachicha, S., Ferchichi, M., Woodward, S., Moncef, N., Cegarra, J., Mechichi, T. (2012). Co-composting of spent coffee ground with olive mill wastewater sludge and poultry manure and effect of *Trametes versicolor* inoculation on the compost maturity. *Chemosphere*, 88(6):677–682.

Haddis, A., Devi, R. (2008). Effect of effluent generated from coffee processing plant on the water bodies and human health in its vicinity. *J. Hazard. Mater.*, 152:259–262.

Hamdouche, Y., Meile, J.C., Nganou, D.N., Durand, N., Teyssier, C., Montet, D. (2016). Discrimination of post-harvest coffee processing methods by microbial ecology analyses. *Food Control*, 65:112–120.

Hoseini, M., Cocco, S., Casucci, C., Cardelli, V., Corti, G. (2021). Coffee by-products derived resources. *A review. Biomass and Bioenergy*, 148:106009.

Janissen, B., Huynh, T. (2018). Chemical composition and value-adding applications of coffee industry by-products: a review. *Resour. Conserv. Recycl.*, 128:110–117, 10.1016/j.resconrec.2017.10.00

Jooste, T., García-Aparicio, M., Brienzo, M., van Zyl, W.H., Görgens, J.F. (2013). Enzymatic hydrolysis of spent coffee ground. *Appl. Biochem. Biotechnol.*, 169:2248–2262.

Kassa, H., Workayehu, T. (2014). Evaluation of some additives on coffee residue (coffee husk and pulp) quality as compost, southern Ethiopia. *Int. Invent. J. Agric. Soil Sci.*, 2:2408–7254. http://internationalinventjournals.org/journals/IIJAS

Kyzas, G.Z. (2012a). Commercial coffee wastes as materials for adsorption of heavy metals from aqueous solutions. *Materials*, 5:1826–1840.

Kyzas, G.Z. (2012b). A decolorization technique with spent "Greek coffee" grounds as zero-cost adsorbents for industrial textile wastewaters. *Materials*, 5:2069–2087.

Kyzas, G.Z., Lazaridis, N.K., Mitropoulos, A.C. (2012). Removal of dyes from aqueous solutions with untreated coffee residues as potential low-cost adsorbents: equilibrium, reuse and thermodynamic approach. *Chem. Engineer. J.*, 189–190:148–159.

Kondamudi, N., Mohapatra, S.K., Misra, M. (2008). Spent coffee grounds as a versatile source of green energy. *J. Agric. Food Chem.*, 56 (24):11757–11760.

Kwon, E.E., Yi, H., Jeon, Y.J. (2013). Sequential co–production of biodiesel and bioethanol with spent coffee grounds. *Bioresour. Technol.*, 136:475–480.

Leifa, F., Pandey, A., Soccol, C.R. (2001). Production of *Flammulina velutipes* on coffee husk and coffee spent-ground. *Brazilian Arch. Biol. Technol.*, 44:205–212.

Liu, K., Price, G.W. (2011). Evaluation of three composting systems for the management of spent coffee grounds. *Bioresour. Technol.*, 102:7966–7974.

Machado, C.M.M., Oliveira, B.H., Pandey, A., Soccol, C.R. (2000). Coffee husk as substrate for the production of gibberellic acid by fermentation. In: T. Sera, C.R. Soccol, A. Pandey, S. Roussos (Eds.), *Coffee Biotechnol. Qual. Proc. 3rd Int. Semin. Biotechnol. Coffee Agro-Industry, Londrina, Brazil*, Springer Netherlands, Dordrecht, pp. 401–408, 10.1007/978-94-017-1068-8_37.

Machado, E.M.S., Rodriguez-Jasso, R.M., Teixeira, J.A., Mussato, S.I. (2012). Growth of fungal strains on coffee industry residues with removal of polyphenolic compounds. *Biochem. Engineer. J.*, 60:87–90.

Martínez-Carrera, D., Aguilar, A., Martínez, W., Bonilla, M., Morales, P., Sobal, M. (2000). Commercial production and marketing of edible mushrooms cultivated on coffee pulp in Mexico. In: T. Sera, C.R. Soccol, A. Pandey, S. Roussos (Eds.), *Coffee Biotechnology and Quality*. Kluwer Academic Publishers, Dordrecht, The Netherlands. pp. 471–488.

Martínez-Carrera, D., Morales, P., Martínez, W., Sobal, M., Aguilar, A. (1996). Large-scale drying of coffee pulp and its potential for mushroom cultivation in Mexico. *Micologia Neotropical Aplicada*, 9:43–52.

Martinez-Saez, N., Ullate, M., Martin-Cabrejas, M.A., Martorell, P., Genovés, S., Ramon, D., Del Castillo, M.D. (2014). A novel antioxidant beverage for body weight control based on coffee silverskin. *Food Chem.*, 150:227–234, 10.1016/j.foodchem.2013.10.100

Manni, A., Haddarb, A.E., Hassanic, I-E.E.A.E., Bouaria, A.E., Sadika, C. (2019). Valorization of coffee waste with Moroccan clayto produce a porous red ceramic (class BIII). *Boletín de la sociedad española de cerâmica y vidrio*, 58:211–220.

Mata, T.M., Martins, A.A., Caetano, N.S. (2018). Bio–refinery approach for spent coffee grounds valorization. *Bioresour. Technol.*, 247:1077–1084.

Matos, A.T. (2003). Tratamento e destinação final dos resíduos gerados no beneficiamento do fruto do cafeeiro. In: L. Zambolin (Ed.), *Produção integrada de café*, Editora UFV, Viçosa, Brazil. pp. 647–708.

Mazzafera, P. (2002). Degradation of caffeine by microorganisms and potential use of decaffeinated husk and pulp in animal feeding. *Scientia Agricola*, 59:821–851.

Mazzafera, P. (1999). Chemical composition of defective coffee beans, *Food Chem.*, 64: 547–554. 10.1016/S0308-8146(98)00167-8

Menezes, E.G.T., Carmo, J.R., Menezes, A.G.T., Alves, J.G.L.F., Pimenta, C.J., Queiroz, F. (2013). Use of different extracts of coffee pulp for the production of bioethanol. *Appl. Biochem. Biotechnol.*, 169:673–687.

Miito, G.J., Banadda, N. (2017). A short review on the potential of coffee husk gasification for sustainable energy in Uganda. *F1000Research*, 6:1–11, 10.12688/f1000research. 10969.1.

Mohapatra, B.R., Harris, N., Nordin, R., Mazumder, A. (2006). Purification and characterization of novel caffeine oxidase from *Alcaligenes* species. *J. Biotechnol.*, 125: 319–332.

Moreira, M.D., Melo, M.M., Coimbra, J.M., Reis, K.C., Schwan, R.F., Silva, C.F. (2018). Solid coffee waste as alternative to produce carotenoids with antioxidant and antimicrobial activities. *Waste Management*, 82:93–99.

Murthy, P.S., Naidu, M.M. (2012a). *Sustainable Management of Coffee Industry By- Products and Value Addition - A Review*, Elsevier B.V, 10.1016/j.resconrec.2012.06.005

Murthy, P.S., Naidu, M.M. (2012b) Recovery of phenolic antioxidants and functional compounds from coffee industry by-products. *Food Bioprocess Technol.*, 5:897–903.

Murthy, P.S., Madhava Naidu, M. (2012c) Production and application of xylanase from *Penicillium* sp. utilizing coffee by-products. *Food Bioprocess Technol.*, 5:657–664.

Murthy, P.S., Manjunatha, M.R., Sulochannama, G., Naidu, M.M. (2012). Extraction, characterization and bioactivity of coffee anthocyanins. *Eur. J. Biol. Sci.*, 4:13–19.

Murthy, P.S., Naidu, M. (2010). Protease production by *Aspergillus oryzae* in solid-state fermentation utilizing coffe by-products. *World Appl. Sci. J.* 8: 199–205.

Murthy, P.S., Naidu, M.M., Srinivas, P. (2009). Production of α-amylase under solid-state fermentation utilizing coffee waste. *J. Chem. Technol. Biotechnol.*, 84:1246–1249.

Mussatto, S.I., Ercília, E.M.S., Martins, S., Teixeira, J.A. (2011). Production, composition, and application of coffee and its industrial residues. *Food Bioprocess Technol.*, 4:661–672.

Mussatto, S.I., Teixeira, J.A. (2010). Increase in the fructooligosaccharides yield and productivity by solid-state fermentation with *Aspergillus japonicus* using agro-industrial residues as support and nutrient source. *Biochem. Engineer. J.*, 53:154–157.

Nakamura, T., Hirata, M., Kawasaki, N., Tanada, S., Tamuro, T., Nakahori, Y. (2009). Decolorization of indigo carmine by charcoal from extracted residue of coffee beans. *J. Environ. Sci. Health Part A*, 3:555–562.

Napolitano, A., Fogliano, V., Tafuri, A., Ritieni, A. (2007). Natural occurrence of ochratoxin A and antioxidant activities of green and roasted coffees and corresponding by-products. *J. Agricult. Food Chem.*, 55:10499–10504.

Namane, A., Mekarzia, A., Benrachedi, K., Belhaneche-Bensemra, N., Hellal, A. (2005). Determination of the adsorption capacity of activated carbon made from coffee grounds by chemical activation with $ZnCl_2$ and H_3PO_4. *J. Hazard. Mater.*, 119:189–194.

Navya, P.N., Pushpa, S.M. (2013). Production, statistical optimization and application of endoglucanase from *Rhizopus stolonifer* utilizing coffee husk. *Bioproc. Biosyst. Eng.* 36:1115–1123, 10.1007/s00449-012-0865-3

Nayak, S., Harshitha, M.J., Maithili, S.C., Anilkumar, H.S., Rao, C.V. (2012). Isolation and characterization of caffeine degrading bacteria from coffee pulp. *Indian J. Biotechnol.*, 11:86–91.

Nyangito, H. (2000). *Delivery of Services to Smallholder Coffee Farmers and Impacts on Production under Liberalization in Kenya*, Kenya Institute for Public Policy Research and Analysis, 2000.

Nuamsrinuan N., Naemchanthara P., Limsuwan P., Naemchanthara K. (2019). Fabrication and characterization of particle board from coffee husk waste. *Appl. Mech. Mater.*, 891: 111–116. 10.4028/www.scientific.net/AMM.891.111

Orozco, A.L., Pérez, M.I., Guevara, O., Rodrígues, J., Hernández, M., González-Vila, F.J., Polvillo, O., Arias, M.E. (2008). Biotechnological enhancement of coffee pulp residues by solid-state fermentation with *Streptomyces*. Py–GC/MS analysis. *J. Anal. Appl. Pyrolysis*, 81:247–252.

Pandey, A., Soccol, C.R., Nigam, P., Brand, D., Mohan, R., Roussos, S. (2000). Biotechnological potential of coffee pulp and coffee husk for bioprocesses. *Biochemical Engineering Journal*, 6:153–162.

Passos, C.P., Coimbra, M.A. (2013). Microwave superheated water extraction of polysaccharides from spent coffee grounds. *Carbohydrate Polymers*, 94:626–633.

Pflunger, R.A. (1975). Soluble coffee processing. In: C. Mantell (Ed.), *Solid Wastes: Origin, Collection, Processing, and Disposal*. John Wiley & Sons Inc, New York.

Plaza, M.G., González, A.S., Pevida, C., Pis, J.J., Rubiera, F. (2012). Valorisation of spent coffee grounds as CO2 adsorbents for postcombustion capture applications. *Appl. Energy*, 99:272–279.

Poltronieri, P., Rossi, R. (2016). Challenges in specialty coffee processing and quality assurance. *Challenges*, 7:19, 10.3390/challe7020019

Prata, E.R.B.A., Oliveira, L.S. (2007). Fresh coffee husks as potential sources of anthocyanins. *LWT - Food Sci. Technol.*, 40: 1555–1560, 10.1016/j.lwt.2006.10.003

Protásio, T.P., Bufalino, L., Tonoli, G.H.D., Guimarães Junior, M., Trugilho, P.F., Mendes, L.M. (2013). Brazilian lignocellulosic wastes for bioenergy production: Characterization and comparison with fossil fuels. *BioResources*, 8:1166–1185.

Ramirez-Coronel, M.A., Marnet, N., Kolli, V.S.K., Roussos, S., Guyot, S., Augur, C. (2004). Characterization and estimation of proanthocyanidins and other phenolics in coffee pulp (Coffea arabica) by thiolysis-high-performance liquid chromatography. *J. Agric. Food Chem.*, 52 (2004): 1344–1349, 10.1021/jf035208t

Rodrigues, F., Pereira, C., Pimentel, F.B., Alves, R.C., Ferreira, M., Sarmento, B., Amaral, M.H., Oliveira, M.B.P.P. (2015). Are coffee silverskin extracts safe for topical use? An *in vitro* and *in vivo* approach. *Ind. Crop. Prod.,* 63:167–174, 10.1016/j.indcrop.2014.10.014.

Rodríguez, V.N., Zuluaga, V.J. (1994). Cultivo de *Pleurotus pulmonarius* en pulpa de café. *Cenicafé*, 45:81–92.

Roh, J.R., Umh, H.N., Yoo, C.M., Rengaraj, S., Lee, B., Kim, Y. (2012). Waste coffee-grounds as potential biosorbents for removal of acid dye 44 from aqueous solution. *Korean J. Chem. Engineer.*, 29:903–907.

Roussos, S., Aquiahuatl, M.D., Trejohernandez, M.D., Perraud, I.G., Favela, E., Ramakrishna, M., …Viniegragonzalez, G. (1995). Biotechnological management of coffee pulp – isolation, screening characterization, selection of caffeine degrading fungi and natural microflora present in coffee pulp and husk. *Appl. Microbiol. Biotechnol.* 42(5):756–762.

Saenger, M., Hartge, E.-U., Werther, J., Ogada., T., Siagi, Z. (2001). Combustion of coffee husks. *Renew. Energy*, 23:103–121.

Safarik, I., Horska, K., Svobodova, B., Safarikova, M. (2012). Magnetically modified spent coffee grounds for dyes removal. *Eur. Food Res. Technol.*, 234: 345–350.

Saratale, G.D., Bhosale, R., Shobana, S., Banu, J.R., Pugazhendhi, A., Mahmoud, E., Sirohi, R., Bhatia, S.K., Atabani, A.E., Mulone, V., Yoon, J-J., Shin, H.S., Kumar, G. (2020). A review on valorization of spent coffee grounds (SCG) towards biopolymers and biocatalysts production. *Bioresour. Technol.*, 314:123800.

Schwan, R.F., Silva, C.F., Batista, L.R., Dias, D.R. (2014). Coffee. In: C.W. Bamforth, R.E. Ward. (Eds.), *Oxford Handbook of Food Fermentations*. London: Oxford, pp. 17–30.

Schwan, R.F., Silva, C.F., Batista, L.R. (2012). Coffee fermentation. In: Y.H. Hui, E. Özgül Evranuz (Eds.), *Handbook of Plant-Based Fermented Food and Beverage Technology*. Boca Raton: CRC Press, pp. 677–690.

Sekhar D., Kumar P.B.P., Rao K.T. (2014). Effect of coffee husk compost on growth and yield of paddy. *J. Acad. Ind. Res.*, 3:195–197. http://jairjp.com/SEPTEMBER2014/09 SEKHAR.pdf

Shankaranand V.S., Lonsane B.K. (1994). Coffee husk: an inexpensive substrate for production of citric acid by *Aspergillus Niger* in a solid-state fermentation system. *World J. Microbiol. Biotechnol.*, 10:165–168, 10.1007/BF00360879

Shemekite, F., Gómez-Brandón, M., Franke-Whittle, I.H., Praehauser, B., Insam, H., Assefa, F. (2014). Coffee husk composting: an investigation of the process using molecular and non-molecular tools. *Waste Manag.*, 34:642–652, 10.1016/j.wasman.2013.11.010

Shenoy, D., Pai, A., Vikas, R.K., Neeraja, H.S., Deeksha, J.S., Nayak, C., Rao, C.V. (2011). A study on bioethanol production from cashew apple pulp and coffee pulp waste. *Biomass Bioenergy*, 35:4107–4111.

Sikka, S.S., Chawla, J.S. (1986). Effect of feeding spent coffee grounds on the feedlot performance and carcass quality of fattening pigs. *Agricult. Wastes*, 18(1986): 305–308.

Sikka, S.S., Bakshi, M.P.S., Ichhponani, J.S. (1985). Evaluation in vitro of spent coffee grounds as a livestock feed. *Agricult. Wastes*, 13:315–317.

Silva, M.A., Nebra, S.A., Machado Silva, M.J., Sanchez, C.G. (1998). The use of biomass residues in the Brazilian soluble coffee industry. *Biomass Bioenergy*, 14:457–467.

Soares, M., Christen, P., Pandey, A., Soccol, C.R. (2000). Fruity flavour production by *Ceratocystis fimbriata* grown on coffee husk in solid-state fermentation. *Process Biochem.*, 35: 857–861, 10.1016/S0032-9592(99)00144-2

Somashekar, K.L., Appaiah, K.A.A. (2013). Coffee cherry husk – A potential feed stock for alcohol production. *Int. J. Environ. Waste Manag.*, 11 (2013): 410–419, 10.1504/IJEWM.2013.054242

Souza, A.L., Garcia, R., Bernardino, F.S., Rocha, F.C., Valadares Filho, S.C., Pereira, O.G., Pires, A.J.V. (2004). Coffee hulls in the diet of sheep: intake and apparent digestibility. *Brazilian J. Animal Sci.*, 33 (Suppl. 2): 2170–2176.

Suarez, J.A., Luengo, C.A. (2003). Coffee husk briquettes: a new renewable energy source. *Energy Sources*, 25:961–967.

Tuan, B. (2005). Efficience of using coffee husk to applied for Robusta coffee in Central Highland. *J. Soil Sci.*, 22:10–15.

Todaka, M., Kowhakul, W., Masamoto, H., Shigematsu, M. (2018). Improvement of oxidation stability of biodiesel by an antioxidant component contained in spent coffee grounds. *Biofuels* 1–9, 10.1080/17597269.2018.1468977

Tokimoto, T., Kawasaki, N., Nakamura, T., Akutagawa, J., Tanada, S. (2005). Removal of lead ions in drinking water by coffee grounds as vegetable biomass. *J. Colloid. Interface Sci.*, 281:56–61.

Tuntiwiwattanapun, N., Monono, E., Wiesenborn, D., Tongcumpou, C. (2017). *In-situ* transesterification process for biodiesel production using spent coffee grounds from the instant coffee industry. *Ind. Crop. Prod.*, 102:23–31.

Vardon, D.R., Moser, B.R., Zheng, W., Witkin, K., Evangelista, R.L., Strathmann, T.J., Rajagopalan, K., Sharma, B.K. (2013). Complete utilization of spent coffee grounds to produce biodiesel, bio-oil, and biochar. ACS Sustain. *Chem. Eng.*, 1:1286–1294.

Vélez, J., Chejne, F., Valdés, C., Emery, E., Londöo, C. (2009). Co-gasification of Colombian coal and biomass in fluidized bed: an experimental study. *Fuel*, 88:424–430.

Wilson, L., John, G.R., Mhilu, C.F., Yang, W., Blasiak, W. (2010). Coffee husks gasification using high temperature air/steam agent. *Fuel Process. Technol.*, 91:1330–1337, 10.1016/j.fuproc.2010.05.003.

Wu, H., Hu, W., Zhang, Y., Huang, L., Zhang, J., Tan, S., Cai, X., Liao, X. (2016). Effect of oil extraction on properties of spent coffee ground-plastic composites. *J. Mater. Sci.*, 51: 10205–10214.

Yanagimoto, K., Ochi, H., Lee, K.-G., Shibamoto, T.J. (2004). Anti-oxidative activities of fractions obtained from brewed coffee. *J. Agric. Food Chem.*, 52 (3):592–596.

Zambrano, F.D.A., Cárdenas, C.J. (2000). Manejo y tratamiento primario de lixiviados producidos en la tecnología BECOLSUB. *Chinchiná, Cenicafé.* 8 pp. (Avances Técnicos N° 280).

Zambrano, F.D.A., Isaza, H.J.D. (1994). Lavado del café en los tanques de fermentación. *Cenicafé*, 45:106–118.

Zayas, T.P., Geissler, G., Hernandez, F. (2007). Chemical oxygen demand reduction in coffee wastewater through chemical flocculation and advanced oxidation processes. *J. Environ. Sci.*, 19:300–305.

Zuorro, A., Lavecchia, R. (2012). Spent coffee grounds as a valuable source of phenolic compounds and bioenergy. *J. Cleaner Prod.*, 34:49–56.

9 Coffee Husk Waste Valorization Using Thermal Pretreatment Associated to Bioprocess to Produce Bioproducts: Characterization, Kinetic, Economic Assessment, and Challenges

Nayara Clarisse Soares Silva,
Yasmim Arantes da Fonseca,
Aline Gomes de Oliveira Paranhos,
Oscar Fernando Herrera Adarme,
Sérgio Francisco de Aquino, and
Leandro Vinícius Alves Gurgel
Laboratory of Technological and Environmental Chemistry, Department of Chemistry, Institute of Exact and Biological Sciences (ICEB), Federal University of Ouro Preto, Campus Universitário Morro do Cruzeiro, Bauxita, Ouro Preto, Minas Gerais, Brazil

Juan Daniel Valderrama Rincón
Grupo GRESIA, Department of Environmental Engineering, Universidad Antonio Nariño, Bogotá, Colombia

Héctor Javier Luna Wandurraga
Postgraduate Program in Environmental Engineering, Federal University of Ouro Preto, Ouro Preto, Minas Gerais, Brazil

DOI: 10.1201/9781003128977-9

Bruno Eduardo Lobo Baêta
Laboratory of Technological and Environmental Chemistry, Department of Chemistry, Institute of Exact and Biological Sciences (ICEB), Federal University of Ouro Preto, Campus Universitário Morro do Cruzeiro, Bauxita, Ouro Preto, Minas Gerais, Brazil

CONTENTS

9.1 INTRODUCTION

The environmental problems derived from the fossil fuel usage have served as motivation in the search for energy production and other bioproducts from renewable sources. In this order of ideas, the agro-industrial residues became promissory raw materials given their relative abundance, specifics characteristics and low cost. Several studies support the use of lignocellulosic residues as substrate for the obtention of high added-value products such as ethanol, lactic acid, butanol, polyhydroxyalkanoates, and biogas (Amiri and Karimi 2015; Mesquita, Ferraz, and Aguiar 2016; Rios-González et al. 2017).

In the context of agriculture, coffee is the second most commercialized commodity in the world, with a production of 174.6 million sacks (60 kg) for the 2018/2019 harvest (Mudgil and Barak 2018; USDA 2019). One important residue from the agro-industrial coffee production is the coffee husk waste (CHW), which is rich in compounds like caffeine, phenols, tannins, and carbohydrates. However, due to its low-density content in the lignocellulosic biomass, the applications of CHW are usually limited to animal feeding supplements and composting (Baêta et al. 2017a; Pandey et al. 2000); hence, the possibility of using it under a biorefinery approach becomes an interesting alternative for the concomitant valorization and sustainable disposition of this kind of residues.

Considering that CHW has a relatively high content of holocellulose (53.67–58.13%) (Bekalo and Reinhardt 2010; Pandey et al. 2000; Baêta et al. 2017a), some studies address the possibility of using that biomass as raw material for bioprocesses. Among those, anaerobic digestion (AD) has been considered as one of the CHW valorization alternatives, obtaining energy as the main product. Respect to this, it has been shown that higher biogas yields can be obtained when the biomass is subject to a pretreatment (Zheng et al. 2014). Hydrothermal liquefaction, also known as liquid hot water (LHW) pretreatment, is one of the suggested pretreatments for a better exploitation of biomass in the context of a lignocellulosic biorefinery; acting on the decomposition of the complex lignin-carbohydrate structure, specifically by hemicelluloses solubilization (Wang et al. 2018).

The pretreatment by LHW generates two process currents: a solid fraction rich in cellulose and a liquid phase (hydrolysate) containing oligosaccharides, organic acids, 2-furfuraldehyde (FF), 5-hydroxymethyl-2-furfuraldehyde (HMF), and other phenolic compounds (PCs) derived from lignin (Ahmad, Silva, and Varesche 2018). These solid and liquid fractions can be further processed together or separated using diverse technological routes. However, even though LHW is a widely reported technology in literature, it usually presents the high energy requirement for water heating as a disadvantage and, hence, the energy balances are mandatory for the techno-economic assessment of processes involving this technology (Momayez, Karimi, and Sárvári 2018). However, when consider the utilization of hydrolysate to produce biogas and bioenergy the balance energy of the process can be improved.

Given the above, this research presents the influence of the LHW pretreatment severity on the CHW fractionation and further biogas production by AD in liquid phase (L-D), solid phase (S-D), and semi-solid phase (SL-D). Additionally, to the best of our knowledge, this is the first study that makes an energetic-economic estimation of the CHW fractionation, after pretreatment, when considering two possible biorefinery technological routes: ethanol production and biogas production.

9.2 COFFEE HUSKS WASTE CHARACTERIZATION

CHW was kindly provided by a farm "Jangada (Criminoso district, Lavras, MG, Brazil)" and was collected in the 2015/2016 harvest season. The raw coffee husks were dried under sunlight until a less than 10% moisture content was achieved. After this procedure, they were stored in light absence at room temperature (~22°C) prior to use.

The dry basis chemical composition of CHW was (%wt) 36.0 ± 0.9% cellulose, 20.4 ± 1.3% de hemicelluloses, 31.1 ± 1.2% lignin, and 16.6 ± 0.9% extractables. These results are in agreement with other studies (Baêta et al. 2017b; Kumari and Singh 2018; Pandey et al. 2000; Morales-Martínez et al. 2020).

9.3 EXPERIMENTAL DESIGN OF THE PRETREATMENT

The LHW pretreatment experiments were performed according to a 2^3 Box-Behnken design to evaluate the influence of the independent variables: X_1 (temperature – T, °C), X_2 (time – t, min), and X_3 (liquid-solid ratio – LSR, mL g^{-1}), on the dependent variables (response): Y_1 (reductive sugars concentration – RS, g L^{-1}), and Y_2 (phenolic compounds concentration – PCs, g L^{-1}) for the hydrolyzate. The matrix of experiments with coded and uncoded levels of the independent variables is shown in Table 9.1.

The selection of the response variables obeys first to the relative easiness of the carbohydrates' degradation by AD and, second, to the inhibitory characteristics of the phenolic compounds on biogas production by AD when using agro-industrial residues (Monlau et al. 2014; Caroca et al. 2021). Considering that CHW has an elevated content of lignin (22–31%) (Baêta et al. 2017a) relatively high quantities of PCs are solubilized during the LHW pretreatment of the biomass. Although several authors report the effect of these compounds on AD for other kinds of residual biomass like fruits residues, sugarcane bagasse, corn residues, and olive mill solid waste (Wikandari et al. 2015; Battista, Fino, and Ruggeri 2014; Michelin et al. 2015; Chen et al. 2020; Caroca et al. 2021), the effect of PCs on the biomethanization of CHW has not been reported yet.

9.4 LIQUID HOT WATER PRETREATMENT OF CHW

The CHW pretreatment experiments by LHW were done in autoclave-like stainless steel (316 L) cylindrical reactors (16.6 cm height × 6.0 cm internal diameter × 8.0 cm external diameter). For all the experiments, the reactors were fed with 30.0 g of CHW (dry basis) and distilled water in LSR proportions as described for the experimental conditions in (Table 9.1). The reactors were heated using a glycerin bath (Marconi, model MA 159) provided with a thermocouple temperature control. After the reaction time was completed, the reactors were immediately cooled down using an ice bath before opening them. After that, the solid and liquid fractions were separated by vacuum filtration using a Büchner funnel before chemical characterization.

The effect of temperature and time under every pretreatment experimental condition was evaluated according to the severity factor (R_0) (Eq. 9.1):

$$\log R_0 = \log\left[t \exp\left(\frac{T - 100}{14.75}\right)\right] \tag{9.1}$$

where t is the reaction time (min) and T is the pretreatment temperature (°C).

TABLE 9.1
The Matrix of Experiments with Coded and Uncoded Levels of the Independent Variables

Experiments	Independent Variables (Factors)			Dependent Variables (Responses)		Hydrolyzate Characterization		
	Temperature (X_1, °C)	Time (X_2, min)	*LSR* (X_3, mL g^{-1})	Reducing Sugars, RS (Y1, g L^{-1})	Phenolic Compounds, PC (Y2, g L^{-1})	COD (g O_2 L^{-1})	Caffeine (g L^{-1})	Inhibitors (HMF + FF) (g L^{-1})
1	120 (−1)	20 (−1)	7.5 (0)	1.1 ± 0.2	1.0 ± 0.4	9.7	0.22	0.00
2	200 (+1)	20 (−1)	7.5 (0)	5.3 ± 0.0	3.6 ± 0.1	36.9	0.29	0.04
3	120 (−1)	90 (+1)	7.5 (0)	2.7 ± 0.2	1.6 ± 0.1	18.8	0.24	0.00
4	200 (+1)	90 (+1)	7.5 (0)	6.6 ± 0.6	4.8 ± 0.4	25.0	0.21	0.38
5	120 (−1)	55 (0)	5 (−1)	4.1 ± 0.3	1.5 ± 0.4	30.8	0.33	0.02
6	200 (+1)	55 (0)	5 (−1)	12.7 ± 0.1	4.5 ± 0.1	38.4	0.40	0.28
7	120 (−1)	55 (0)	10 (+1)	0.7 ± 0.2	2.0 ± 0.1	12.7	0.57	0.00
8	200 (+1)	55 (0)	10 (+1)	6.9 ± 0.3	2.2 ± 0.0	16.4	0.25	0.18
9	160 (0)	20 (−1)	5 (−1)	5.5 ± 0,3	2.5 ± 0.2	32.9	0.34	0.03
10	160 (0)	90 (+1)	5 (−1)	5.2 ± 0.2	2.4 ± 0.1	36.2	0.24	0.04
11	160 (0)	20 (−1)	10 (+1)	1.2 ± 0.2	1.7 ± 0.2	17.4	0.20	0.00
12	160 (0)	90 (+1)	10 (+1)	1.5 ± 0.0	1.7 ± 0.1	29.2	0.15	0.03
13(CP)	160 (0)	55 (0)	7.5 (0)	4.2 ± 0.7	3.9 ± 0.2	29.4	0.22	0.03
14(CP)	160 (0)	55 (0)	7.5 (0)	3.7 ± 0.3	4.1 ± 0.1	33.0	0.23	0.03
15 (CP)	160 (0)	55 (0)	7.5 (0)	3.3 ± 0.2	3.8 ± 0.5	29.4	0.23	0.03

9.4.1 Effect of the LHW Pretreatment on the Hemicellulosic Hydrolyzate (Liquid Fraction)

The results obtained for the experimental conditions, described by the Box-Behnken matrix, on the evaluated response variables (RS and PCs), are presented in Table 9.1. The ANOVA results (Table 9.2) show that the generated models for RS and PCs have correlation coefficients (R^2) higher than 0.93 and the lack of fit was not significative (*p-valor* > 0.05) suggest that the models are adjusted for the main linear and quadratic effects but do not have into account the effect of interactions. It is important to mention that the Behnken Box approach is designed for the obtention of first- and second-order coefficients.

The analysis of the standardized effects of the independent variables (T and LSR) showed that these effects were significative for both of the evaluated response variables, as shown in Figure 9.1. With respect to the hydrolysate RS content, *T* has a positive (18.5103) (Figure 9.1) and significative effect (*p* valor = 0.003 < 0.005) (Table 9.2), indicating that the higher the temperature the higher the RS content generated by LHW. According to other authors, the removal of 80% or more hemicellulose can be achieved at 200°C, which resulted similar to the observations in this study for temperatures equal and over 200°C (Table 9.1; experimental conditions 2, 4, 6, and 8), resulting in higher RS content in the hydrolysate.

The effect of temperature on the liquid fraction RS content was also observed by Wang et al. (2018) for the hydrothermal pretreatment of rice straw (LSR = 9 mL g^{-1} and 15 min). The authors observed an increase in the glucose content from 1.3 g L^{-1} to 1.9 g L^{-1} and an increase in the xylose content from 0 g L^{-1} to 0.3 g L^{-1} when the pretreatment temperature was increased from 90°C to 210°C, respectively; indicating that sugars hydrolysis increased according to the pretreatment severity.

Similar to what happened with RS, the independent variable *T* also presented a significate (*p* valor = 0.002 < 0.05) (Table 2) and positive (19,879) (Figure 1) effect on PCs. For higher temperatures, the deacetylation of the hemicelluloses promotes an increase in the acetic acid concentration in the reactive mixture, with a consequent pH decrease, favoring the hydrolysis of the α-O-4 and β- O- 4 bonds in lignin and solubilizing the phenolic units. As shown in Table 1, when the temperature was increased from 120°C (condition 5–1.51 g PC L^{-1}) to 200 °C (condition 8–2.22 g PC L^{-1}), PCs content increased in 47%, even though condition 8 resulted in a higher dilution. Similarly, conditions 2, 4, and 6 (*T* = 200°C) resulted in higher PCs content independently from time and LSR.

9.4.2 Response Surfaces

The response surfaces were generated using Eq. (9.2):

$$Y = \beta_0 + \sum \beta_i x_i + \sum \beta_{ii} x_i^2 + \sum \beta_{ij} x_i x_j \tag{9.2}$$

where Y is the response variable, β_0 is a model constant, β_i is the coefficient of the linear term, β_{ii} is the coefficient of the quadratic term, β_{ij} is the coefficient of the

TABLE 9.2

Estimated Effects and Variance Analysis (ANOVA) for Reducing Sugars and Phenolic Compounds in the LHW Pretreatment of Coffee Husk

Factors	Response: Y_1 – Reducing Sugars (g L^{-1})						Response: Y_2 – Phenolic compounds (g L^{-1})					
	Effects	SS	DF	MS	*F* value	*p* value	Effects	SS	DF	MS	*F* value	*p* value
Temperature (°C) (*L*)	4.45833	65.5512	1	65.55125	322.3832	0.003088*	2.25000	10.12500	1	10.12500	433.9286	0.002297*
Temperature (°C) (Q)	5.72500	7.9878	1	7.98776	39.2840	0.024523*	0.35417	0.46314	1	0.46314	19.8489	0.046867*
Time (min) (*L*)	−1.47083	1.0513	1	1.05125	5.1701	0.150846	0.42500	0.36125	1	0.36125	15.4821	0.058938
Time (min) (*Q*)	0.72500	6.0416	1	6.04160	29.7128	0.032046*	0.82917	2.53853	1	2.53853	108.7940	0.009067*
LSR (mL.g^{-1}) (*L*)	1.27917	36.9800	1	36.98000	181.8689	0.005454*	−0.82500	1.36125	1	1.36125	58.3393	0.016713*
LSR (mL.g^{-1}) (*Q*)	−4.30000	2.9631	1	2.96314	14.5728	0.062279	1.02917	3.91083	1	3.91083	167.6071	0.005913*
$T \times t$	−0.89583	0.0225	1	0.02250	0.1107	0.771030	0.30000	0.09000	1	0.09000	3.8571	0.188497
$T \times$ RLS	−0.15000	1.4400	1	1.44000	7.0820	0.116947	−1.40000	1.96000	1	1.96000	84.0000	0.011696
$t \times$ RLS	−1.20000	0.0900	1	0.09000	0.4426	0.574315	0.05000	0.00250	1	0.00250	0.1071	0.774506
Lack of fit		6.8425	3	2.28083	11.2172	0.082956		1.30750	3	0.43583	18.6786	0.051244
Pure Error		0.4067	2	0.20333				0.04667	2	0.02333		
Total		130.4773	14					21.43733	14			

L, linear term; *Q*, quadratic term; SS, sum of squares; DF, degree of freedom; MS, mean squares. Y_1) $R^2 = 0.9444$; Y_2) $R^2 = 0.9368$.

Note

* *p* values less than 0.05 indicated that model terms are significant.

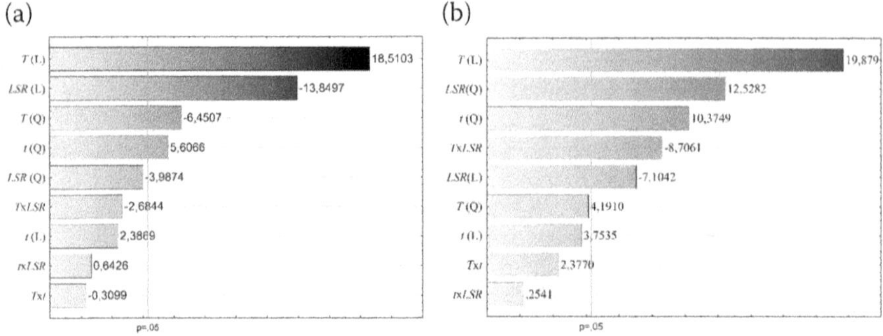

FIGURE 9.1 Pareto chart with the standardized effects for the independent variables: (a) RS concentration (g L^{-1}) and (b) PC concentration (g L^{-1}) in the hydrolysate.

interaction of the terms, x_i and x_j are the coded levels of the variables X_i and X_j, respectively. The variables investigated were coded using Eq. (9.3).

$$x_i = \left(\frac{X_i - X_0}{\Delta X_i} \right) \tag{9.3}$$

where x_i is the coded value of the independent variables under study, X_i is the uncoded value of the independent variable, X_0 is the uncoded value of the center point, and ΔX_i is the uncoded value of the difference between X_i and X_0.

The response surfaces and the statistical analyses (F test and analysis of variance (ANOVA)) were performed using the Statistica® software (Statsoft Inc, version 12, USA). ANOVA was performed to test the significance of the responses (RS and PC) and to evaluate the adjustment of the experimental data to the model equation (Eq. 9.2). The probability values (p values) were used to assess the significance of each model coefficient and the contribution of each independent variables to the response evaluated. Model parameters with p values < 0.05 (95% confidence level) were considered as significant.

The second-order polynomial model equations were obtained after a multivariate regression analysis for the responses RS (Eq. 9.4) and PC (Eq. 9.5), in terms of the decoded values:

$$\begin{aligned} \text{RS/g.L}^{-1} = {} & 4.458 + 2.863T - 0.735T^2 + 0.363t + 0.640t^2 - 2.150LSR \\ & -0.448LSR^2 - 0.075T * t - 0.600T * LSR + 0.150t * LSR \end{aligned} \tag{9.4}$$

$$\begin{aligned} \text{PCs/g.L}^{-1} = {} & 2.458 + 1.125T + 0.177T^2 + 0.213t + 0.415t^2 \\ & - 0.413LSR + 0.515LSR^2 + 0.150T * t - 0.700T * t \\ & + 0.025t * SLR \end{aligned} \tag{9.5}$$

The response surfaces for the dependent variables RS and PCs are shown in Figure 9.2. Given t presented the lowest significance statistical effect, that variable

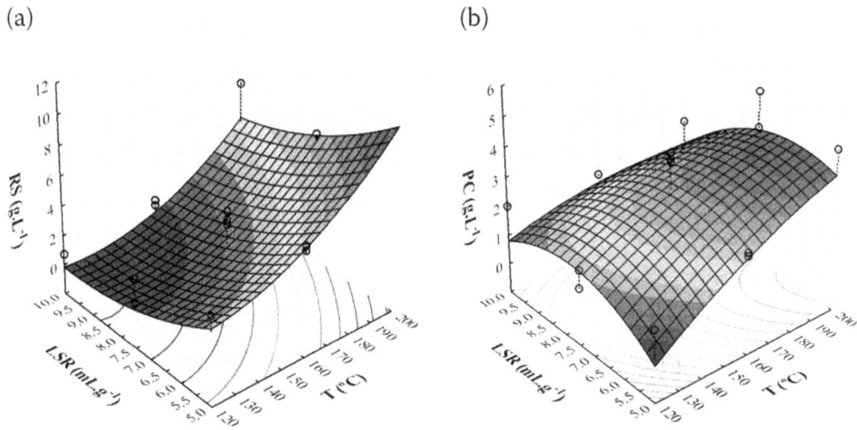

FIGURE 9.2 (a) Response surface for RS and (b) response surface for PCs.

was fixed to its lowest level (20 min) and only the variables *T* and SLR were plotted. Figure 9.2a shows that the conditions resulting in the largest RS liberation to the hydrolysate, correspond to the high temperatures (T>170°C) and low LSR (< 7.0 mL g^{-1}) region. On the other hand, as shown in Figure 9.2b, the conditions that permit the highest RS release also permit the highest PCs release to the hydrolysate, which indicates that a more detailed evaluation is necessary to establish the best condition with respect to the viability of the biomethanization of pretreated CHW.

9.4.3 Characterization of Solid Fraction Generated after the Liquid Hot Water Pretreatment

The raw CHW and the pretreated solid fractions were milled using a Willey mill (Marconi, model MA048). The fraction of material passing the 20 mesh was used for characterization. The cellulose, hemicellulose, and lignin contents were quantified according to the protocols described by NREL (National Renewable Energy Laboratory) and by TAPPI (Technical Association of the Pulp and Paper Industry) (Sluiter et al. 2004; Golschimdt 1971). The sugars concentration (cellobiose, glucose, xylose, and arabinose), organic acids (formic and acetic), and aldehydes (FF and HMF) in the liquid fraction, generated by using the Klason method (according to a modified TAPPI T222 om-2), was determined using high-efficiency liquid chromatography (HPLC) in a Shimadzu® chromatograph, equipped with an Aminex HPX 87H (300 × 7.8 mm BIO-RAD) column heated at 55°C. The mobile phase was sulfuric acid (5 mmol L^{-1}) at a 0.6 mL min^{-1} flow rate. Sugars were identified and quantified using a refraction index detector (model RID-6A), while for the organic acids and aldehydes, a UV–vis detector (model SPD-20A) was used at wavelengths of 210–274 nm, respectively (Baêta et al. 2016).

The raw CHW and the pretreated liquid and solid fractions chemical oxygen demand (COD) was determined according to the standard method UNE 77004:2002 (Baêta et al. 2017a). Total (TS) and volatile (VS) solids were measured according to the Standard Methods for the Examination of Water and Wastewater. The PCs quantification was done by a colorimetric method described by Mussatto et al. (2011), with modifications: the extraction was made with acetone (70%) and sonication, using a sample:solvent ratio of 1:10 (mL:mL or g:mL). After centrifugation (250 × *g*, 10 min), 50 µL of the filtered extract was mixed with 450 µL of distilled water, 1.25 mL of sodium carbonate (7.5% w/v), and 520 µL of the Folin-Ciocalteu reagent. The tubes containing the samples were protected from light and mixed by vortexing for 3 min. Absorbance was measured after 35 min at 725 nm. The total content of PCs was expressed in grams of tannic acid per liter.

9.4.4 Characterization of Liquid Fraction Generated after the Liquid Hot Water Pretreatment

The RSs present in the liquid fraction, after the LHW pretreatment, were quantified according to the 3,5-dinitrosalicylic (DNS) method reported by Miller (1959). To determine the caffeine content, the hydrolysates were diluted 10 times. The analysis was performed in a gas chromatograph coupled to a mass spectrometer (GC-MS) Shimadzu model GCMS-QP2010 plus, according to the methodology described by Rodrigues et al. (2014).

9.5 EXPERIMENTAL CONDITIONS FOR THE CHW PRETREATMENT BY LHW

Starting from the previously generated quadratic models for the response variables, a global desirability function was used to generate three LHW pretreatment conditions to be evaluated respect to the biomethane production (BP) using the pretreated CHW. The input values for the calibration of the desirability function and the independent variable levels generated by the prediction tool available in the software Statistica® (Statsoft Inc., version 12) are shown in Table 9.3. The three generated desirability conditions: BD1 (120°C; 20 min; SLR 5 mL g^{-1}), BD2 (189°C; 90 min; SLR 5 mL g^{-1}), and BD3 (200°C; 61 min; SLR 5 mL g^{-1}), represent scenarios with high RSs concentrations (> 4.6 g L^{-1}) and a variety of PCs content (Table 9.3); low (2.0 g L^{-1}), medium (4.6 g L^{-1}), and high (6.3 g L^{-1}), respectively.

Additionally, the pretreatment was also performed under a desirability condition suitable for ethanol production (condition ED), which only differs from condition BD3 by using a reaction time 20 min shorter. The hydrolysate obtained under the ED condition was also tested for biomethane production to take into account alternative technological routes under the biorefinery context.

9.5.1 Analysis of the Results of Liquid Fraction after Pretreatment

Table 9.1 shows the characterization of the hydrolysates obtained under the experimental conditions established in the Box-Behnken matrix, while the characterization for the desirability conditions established for biogas production (BD) and ethanol production (ED) are shown in Table 9.3.

TABLE 9.3
Desirability Conditions for BD and ED

Fractions	Parameters	BD1	BD2	BD3	ED
		T = 120 ± 2°C / t = 20 ± 1 min / LSR = 5 mL g^{-1} / log R_0 = 1.89	T = 189 ± 2°C / t = 90 ± 1 min / LSR = 5 mL g^{-1} / log R_0 = 4.57	T = 200 ± 2°C / t = 61 ± 1 min / LSR = 5 mL g^{-1} / log R_0 = 4.73	T = 200 ± 2°C / t = 41 ± 1 min / LSR = 5 mL g^{-1} / log R_0 = 4.56
Liquid (hydrolysate)	Reducing sugars (*RS*, g L^{-1})	4.6	8.2	11.0	8.7
	Phenolic compounds (*PC*, g L^{-1})	2.0	4.6	6.3	5.5
	COD (g O2 L^{-1})	18.9	23.3	31.0	47.7
	Caffeine (g L^{-1})	0.3	0.2	0.2	0.2
	Inhibitors (HMF + FF) (g L^{-1})	ND	0.4	0.3	0.9
Solid	% Cellulose (% Cel. Removal)	30.6 ± 0.2 (31.77)	36.6 ± 1.7 (35.86)	48.1 ± 2.1 (38.12)	47.9 ± 1.7 (38.12)
	% Hemicellulose (% Hemi. Removal)	21.0 ± 5.0 (28.27)	11.1 ± 1.2 (70.20)	7.8 ± 0.2 (84.63)	5.8 ± 0.4 (89.5)
	% Lignin (% Lig. Removal)	37.5 ± 0.5 (2.90)	38.1 ± 4.0 (22.47)	40.7 ± 3.8 (39.20)	38.2 ± 1.2 (47.4)
	% Balance	89.15	85.80	96.60	93.3

The results (Table 9.1) suggest that the LHW pretreatment produced a low COD solubilization. For instance, under condition 12 (160°C, 90 min, and 10 mL g^{-1}), only 13.6% of the COD available in the raw CHW (1.29 g COD per g of raw CHW) was solubilized. On the other hand, under condition 1 (120°C, 20 min, and 7.5 mL g^{-1}) that value is reduced to 4.6%. These indicate that, in general, the LHW pretreatment is not amazingly effective with respect to COD solubilization, and a good part of organic matter continues in the solid fraction after the pretreatment.

With respect to caffeine, the highest concentration of this compound in the hydrolysates was 0.57 g L^{-1} when operating under condition 7 (120°C, 55 min, and 10 mL g^{-1}). This is important because, according to Widjaja et al. (2019), caffeine concentrations over 2.5 g L^{-1} may inhibit growth for some bacterial species, provoking negative effects on biogas production. When considering the evaluated desirability conditions (Table 9.3), a decrease in the caffeine content was observed when the LHW severity was increased by using temperatures over 120°C. A possible explanation for this behavior is related to the boiling point of caffeine, which is 178°C; hence, under conditions like 8 (200°C), part of the caffeine might be lost as vapor besides being degraded into other volatile compounds which escape from the reactor right after opening it.

From Table 9.1, it is possible to conclude that increasing the LHW severity promotes the conversion of pentoses and hexoses into inhibition products (FF and HMF). A similar behavior was observed by Baêta et al. (2016) and Batista et al. (2019), who studied the sugarcane bagasse and sugarcane straw auto-hydrolysis. Additionally, in a similar way to what happens with caffeine, FF and HMF result toxic to the fermentative processes (Monlau et al. 2014). However, the concentrations of the furanic compounds in the hydrolysates generated from the CHW were always lower than 0.38 g L^{-1}. Furthermore, Quéméneur et al. (2012) showed that FF and HMF were not toxic during methane production from xylose at concentrations over 1.0 g L^{-1}. In another study, Park et al. (2012) reported that inhibition of methane production from acetate was only happening for HMF concentrations over 5.0 g L^{-1}.

9.5.2 Analysis of the Results of Solid Fraction after Pretreatment

The analysis of the solid fractions obtained after pretreatment under several desirability conditions is presented in Table 9.3. The results show that hemicellulose was the most affected fraction after pretreatment, a characteristic behavior when doing thermal liquefaction. The most severe pretreatment condition ($\log R_0$ = 4.73) promoted a high hemicellulose removal (>84%), generating a cellulose-rich (48%) and lignin-rich (41%) solid fraction. In comparison, raw CHW contains 36% cellulose and 31% lignin.

Importantly, these results are in accordance with the hydrolysates characterization. Under the most severe conditions, there are more carbohydrates being removed from the solid phase while generating a liquid fraction enriched in RSs and furanic compounds (FF and HMF). On the other hand, besides the effect caused by the hemicellulose removal from the solid phase, lignin removal would also result in higher concentrations of PCs in the hydrolysates.

9.6 BIOGAS PRODUCTION BY ANAEROBIC DIGESTION

BMP assays were done for all the three desirability conditions for biogas production and the one for ethanol production (Table 9.3). Additionally, the AD assays were done in solid phase (S-D; ST ≥ 15%), liquid phase (L-D; ST < 10%), and semi-solid phase (SL-D; 10% ≤ ST < 15%). It is worth to make clear that for the L-D condition, the pretreated hydrolysate was used, while for the S-D condition, the pretreated solid phase was used. For the SL-D condition, a mixture of both phases was used.

All the assays were done using glass bottles which served as batch reactors with 120 mL of the working volume and 60 mL of headspace. For the L-D assays, the bottles were filled with 6 mL of hydrolysate and microbial inoculum such that the substrate/microorganisms ratio (F/M) was 0.4 g COD g VS-inoculum^{-1}, according to (B. Baêta et al. 2016). For the S-D and SL-D assays, the TS content was 15 and 12.5%, respectively; and the F/M was 2 g COD g VS-inoculum^{-1} (Lima et al. 2018).

The inoculum used for the BMP assays was obtained from a pilot-scale UASB reactor fed with wastewater at Center of Research and Training in Sanitation (*CePTS UFMG-COPASA*). The inoculum was mixed with cow manure in a proportion equal to 1 g VS-sludge:1 g VS-manure (Lima et al. 2018). The BMP assays were done in duplicate and a bottle without added substrate was used as control under every condition, in order to quantify the endogenous methane production of the inoculum.

For all bottles, pH was adjusted to 7.5 ± 0.2 using HCl (2 mol L^{-1}) and NaOH (0.1 mol L^{-1}). After that, powdered activated carbon (PAC) was added to the bottles (4 g L^{-1}), to control the accumulation of VFAs, facilitating the pH buffering (dos Santos et al. 2018). After PAC addition, the bottles were purged with $N_{2(g)}$ for 5 min, to ensure anaerobic conditions before sealing them. The bottles were kept in an incubator (Thoth®, model 6440), with agitation (180 rpm) at a controlled temperature of 35 ± 2°C to favor the growth of methanogenic microorganisms.

Biogas production was measured daily using the accumulated pressure in the bottle tanks to a differential manometer (PM 9100-HA). A gas chromatograph (Shimadzu Model 2014/TCD) was used to establish the methane content in the biogas (%, v v^{-1}), using operation conditions as described by (Lima, Adarme, Baêta, Gurgel, and de Aquino 2018).

The assays were maintained in operation until the variation in the biogas production was lower than 5% during a time-lapse of 24 hours. The methane production results for all assays were presented as Nm^3CH_4 kg COD^{-1} measured under normal conditions of temperature and pressure (CNTP: 273 K; 101315 Pa), as recommended by IUPAC.

9.6.1 Biochemical Methane Potential of the Raw CHW and the Liquid and Solid Fractions Obtained after Pretreatment

Table 9.4 presents the characterization analysis for selected compounds at the beginning of the BMP assays. It also presents the accumulated methane production under all the evaluated conditions.

TABLE 9.4
Results of Biogas Assay

Reactor (conditions)	F/I ratio gCOD (gSSV inóculo^{-1})	TS (%)	Phenolic Compounds (g L^{-1})	Holocellulose (g L^{-1})	Reducing Sugars (g L^{-1})	Total Sugars (g L^{-1})	S/Ph Ratio	Initial pH	Final pH	Methane Production (NmL CH_4 gCOD^{-1})	HAc (g L^{-1})	HPr (g L^{-1})	HBu (iso) (g L^{-1})	HBu (g L^{-1})	TVFAs (g L^{-1})
Raw CH	2.8	15	0.2	52.1	–	52.1	260.5	7.57	5.98	6.7 ± 0.5	1.55	2.94	ND	ND	4.5
L-D1	0.4	≤10	0.2	–	0.5	0.5	2.5	7.55	6.23	130.2 ± 13.8	ND	ND	ND	ND	ND
L-D2	0.4	≤10	0.4	–	0.8	0.8	2	7.56	6.34	133.5 ± 8.2	0.30	0.80	ND	ND	0.1
L-D3	0.4	≤10	0.6	–	1.1	1.1	1.8	7.61	6.57	225.9 ± 13.9	0.27	0.57	ND	ND	0.9
SL-D1	3.4	12.5	0.4	17.7	1.1	18.8	47	7.62	6.22	22.9 ± 0.6	2.84	4.09	0.19	ND	7.1
SL-D2	3.1	12.5	0.8	13.6	2.9	16.5	20.6	7.60	5.94	16.5 ± 2.3	4.46	2.98	ND	ND	7.5
SL-D3	3.4	12.5	1.7	19.1	4.2	23.3	13.7	7.56	5.86	ND	10.33	1.15	0.33	0.97	12.8
S-D1	3.3	15	0.2	24.4	–	24.4	122	7.55	6.34	23.9 ± 2.1	0.99	6.44	ND	ND	7.5
S-D2	3.0	15	1.1	20.6	–	20.6	18.72	7.52	6.13	17.8 ± 1.2	ND	5.26	ND	ND	7.3
S-D3	3.2	15	4.6	28.7	–	28.7	6.2	7.54	5.96	ND	7.15	0.13	ND	1.07	8.3
ED	0.4	≤10	0.8	–	8.7	8.7	10.1	7.55	6.41	127.3	ND	ND	ND	ND	ND

F/I, food-to-inoculum ratio; TS, total solids; total sugars, Holocellulose + reducing sugars; S/Ph, sugar-to-phenol ratio; COD, chemical oxygen demand; HAc, acetic acid; HPr, propionic acid; HBu (iso), isobutyric acid; HBu, butyric acid; TVFAs, total volatile fatty acids; ND, not detected.

In general, the pretreatment resulted to be efficient for increasing the biodegradability of raw CHW, as shown in Table 9.4. Except for conditions SL-D3 and S-D3, all the studied scenarios permitted an accumulated methane production higher than what was obtained using just raw CHW. When comparing substrates, the liquid fractions obtained from pretreatment resulted in higher biogas production, which can be explained by the higher mass transfer expected for systems with lower heterogeneity (lower content of total suspended solids), as shown in Table 9.4.

In contrast with what was observed for the other AD systems, the most severe pretreatment condition conduced to the highest accumulated methane production (225.9 ± 13.9 NmL CH_4 $gCOD^{-1}$), with a biodegradability of 64%, when considering that the maximum theoretical production should be 350 NmL CH_4 $gCOD^{-1}$. This value was higher than the one obtained by (dos Santos et al. 2018) (218.2 NmLCH_4 $gCOD^{-1}$), who evaluated the biogas production from the liquid fraction obtained by oxidative pretreatment of CHW (LSR 10 ml g^{-1}, pH 11, and specific applied ozone load (*SAOL*) 18,50 mg O_3 g CHW^{-1}). The BMP assays in this study also had PAC at a concentration of 4 g L^{-1}, similarly to the present study.

The methane production from the liquid fraction under the ED condition was 127.29 NmL g COD^{-1}. This production was lower, possibly because the RSs:PCs ratio was lower than what was used for the other conditions. However, it presents the best enzymatic hydrolysis yields for the production of fermentable sugars from pretreated CHW. In the end, the results showed above suggest that high-severity conditions in the pretreatment are useful for the production of two biofuels: methane and ethanol.

With respect to the S-D and SL-D systems, it is possible to infer that conditions BD1 and BD2 present the highest biogas production. This can be explained by the deconstruction of the complex structure of the biomass, enriching the solid fractions in cellulose. Higher lignin content was also observed for BD3, as shown in Table 9.3. It can also be suggested that repolymerization of the phenolic fragments may act as a physical barrier during the pretreated CHW microbial degradation. Figure 9.3 shows a relationship between PCs content and biogas production for S-D and SL-S, following and inversely proportional linear tendency, that is, when the severity of the pretreatment increases, the PCs content also increases and biogas production decreases. According to Jönsson and Martín (2016) and Monlau et al. (2014), PCs can cause damage on the microbial cells related to AD, altering the cellular membrane permeability in a selective manner and causing the inactivation of essential enzymatic machinery.

The methane production results shown in this study for SL-D and S-D were lower than the results obtained by Passos et al. (2018) (144.96 NmL CH_4 $gCOD^{-1}$) for CHW pretreated by steam explosion (TS = 15%, 120°C, and 60 min). One possible hypothesis explaining the higher degradability of CHW after that pretreatment is that steam explosion tends to generate a higher fragmentation of the biomass when compared to conventional hydrothermal liquefaction, making carbohydrates more available for microorganisms. The steam explosion also generates lower amounts of PCs as byproducts, taking into account that during the decompression of the system, a fraction of the volatile compounds escape alongside water vapor (flash).

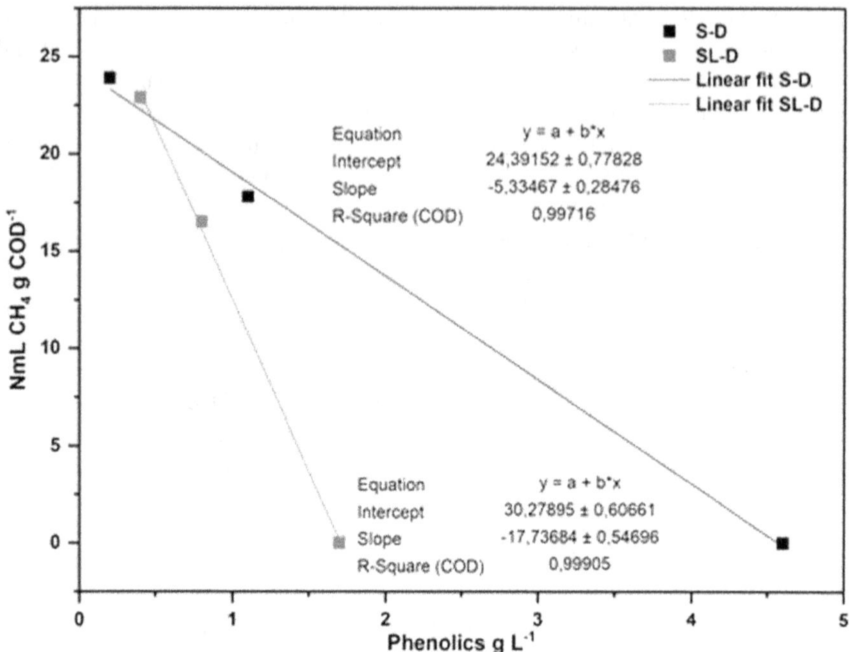

FIGURE 9.3 Interaction between phenolic compounds during the anaerobic digestion and the methane production.

9.6.2 Effect of VFAs Accumulation on Biogas Production

In the same way, PCs can act as AD inhibitors, the accumulation of VFAs in the reactors can induce kinetic and thermodynamic failure of methanogenic systems (De Aquino and Chernicharo 2005). Table 9.4 shows the VFAs concentration by the end of the BMP assays. It can be seen that L-D presented the lowest VFAs concentration probably because it was the condition with the lowest organic load (0.4 gCOD gSSVinóculo^{-1}).

The S-D3 and SL-D3 conditions resulted in 7.15 and 10.33 g L^{-1} of acetic acid, respectively (Table 9.4). On the other hand, under the S-D1 and SL-D1 conditions, the propionic acid concentrations were 6.44 and 4.09 g L^{-1}, respectively. Previous studies reported that a concentration of propionic acid of 900 mg L^{-1} alongside an acetic acid concentration over 800 mg L^{-1} inhibit the growth of acetoclastic methanogenic microorganisms, disrupting the kinetic and thermodynamic equilibrium of the system (Zeng et al. 2013). Given the above, the accumulation of acids in the S-D and SL-D systems may have been caused by relatively high concentrations of PCs.

The substrates that were liquefied under the most severe conditions (condition BD3) conduced to a lower production of propionic acid in the fermentative mixture for the S-D and SL-D systems. It is also important to mention that four carbon acids (butyric and isobutyric) were only detected under one pretreatment condition (BD3), for the S-D and SL-D systems. In agreement with da Fonseca et al. (2021), the higher solids contents in AD (12–15%) that are also associated with higher PCs contents, provoke the deviation of some microbial metabolic routes as a way to maintain their enzymatic systems functioning and permitting their survival despite the fermentative medium toxicity.

The holocellulose concentration (Table 9.4) for the raw CHW and the SL-D and S-D evaluated conditions, can be associated to lower methane production and higher VFAs accumulation when compared to the L-D and ED conditions. From Table 9.4, the highest VFAs concentrations were found under the conditions SL-D3 (12.8 g L^{-1}) and S-D3 (8.3 g L^{-1}). The holocellulose concentration is increased according to the pretreatment severity for the SL-D and S-D systems as well as the PCs concentration. The highest concentration of PCs occurred also under the conditions SL-D3 (1.7 g L^{-1}) and S-D3 (4.6 g L^{-1}). This can be explained based on the lignin degradation increment in the solid phase (Table 9.3) and previous reports by Navarro et al. (2020) where it was found that holocellulose favors an increase in acidogenesis. In this order of ideas, the presence of holocellulose may act as a predictor of acidification and methane production inhibition.

Taking into account that the methane production for the S-D and SL-D systems were relatively low, a technical alternative is to recover the excess of VFAs produced under the condition B3 by using an eventual separation operation. In the work by Tonucci et al. (2020), the authors reported high VFAs recovery efficiencies when using selective adsorbents, like molecularly imprinted polymers (MIPs), in high VFAs concentration environments, reaching butyric acid removals over 60% from an AD effluent. Hence, the production of VFAs from CHW may constitute a promissory technological route to include an additional added-value bioproduct to a coffee industry-associated biorefinery.

9.6.3 Methane Production Kinetics Modeling

The modeling of the methane production kinetics by AD was done according to the modified Gompertz Eq. (9.6) and Groots Eq. (9.7) models.

$$P = P_0 exp\left\{-exp\left[\frac{\mu_m e}{P_0}(\lambda - t) + 1\right]\right\} \tag{9.6}$$

where P is the accumulated methane yield (NmL g COD^{-1}), P_0 is the maximum methane yield (NmL $gCOD^{-1}$), t is the incubation time (days), λ is the lag phase time (days), μ_m is the maximum methane production rate (NmL $gCOD^{-1}$ d^{-1}), and e is the Euler constant (2.71828).

$$P = \sum_{i=1}^{n} \frac{A_i}{1 + \frac{B_i^{C^i}}{t^{C^i}}} \tag{9.7}$$

where P is the accumulated methane yield (NmL g COD^{-1}), t is the incubation time (days), A_i (NmL g COD^{-1}) is the asymptotic methane production for phase i, B_i

(days) is the incubation time for reaching half of the asymptotic methane production for phase i, and C_i is a constant that determines the characteristic curvature profile of the phase i. The value of i indicates the number of phases (1 to n). Data processing was done using Statistica® (Statsoft Inc., version 12). The validation of the adjusted models was done based on the coefficient of determination (R2), the root Mean Square Error (RMSE), and the Akaike information criterion (AIC) (Adarme et al. 2019; Paranhos et al. 2020; Lima et al. 2018).

9.6.4 Results of Methane Production Kinetics Modeling

A kinetic analysis and modeling were done for the experiments that presented the highest accumulated methane productions (L-D3, S-D1, and ED) and the results were compared with the methane production experiments using raw CHW.

The Gompertz and Groot's kinetic models were used for adjustment of the methane production data as shown in Figure 9.4 and the obtained kinetic parameters are shown in Table 9.5. From the results, it is possible to conclude that the model that better adjusted to the methane production kinetics is the Groot's model, considering that it resulted in the highest R^2 values and the lowest values for RMSE, NRMSE, and AIC.

When considering the Groot's model, the values obtained for methane production during phase 1 (A1) when using raw CHW and conditions S-D1, L-D3, and

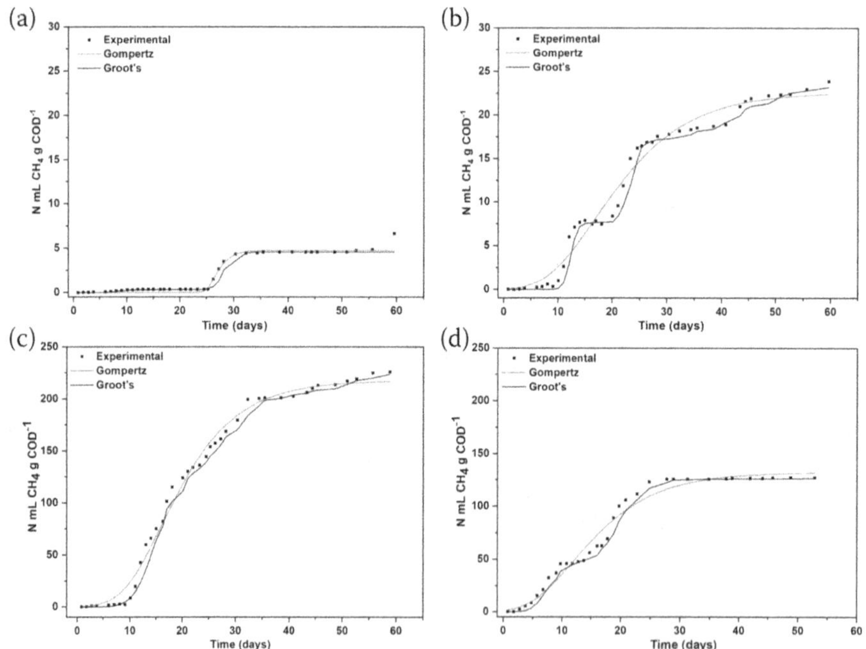

FIGURE 9.4 Raw and pretreated CHW AD accumulated methane production data adjusted to the modified Gompertz model and to Groot's model. (a) raw CHW, (b) S-D1, (c) L-D3, (d) ED.

TABLE 9.5
Kinetic Parameters for Anaerobic Digestion Assays

Model	Parameters	Raw CHW	S-D 1	S-D 2	SL-D 1	SL-D 2	L-D 1	L-D 2	L-D 3	ED
Gompertz	P_0	4.757	22.603	14.399	22.885	14.917	145.860	133.561	217.649	132.265
	R	1.111	0.903	0.740	1.671	0.668	3.179	5.691	10.201	6.392
	λ	24.944	7.920	10.090	16.272	21.917	4.023	7.757	8.378	4.416
	RMSE	0.391	1.280	1.238	1.261	0.999	7.768	4.576	7.742	7.310
	NRMSE	5.875	5.367	6.961	5.496	6.043	5.967	3.429	3.427	5.743
	AIC	−71.230	23.698	21.016	22.503	3.907	167.945	125.622	167.682	139.206
	R^2	0.968	0.977	0.955	0.984	0.969	0.966	0.992	0.992	0.977
Groot	A_1	0.332	7.660	5.204	1.964	0	42.760	59.701	140.514	51.903
	B_1	7.660	11.343	12.840	10.912	15.318	8.671	12.163	14.481	7.002
	C_1	6.786	19.181	11.259	33.100	32.690	7.819	6.718	58.334	4.229
	A_2	4.236	9.376	7.086	12.380	8.390	64.757	66.848	64.755	74.201
	B_2	27.227	22.239	22.293	23.679	27.248	27.976	25.684	28.188	18.972
	C_2	33.285	24.293	34.681	21.218	22.187	10.937	8.324	11.617	11.370
	A_3	–	6.986	5.596	9.464	11.741	26.074	6.830	23.784	22.911
	B_3	–	42.668	52.112	27.483	52.488	49.845	48.423	50.436	64.227
	C_3	–	7.553	280.843	3.489	7.124	12.587	20.136	12.866	12.839
	RMSE	**0.091**	**0.311**	**0.237**	**0.243**	**0.350**	**2.691**	**2.447**	**3.753**	**2.039**
	NRMSE	**1.869**	**1.304**	**1.332**	**1.058**	**2.119**	**2.067**	**1.833**	**1.661**	**1.602**
	AIC	**−180.157**	**−777.729**	**−100.087**	**−98.035**	**−67.970**	**99.182**	**91.370**	**126.449**	**67.886**
	R^2	**0.998**	**0.998**	**0.998**	**0.999**	**0.996**	**0.996**	**0.998**	**0.998**	**0.998**

ED, were in general lower (Table 9.5, 0.332, 7.660, 140.514, and 51.903 NmLCH_4 gCOD^{-1}) than the values obtained during phase 2 (A2) (Table 9.5, 4.236, 9.376, 64.755, and 74.201 NmLCH_4 gCOD^{-1}). Except for L-D3, this behavior may be explained based on each of the substrate's fermentation velocity during each phase. In the L-D3 case, the first phase (A1) is related to the rapid autonomous fermentation of each substrate (easily fermentable soluble fraction) (Cone, van Gelder, and Driehuis 1997; Rodrigues et al. 2014). In this sense, the higher severity under condition D3 made it possible to release higher amounts of fermentable sugars towards the liquid phase (Table 9.3), which were promptly consumed by microorganisms, explaining a higher methane production velocity for this case. For the second phase (A2), a structural degradation of the residual carbohydrates is expected and it depends on their specific chemical characteristics so, the net methane production is the sum of the methane production during both phases (Groot et al. 1996).

The S-D digestions (raw CHW, S-D1, and ED), on one hand follow a sequential process where the first phase (A1) involves an extracellular hydrolysis of the lignocellulosic polysaccharides to obtain monosaccharides. Later, the produced monosaccharides are fermented by the anaerobic consortium to produce short-chain VFAs that, then, conduce to the production of more methane (López, Dijkstra, Dhanoa, and Bannink 2011). This proposed mechanism explains the higher methane production velocities for phase A2 in some experiments. A similar observation was reported by (Paranhos et al. 2020), who evaluated the co-digestion of lignocellulosic residues with poultry manure using S-D systems. The presence of multiple slopes in the methane production curves can also be related with the difficulties experienced by the microbial consortium to adapt to inhibitory compounds such as the lignin fractions, the furanes, and the PCs.

The slopes and phases for the Groot's model can be seen more easily when analyzing the solid and semi-solid systems, as presented in Figure 9.4. This is consistent with the fact that both raw CHW and pretreated CHW are complex biomasses showing a difficult biodegradation. These lignocellulosic residues are recalcitrant in part because their chemical and structural characteristics can affect the process negatively. One of the chemical barriers is the toxic effect of the PCs from the cell walls, while the structural barriers are the crystallinity and low accessibility of cellulose, as well as the low digestibility of some vegetal tissues such as the cuticle and the warty layer (Groot et al. 1996).

Even though the Groot's model shows the best adjustment, the Gompertz model permits to obtain some important parameters from the analytical point of view, such as the lag phase time for methane production. From Table 9.5 and Figure 9.4(a-c), it can be concluded that raw CHW was the substrate that required the longest adaptation time, with a lag phase over 24 days. This observation is explained because of the higher structural complexity of the raw CHW in relation to the lignin-carbohydrate complex, which makes more difficult the access of the enzymatic machinery.

A comparison between the S-D and the SL-D systems shows that the higher severity pretreated substrates presented longer lag phases, similar to what was observed for L-D (Table 9.5, Values of λ). The most severe LHW conditions promoted a higher sugar solubilization, resulting in a lignin-richer solid phase which is consequently more recalcitrant (Table 9.3). This also generates liquid fractions richer in PCs (Table 9.1). It is well known that applying pretreatments to lignocellulosic biomass can promote the formation of specific lignin degradation byproducts, mainly those made of PCs, which can inhibit or reduce the AD velocity (Poirier et al. 2016; Monlau et al. 2014). Schroyen et al. (2018) studied the impact of PCs release during aggressive pretreatments by adding phenolic acids to the medium. The authors observed that adding 2000 mg L^{-1} of ferulic acid and 4-hydroxybenzoic acid, inhibited hydrolysis in 22%. Another study by Barakat et al. (2012) found that the presence of syringaldehyde and vanillin (1 g L^{-1}) increased the microorganisms adaptation time (lag phase) when digesting xylose. These studies further support and help to explain the long lag phases observed under conditions of high-severity pretreatment and higher initial PCs concentration in the system, as shown in Table 9.4.

The shorter lag phases observed for the L-D experiments, when compared with the solid and semi-solid systems, can be attributed mainly to the availability of hydrolyzed and partially hydrolyzed sugars which are readily available for the microorganisms in the liquid fraction. On the other hand, it is also due to the higher toxicity observed when some inhibitory compounds are present in fermentative media with higher contents of total solids (15 and 12%, respectively), as a consequence of the higher resistance to the molecular diffusion phenomena. This is consistent with how the biological processes, like AD, are complex and dependent on a variety of factors such as the substrate characteristics and the operation conditions of the process (pH, F/M ratio, load, TS, etc.).

9.7 VIABILITY ASSESSMENT OF THE USE OF LHW AS AN ALTERNATIVE FOR THE VALORIZATION OF CHW

To assess the energetic viability of the process, an energy balance was done considering the energy invested in the LHW pretreatment (E(−), Eq. 9.8) and the energy obtained (E(+), Eq. 9.9) from the combustion of the methane generated by AD in an energy and heat cogeneration system (CHP).

$$E(-)/MJ\ kg_{Casca\ bruta}{}^{-1} = Cp_{H_2O}m(T_f - 100\ °C) \tag{9.8}$$

where C_p is the average heat capacity for water at the operation temperatures (4.19×10^{-3} MJ/kg°C), m is the mass of water used (kg), and T_f is the final operation temperature (°C).

$$E(+)(\text{MJ kg}_{\text{CB}}{}^{-1}) = \frac{NmLCH_4}{gDQO_{entrada}} \times \frac{gDQO}{gmass_{(HH\ ou\ CPT)}} \times \frac{gmass_{(HH\ ou\ CPT)}}{g_{CB}} \times \frac{1000g}{\text{kg}_{CB}} \times \frac{10^{-6}\text{m}^3}{\text{mL}} \times \frac{34.5\text{MJ}}{\text{Nm}^3\text{CH}_4} \tag{9.9}$$

where $gmass_{(HH)}$ is the hydrolysate mass recovered from the hydrothermal pretreatment (g) and $gmass_{(CPT)}$ is the pretreated CHW mass, considering only the solid fraction for solid digestion and the solid fraction with the hydrolysate for the semi-solid digestion.

For the $E(-)$ calculations, as described in Eq. 9.8, the initial water temperature was considered as 100°C in vapor phase, given that it is possible to use the water in the CHP cooling system, by mean of a heat exchanger, for heating and vaporizing the water fed to the LHW pretreatment. The calculation of the energy $E(-)$ used during LHW was simplified in such a way that only water was taken into account, given the solids content of the system was always below 20%.

On the other hand, the calculation of $E(+)$, as described in 9.9, considered the combustion $\Delta H°$ for methane as 34.5 MJ $(Nm^3CH_4)^{-1}$. The yields and efficiencies for the CHP system were taken from (Cano, Pérez-Elvira, and fdz-polanco 2015). According to this author, for a conventional CHP system, the combustion efficiency is 85%, where the conversion of chemical to thermal energy corresponds to 65% and the conversion to electrical energy corresponds to 35%.

The energy gain is given by the difference between the generated energy $E(+)$ and the consumed energy for LHW, $E(-)$. The economic rentability is given based on the possibility of selling the generated electrical energy using values from the Brazilian ministry of mines and energy US$ 0.0676/kWh (R$ 0.35/KWh *ordinance n°65 de 2018,* Ministry of Mines and Energy of Brazil). The energetic-economical estimation evaluated for methane production from the liquid fraction and ethanol production from a pretreated solid fraction, took into account the glucose mass generated by the enzymatic hydrolysis of CHW and the theoretical yield of ethanol from a given mass of glucose (51%, 0.51 gEtOH $gGlu^{-1}$), as described in Eq. 9.10.

$$g_{etanol}/g_{casca\ bruta} = \frac{gGlicose_R}{g\ Casca_{PT}} \times \frac{g\ Casca_{PT}}{g\ Casca_{bruta}} \times \frac{0,51\ g_{etanol}}{gGlicose_R} \qquad (9.10)$$

The economic gain generated from selling the ethanol was calculated from the price of a liter of ethanol (US$ 0.3140. L $EtOH^{-1}$) established for the local union of bioenergy producers on April 17, 2020 (https://www.udop.com.br/). In the end, all the presented calculations for estimating the economic viability of the system were done using 1 ton of raw CHW as the basis of calculation.

To assess the viability of using CHW as raw material for the production of biofuels, an energetical-economic analysis was done, considering the energy balance (thermal energy generated by the biogas combustion – thermal energy invested for LHW), as well as the possible gains generated from the selling of products such as electrical energy and ethanol. For the analysis of the estimations, three different scenarios (S1, S2, and S3) were considered: S1, biogas production from raw CHW; S2, biogas production from a pretreated mixture (semi-solid) (BD1); and S3, ethanol production from the solid fraction (SF) and biogas production from the liquid fraction (LF) obtained from the pretreatment of CHW, the configurations are presented in Figure 9.5.

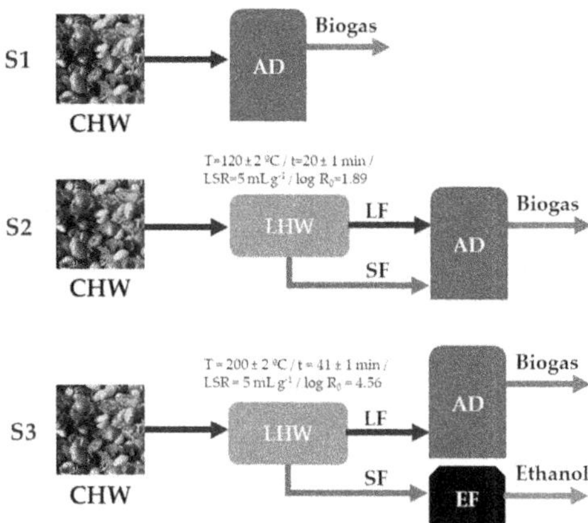

FIGURE 9.5 Assessment energy – economic scenarios.

Figure 9.6 shows the results of the scenarios. The S1 consisted in doing AD to the CHW without using any pretreatment. Even though in such a situation, there is no energy investment, the rentability is considered low. For this condition the only product considered was the electrical energy produced from the CHP system from biogas combustion. AD biogas yield is low in this case (Table 9.4; 6.7 $NmLCH_4$ $gCOD^{-1}$).

On the other hand, as previously observed, the hydrothermal treatment was able to improve the biodegradability of the CHW. Even though AD using conditions BD2 and BD3 resulted in higher methane production, only condition BD1 resulted in a positive energy balance, a reason why it was selected for the viability assessment. Given the positive balance for condition BD1, using the lowest possible temperature (120°C) was justified. This is related to the fact that hydrothermal pretreatments have the energy consumption as a limiting factor, given the high energetic demand associated to heating up the water used in the process. The higher the process temperature, the more difficult it is to obtain a positive energy balance.

According to the data presented in Figure 9.6a the thermal energy generated in the S2 is enough to cover the energy invested during pretreatment and there is a surplus that could be used in other stages of the coffee processing industry, such as the roasting of the grains. Furthermore, the electrical energy generated by the system can be sold to the company in charge of the local grid or used back in the processing plant, contributing to increasing the rentability.

For the S3, the viability assessment took into account the desirability condition for ethanol production (ED-200°C, 41 min and *LSR* de 5 mL g^{-1}), where the LF is used for producing methane and the SF is used for producing ethanol. This scenario resulted to be the most advantageous from the point of view of rentability, as shown in Figure 9.6b. However, it is important to make it clear that under this condition the energetic demand of the pretreatment cannot be covered just by using the produced biogas and a deficit of 1611.19 MJ per ton of CHW is generated.

Given the above, an alternative to increase the thermal viability of the process for S3 is by using part of the raw CHW as fuel for direct combustion and thermal

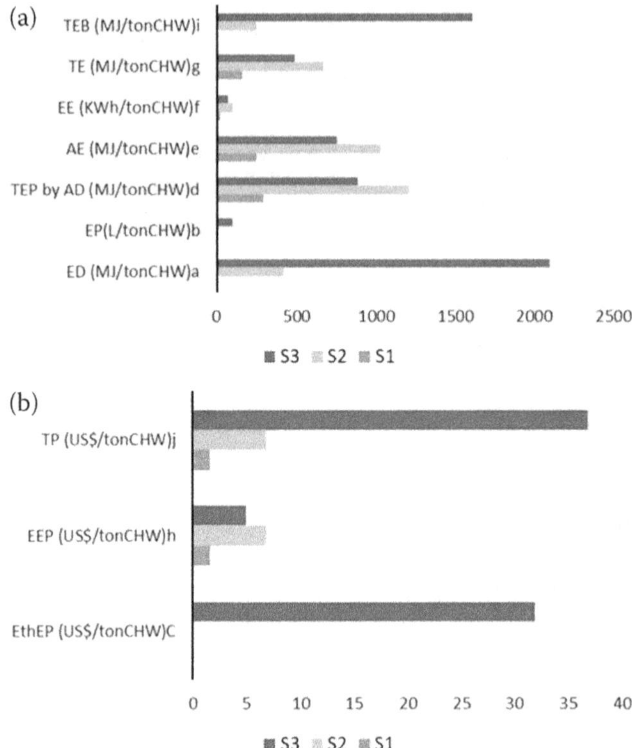

FIGURE 9.6 Energy (a) and economic (b) estimation proposed scenarios. a – ED: Energy demand of LWH to treat 1 ton of Raw CHW. b – EP: Ethanol production considering a conversion of recovery glucose of 0.51 for ED condition (enzymatic hydrolysis process with a load of 40FPU g^{-1}. c – EthEP: Ethanol economic profitability considering o market value of US$ 0.3140/L (UDOP, 2020). d – TEP by AD: Energy production per ton of Raw CHW from AD. e – AE: Available energy considering a CHP system. f and g – EE and TE: Electric energy and thermic energy considering a conversion of available energy of 35 and 65%, respectively. h – EEP: Electrical energy profitability considering US$ 0.0676(KWh (R$ 0.35/kWh, Portaria n° 65 de 2018. Ministério de Minas e Energia). i – TEB: Total energy balance. j – TP: Total profitability.

generation. Considering a heat of combustion of 16.1 MJ/kg for the raw CHW (Saenger et al. 2001) and a thermal efficiency of 80%, approximately 125 kg of raw CHW would be required. Thanks to this modification, the system would become energetically sustainable a gain of US$ 30 could be generated. Because of that, for the scope of this research, the best option for the CHW management is to pretreat it using LHW and to separate the resulting currents so the SF is used to produce ethanol and the LF is used to produce biogas.

Although S3 has been considered as the best technological route for the coffee production industry, the enzymatic mixtures required for the enzymatic hydrolysis represent a considerable drop in the economic gains. Additionally, the fixed costs associated to maintenance and the unit operations required for the separations, represent a techno-economical challenge that may limit the possibilities of an industrial-scale implementation. For this reason, a future more robust analysis is necessary, including the impact of the mentioned factors on the viability assessment of the process.

9.8 CONCLUDING REMARKS

This chapter presents some alternatives for the valorization of CHW including an LHW pretreatment followed by anaerobic digestion. Additionally, a technological route was proposed which includes the association of production of ethanol from solid fraction generated after CHW pretreatment and biogas from liquid fraction showing that it is possible to add value to CHW under a biorefinery approach. For this technological route, an LHW pretreatment of CHW at 200°C, 41 min and LSR de 5 mL g^{-1} was applied. In this scenario, the total profitability considering the sale of EE generated by burning biogas in a CHP system and the sale of ethanol was of approximately US$ 35 per ton of CHW. However, for this condition to be applied, as the energy balance was not positive in the function of necessity higher temperature in the pretreatment, a part of the coffee husk must be used as fuel to complement the thermal energy necessary for the pretreatment. If the proposal is to use the coffee husk waste to produce only biogas, the study allows to conclude that the pretreatment of CHW was efficient and the anaerobic digestion must be carried out with the mixture of solid and liquid fraction generated after the pretreatment, however, the pretreatment must be carried out in low severity conditions (120°C), since the higher severity of CHW pretreatment generating a higher release of toxic compounds hampered the anaerobic digestion.

ACKNOWLEDGMENTS

The authors would like to thank the Brazilian National Council for Scientific and Technological Development (CNPq grant number 438527/2018-6) and Federal University of Ouro Preto (UFOP) for the financial support. This study was also financed in part by the Coordenação de Aperfeiçoamento de Pessoal de Nível Superior – Brasil (CAPES) – Finance Code 001. CNPq – grant numbers 152180/2019-2.

REFERENCES

Adarme, Oscar Fernando Herrera, Bruno Eduardo Lobo Baêta, Jose Balena Gabriel Filho, Leandro Vinícius Alves Gurgel, and Sérgio Francisco de Aquino. 2019. "Use of Anaerobic Co-Digestion as an Alternative to Add Value to Sugarcane Biorefinery Wastes." *Bioresource Technology* 287 (May): 121443. doi:10.1016/j.biortech.2019.121443

Ahmad, Fiaz, Edson Luiz Silva, and Maria Bernadete Amâncio Varesche. 2018. "Hydrothermal Processing of Biomass for Anaerobic Digestion – A Review." *Renewable and Sustainable Energy Reviews* 98: 108–124. doi:10.1016/j.rser.2018.09.008

Amiri, Hamid, and Keikhosro Karimi. 2015. "Autohydrolysis: A Promising Pretreatment for the Improvement of Acetone, Butanol, and Ethanol Production from Woody Materials." *Chemical Engineering Science* 137: 722–729. doi:10.1016/j.ces.2015.07.020

Baêta, B., D. Lima, O. Adarme, L. Gurgel, and S. Aquino. 2016. "Optimization of Sugarcane Bagasse Autohydrolysis for Methane Production from Hemicellulose Hydrolyzates in a Biorefinery Concept." *Bioresource Technology* 200: 137–146. doi:10.1016/j.biortech.2015.10.003

Baêta, Bruno Eduardo Lobo, Paulo Henrique de Miranda Cordeiro, Fabiana Passos, Leandro Vinícius Alves Gurgel, Sérgio Francisco de Aquino, and Fernando Fdz-Polanco. 2017a. "Steam Explosion Pretreatment Improved the Biomethanization of Coffee Husks." *Bioresource Technology* 245 (August): 66–72. doi:10.1016/j.biortech.2017.08.110

Baêta, Bruno Eduardo Lobo, Paulo Henrique de Miranda Cordeiro, Fabiana Passos, Leandro Vinícius Alves Gurgel, Sérgio Francisco de Aquino, and Fernando Fdz-Polanco. 2017b. "Steam Explosion Pretreatment Improved the Biomethanization of Coffee Husks." *Bioresource Technology* 245: 66–72. doi:10.1016/j.biortech.2017.08.110

Barakat, Abdellatif, Florian Monlau, Jean Philippe Steyer, and Hélène Carrere. 2012. "Effect of Lignin-Derived and Furan Compounds Found in Lignocellulosic Hydrolysates on Biomethane Production." *Bioresource Technology* 104: 90–99. doi:10.1016/j.biortech.2011.10.060

Batista, Gustavo, Renata B.A. Souza, Bruna Pratto, Martha S.R. dos Santos-Rocha, and Antonio J.G. Cruz. 2019. "Effect of Severity Factor on the Hydrothermal Pretreatment of Sugarcane Straw." *Bioresource Technology* 275 (December 2018): 321–327. doi:10.1016/j.biortech.2018.12.073

Battista, Federico, Debora Fino, and Bernardo Ruggeri. 2014. "Polyphenols Concentration's Effect on the Biogas Production by Wastes Derived from Olive Oil Production." *Chemical Engineering Transactions* 38 (June). doi:10.3303/CET1438063

Bekalo, Samson Ayele, and Hans Wolf Reinhardt. 2010. "Fibers of Coffee Husk and Hulls for the Production of Particleboard." *Materials and Structures/Materiaux et Constructions* 43 (8): 1049–1060. doi:10.1617/s11527-009-9565-0

Cano, Raul, S. I. Pérez-Elvira, and Fernando fdz-polanco. 2015. "Energy Feasibility Study of Sludge Pre-Treatments: A Review." *Applied Energy* 149 (July). doi:10.1016/j.apenergy.2015.03.132

Caroca, E., A. Serrano, R. Borja, A. Jiménez, A. Carvajal, A.F.M. Braga, G. Rodriguez-Gutierrez, and F.G. Fermoso. 2021. "Influence of Phenols and Furans Released during Thermal Pretreatment of Olive Mill Solid Waste on Its Anaerobic Digestion." *Waste Management* 120 (February): 202–208. doi:10.1016/j.wasman.2020.11.027

Chen, Xiangxue, Rui Zhai, Ying Li, Xinchuan Yuan, Zhi-Hua Liu, and Mingjie Jin. 2020. "Understanding the Structural Characteristics of Water-Soluble Phenolic Compounds from Four Pretreatments of Corn Stover and Their Inhibitory Effects on Enzymatic Hydrolysis and Fermentation." *Biotechnology for Biofuels* 13 (1): 1–13. doi:10.1186/s13068-020-01686-z

Cone, John W., Antonie H. van Gelder, and Frank Driehuis. 1997. "Description of Gas Production Profiles with a Three-Phasic Model." *Animal Feed Science and Technology* 66 (1): 31–45. doi:10.1016/S0377-8401(96)01147-9

da Fonseca, Yasmim Arantes, Nayara Clarisse Soares Silva, Adonai Bruneli de Camargos, Silvana de Queiroz Silva, Hector Javier Luna Wandurraga, Leandro Vinícius Alves Gurgel, and Bruno Eduardo Lobo Baêta. 2021. "Influence of Hydrothermal Pretreatment Conditions, Typology of Anaerobic Digestion System, and Microbial Profile in the Production of Volatile Fatty Acids from Olive Mill Solid Waste." *Journal of Environmental Chemical Engineering*, January. Elsevier, 105055. doi:10.1016/j.jece.2021.105055

De Aquino, Sérgio F, and Carlos A L Chernicharo. 2005. "Acúmulo de Ácidos Gordos Voláteis (AGVs) Em Reatores Anaeróbis Sob Stress:Causas e Estratégias de Controlo." *Eng. Sanit. Ambient. 10*, 152–161.

dos Santos, Lívia Caroline, Oscar Fernando Herrera Adarme, Bruno Eduardo Lobo Baêta, Leandro Vinícius Alves Gurgel, and Sérgio Francisco de Aquino. 2018. "Production of Biogas (Methane and Hydrogen) from Anaerobic Digestion of Hemicellulosic Hydrolysate Generated in the Oxidative Pretreatment of Coffee Husks." *Bioresource Technology* 263: 601–612. doi:10.1016/j.biortech.2018.05.037

Golschimdt, O. 1971. "Ultraviolet Spectra." *Wiley-Interscience*, 241–266.

Groot, Jeroen C. J., John W. Cone, Barbara A. Williams, Filip M. A. Debersaques, and Egbert A. Lantinga. 1996. "Multiphasic Analysis of Gas Production Kinetics for in Vitro Fermentation of Ruminant Feeds." *Animal Feed Science and Technology* 64 (1): 77–89. doi:10.1016/S0377-8401(96)01012-7

Jönsson, Leif J., and Carlos Martín. 2016. "Pretreatment of Lignocellulose: Formation of Inhibitory by-Products and Strategies for Minimizing Their Effects." *Bioresource Technology* 199: 103–112. doi:10.1016/j.biortech.2015.10.009

Kumari, Dolly, and Radhika Singh. 2018. "Pretreatment of Lignocellulosic Wastes for Biofuel Production: A Critical Review." *Renewable and Sustainable Energy Reviews* 90 (April). Elsevier Ltd: 877–891. doi:10.1016/j.rser.2018.03.111

Lima, Diego Roberto Sousa, Oscar Fernando Herrera Adarme, Bruno Eduardo Lobo Baêta, Leandro Vinícius Alves Gurgel, and Sérgio Francisco de Aquino. 2018. "Influence of Different Thermal Pretreatments and Inoculum Selection on the Biomethanation of Sugarcane Bagasse by Solid-State Anaerobic Digestion: A Kinetic Analysis." *Industrial Crops and Products* 111 (November 2017): 684–693. doi:10.1016/j.indcrop.2017.11.048

Lima, Diego Roberto Sousa, Oscar Fernando Herrera Adarme, Bruno Eduardo Lobo Baêta, Leandro Vinícius Alves Gurgel, Sérgio Francisco de Aquino, and Roberto. 2018. "Influence of Different Thermal Pretreatments and Inoculum Selection on the Biomethanation of Sugarcane Bagasse by Solid-State Anaerobic Digestion: A Kinetic Analysis." *Industrial Crops & Products* 111 (January): 684–693. doi:10.1016/j.indcrop.2017.11.048

López, S., J. Dijkstra, M.S. Dhanoa, A. Bannink, E. Kebreab, and J. France. 2011. "Part 3 – Modelling Fermentation, Digestion and Microbial Interactions in the Gut." In *Modelling Nutrient Digestion and Utilisation in Farm Animals*, edited by D. Sauvant, J. van Milgen, P. Faverdin, and N. Friggens, 1st ed., 139–147. Netherlands: Wageningen Academic Publishers. doi:10.3920/978-90-8686-712-7

Mesquita, Jéssica Faria, André Ferraz, and André Aguiar. 2016. "Alkaline-Sulfite Pretreatment and Use of Surfactants during Enzymatic Hydrolysis to Enhance Ethanol Production from Sugarcane Bagasse." *Bioprocess and Biosystems Engineering* 39 (3): 441–448. doi:10.1007/s00449-015-1527-z

Michelin, M., E. Ximenes, M.d.L. Polizeli, and M.R. Ladisch. 2015. "Effect of Phenolic Compounds from Pretreated Sugarcane Bagasse on Cellulolytic and Hemicellulolytic Activities." *Developmental Biology* 14 (June): 1–29. doi:10.1242

Miller, Gail Lorenz. 1959. "Use of Dinitrosalicylic Acid Reagent for Determination of Reducing Sugar." *Analytical Chemistry* 31 (3): 426–428. doi:10.1021/ac60147a030

Momayez, Forough, Keikhosro Karimi, and Ilona Sárvári. 2018. "Enhancing Ethanol and Methane Production from Rice Straw by Pretreatment with Liquid Waste from Biogas Plant" *Energy Conversion and Management* 178 (August): 290–298.

Monlau, F., C. Sambusiti, A. Barakat, M. Quéméneur, E. Trably, J.-P. Steyer, and H. Carrère. 2014. "Do Furanic and Phenolic Compounds of Lignocellulosic and Algae Biomass Hydrolyzate Inhibit Anaerobic Mixed Cultures? A Comprehensive Review." *Biotechnology Advances* 32 (5): 934–951. doi:10.1016/j.biotechadv.2014.04.007

Morales-Martínez, J. L., M. G. Aguilar-Uscanga, E. Bolaños-Reynoso, and L. López-Zamora. 2020. "Optimization of Chemical Pretreatments Using Response Surface Methodology for Second-Generation Ethanol Production from Coffee Husk Waste." *BioEnergy Research*, October. BioEnergy Research. doi:10.1007/s12155-020-10197-6

Mudgil, Deepak, and Sheweta Barak. 2018. *Beverages: Processing and Technology.* Jodhpur: Scientific Publishers.

Mussatto, Solange I., Lina F. Ballesteros, Silvia Martins, and José A. Teixeira. 2011. "Extraction of Antioxidant Phenolic Compounds from Spent Coffee Grounds." *Separation and Purification Technology* 83 (1): 173–179. doi: 10.1016/j.seppur.2011.09.036

Navarro, Ronald R., Yuichiro Otsuka, Kenji Matsuo, Kei Sasaki, Ken Sasaki, Tomoyuki Hori, Hiroshi Habe, et al. 2020. "Combined Simultaneous Enzymatic Saccharification and Comminution (SESC) and Anaerobic Digestion for Sustainable Biomethane Generation from Wood Lignocellulose and the Biochemical Characterization of Residual Sludge Solid." *Bioresource Technology* 300 (December 2019). Elsevier: 122622. doi:10.1016/j.biortech.2019.122622

Pandey, Ashok, Carlos R. Soccol, Poonam Nigam, Debora Brand, Radjiskumar Mohan, and Sevastianos Roussos. 2000. "Biotechnological Potential of Coffee Pulp and Coffee Husk for Bioprocesses." *Biochemical Engineering Journal* 6 (2): 153–162. doi:10.1016/S1369-703X(00)00084-X

Paranhos, Aline, Oscar Adarme, Gabriela Fernandes, Silvana De Queiroz, Sérgio Francisco, De Aquino, and E. Engineering. 2020. "Methane Production by Co-Digestion of Poultry Manure and Lignocellulosic Biomass: Kinetic and Energy Assessment." *Bioresour. Technol.* 300 (December 2019). doi:10.1016/j.biortech.2019.122588

Park, Jeong Hoon, Jeong Jun Yoon, Hee Deung Park, Dong Jung Lim, and Sang Hyoun Kim. 2012. "Anaerobic Digestibility of Algal Bioethanol Residue." *Bioresource Technology* 113: 78–82. doi:10.1016/j.biortech.2011.12.123

Passos, Fabiana, Paulo Henrique Miranda Cordeiro, Bruno Eduardo Lobo Baeta, Sergio Francisco de Aquino, and Sara Isabel Perez-Elvira. 2018. "Anaerobic Co-Digestion of Coffee Husks and Microalgal Biomass after Thermal Hydrolysis." *Bioresource Technology* 253: 49–54. doi:10.1016/j.biortech.2017.12.071

Poirier, Simon, Ariane Bize, Chrystelle Bureau, Théodore Bouchez, and Olivier Chapleur. 2016. "Community Shifts within Anaerobic Digestion Microbiota Facing Phenol Inhibition: Towards Early Warning Microbial Indicators?" *Water Research* 100: 296–305. doi:10.1016/j.watres.2016.05.041

Quéméneur, Marianne, Jérôme Hamelin, Abdellatif Barakat, Jean Philippe Steyer, Hélne Carrre, and Eric Trably. 2012. "Inhibition of Fermentative Hydrogen Production by Lignocellulose-Derived Compounds in Mixed Cultures." *International Journal of Hydrogen Energy* 37 (4): 3150–3159. doi:10.1016/j.ijhydene.2011.11.033

Rios-González, Leopoldo J., Thelma K. Morales-Martínez, María F. Rodríguez-Flores, José A. Rodríguez-De la Garza, David Castillo-Quiroz, Agustín J Castro-Montoya, and Alfredo Martinez. 2017. "Autohydrolysis Pretreatment Assessment in Ethanol Production from Agave Bagasse." *Bioresource Technology* 242: 184–190. doi:10.1016/j.biortech.2017.03.039

Rodrigues, Keila Letícia Teixeira, Ananda Lima Sanson, Amanda de Vasconcelos Quaresma, Rafaela de Paiva Gomes, Gilmare Antônia da Silva, and Robson José de Cássia Franco Afonso. 2014. "Chemometric Approach to Optimize the Operational Parameters of ESI for the Determination of Contaminants of Emerging Concern in Aqueous Matrices by LC-IT-TOF-HRMS." *Microchemical Journal* 117: 242–249. doi:10.1016/j.microc.2014.06.017

Saenger, M., E. U. Hartge, J. Werther, T. Ogada, and Z. Siagi. 2001. "Combustion of Coffee Husks." *Renewable Energy* 23 (1): 103–121. doi:10.1016/S0960-1481(00)00106-3

Schroyen, Michel, Han Vervaeren, Katleen Raes, and Stijn W H Van Hulle. 2018. "Modelling and Simulation of Anaerobic Digestion of Various Lignocellulosic Substrates in Batch Reactors: Influence of Lignin Content and Phenolic Compounds II." *Biochemical Engineering Journal* 134: 80–87. doi:10.1016/j.bej.2018.03.017

Sluiter, Amie, Bonnie Hames, Raymond O. Ruiz, Christopher Scarlata, Justin Sluiter, David Templeton, and Department of Energy. 2004. "Determination of Structural Carbohydrates and Lignin in Biomass." *NREL (National Renewable Energy Laboratory).*

Tonucci, Marina C., Oscar F.H. Adarme, Sérgio F. de Aquino, Bruno Eduardo L. Baeta, and César Ricardo T. Tarley. 2020. "Synthesis of Hybrid Magnetic Molecularly Imprinted Polymers for the Selective Adsorption of Volatile Fatty Acids from Anaerobic Effluents." *Polymer International* n/a (n/a). John Wiley & Sons, Ltd. doi:10.1002/pi.6026

USDA. 2019. "Coffee: World Markets and Trade." United States Department of Agriculture.

Wang, Dou, Fei Shen, Gang Yang, Yanzong Zhang, Shihuai Deng, and Jing Zhang. 2018. "Bioresource Technology Can Hydrothermal Pretreatment Improve Anaerobic Digestion for Biogas from Lignocellulosic Biomass?" *Bioresource Technology* 249 (August 2017). Elsevier: 117–124. doi:10.1016/j.biortech.2017.09.197

Wang, Dou, Fei Shen, Gang Yang, Yanzong Zhang, Shihuai Deng, Jing Zhang, Yongmei Zeng, Tao Luo, and Zili Mei. 2018. "Can Hydrothermal Pretreatment Improve Anaerobic Digestion for Biogas from Lignocellulosic Biomass?" *Bioresource Technology* 249: 117–124. doi:10.1016/j.biortech.2017.09.197

Widjaja, T., S. Nurkhamidah, A. Altway, T. Iswanto, B. Gusdyarto, and F. F. Ilham. 2019. "Performance of Biogas Production from Coffee Pulp Waste Using Semi-Continuous Anaerobic Reactor." *IOP Conference Series: Materials Science and Engineering* 673 (1). doi:10.1088/1757-899X/673/1/012003

Wikandari, Rachma, Noor Kartika Sari, Qurrotul A'yun, Ria Millati, Muhammad Nur Cahyanto, Claes Niklasson, and Mohammad J. Taherzadeh. 2015. "Effects of Lactone, Ketone, and Phenolic Compounds on Methane Production and Metabolic Intermediates During Anaerobic Digestion." *Applied Biochemistry and Biotechnology* 175 (3): 1651–1663. doi:10.1007/s12010-014-1371-7

Zeng, Zequan, Haikui Zou, Xin Li, Moses Arowo, Baochang Sun, Jianfeng Chen, Guangwen Chu, and Lei Shao. 2013. "Degradation of Phenol by Ozone in the Presence of Fenton Reagent in a Rotating Packed Bed." *Chemical Engineering Journal* 229: 404–411. doi:10.1016/j.cej.2013.06.018

Zheng, Yi, Jia Zhao, Fuqing Xu, and Yebo Li. 2014. "Pretreatment of Lignocellulosic Biomass for Enhanced Biogas Production." *Progress in Energy and Combustion Science*. Elsevier Ltd. doi:10.1016/j.pecs.2014.01.001

10 Valorization of Pectin from Fruit Wastes for Sustainable Production of Value-added Compounds by Actinomycetes

Suneetha Vuppu
School of Biosciences and Technology, VIT University, Vellore India

Praveen Kumar
VIT University, India

Arjun Chinamgari
Department of Industrial Engineering, University of Florida, Gainesville, FL, USA

CONTENTS

DOI: 10.1201/9781003128977-10

10.1 INTRODUCTION

In modern enzyme technology, actually in scientific terms, pectin valorization is the technique of reusing, recycling, or composting pectin-rich fruit waste materials and converting them into more useful products including biomaterials, chemicals, fuels, or other sources of energy. Pectins are the substrates in which pectinases catalyze on it. The pectic substances are complex, negatively charged, acidic molecules present as large constituent as Ca-pectate and Mg-pectate in the plant cell walls. Pectins are diverse molecules present in the primary walls of about woody tissue (5%), grass (2–10%), and other plant cell walls (Neill et al., 1990; Ridley et al., 2001). The pectin constitutes about 0.5–4.0% in the fresh weight of plant material (Kashyap et al., 2001). The galacturonic acid residues are joined by α-1–4 linkage with a leading chain that consists of arabinose, rhamnose, xylose, and galactose to form the substantial pectin complex (Gummadi and Panda, 2003; Caffall and Mohnen, 2009). It maintains the structural integrity by providing physical properties, involuntary support to the primary cell wall and contributes intercellular fixation due to the presence of nonesterified carboxyl groups that are connected through divalent cations such as Mg^{2+} and Ca^{2+}, which causes the formation of pectin to gel (Praveen and Suneetha, 2015). Pectin is mainly comprised of three polymers, namely rhamnogalacturonan (RG I), homogalacturonan (HG), and rhamnogalacturonan (RG-II). Pectin is classified based on the degree of methylation (DM), it is classified as high (DM greater than 50%) and low methoxy pectin (DM less than 50%) (Yapo, 2011). The pectin present in small amount in plant materials that are most difficult to hydrolysis causes a negative impact on its byproducts' production from biomass (Latarullo et al., 2016). Mostly, pectins that are available commercially are produced from citrus fruit peels. Earlier reports showed that the degradation of pectin substance has anticancer, antioxidant, and other therapeutic properties. Enzymes are essential molecules present in nature to replace deleterious chemical reactions in an eco-friendly manner. Pectinases(Ramirez-Tapias et al., 2017) or pectinolytic enzymes that are indulged in the pectin breakdown are enormously distributed in higher plants and microorganisms. There are various types of pectinase enzyme produced by microbes, which uses pectin as sole energy and carbon source. These enzymes catalyze degradation of pectin through depolymerization neither by lyases, hydrolases, and de-esterification (esterases) reactions (Garg et al., 2016). This study deals with identification and characterization of pectin from various natural sources and pectinase from *actinomycetes* for industrial applications. The potential pectinolytic isolate VIT-SP4 was isolated from pectin enriched fruit industrial dump sites in Chittoor, Gudipala, Mudigolam, and Madanapalli of Andhra Pradesh and screened for maximum pectinase activity and it was identified as *Streptomyces fumigatiscleroticus* VIT-SP4. The strain of *Streptomyces lydicus* MTCC 7505 with known activity was taken as control strain. The cheapest pectin substrates (fruit peels) were characterized and investigated for pectinase activity. The result showed that isolated VIT-SP4 produces maximum pectinase activity with orange pectin.

The pectin present in small amount in plant materials that are most difficult to hydrolysis causes a negative impact on its byproducts' production from biomass (Latarullo et al., 2016). Mostly, pectins that are available commercially are produced from citrus fruit peels. Earlier, reports showed that the degradation of pectin substance has anticancer, antioxidant, and other therapeutic properties. Valorization of pectin from fruit wastes plays a significant role in various applications. The quantity of fruit juices was increased with pectinase and efficacy was increased with enzyme mixture. Ultimately, it helps in maximal yield and clarification of fruit juices that can be used in fruit juice industries to enhance the nutritional value of juices. The postconstruction crack repair of concrete block was carried out with potential isolate *S. fumigatiscleroticus* VIT-SP4. The calcium pectinate pellets were prepared and pectinase from *S. fumigatiscleroticus* VIT-SP4 was investigated for controlled drug delivery studies. The *in-vitro* analysis of pectin was carried out by AOAC method after pectinase activity of isolate VIT-SP4 on different fruit peels. The results showed the presence of moisture protein, ash, carbohydrate, crude fiber, and fat in different fruit peels similar to previous studies. The pectinase enzyme promotes *Zinnia elegans* plant growth after a time period of 2 weeks. This pectinolytic isolate VIT-SP4 can be a promising strain in food, pharma, and building construction to increase the strength in civil engineering industries. The characterization of pectin and pectinase enzyme by Fourier transform infrared spectroscopy (FTIR), scanning electron microscopy (SEM), and *in-vitro* analysis will be discussed. Hence, the current research aims to study the pectin substrates and pectinase enzymes for industrial exploitation.

10.2 MATERIALS AND METHODS

10.2.1 Chemicals

All chemicals used in this study are from Himedia (AR) grade.

10.2.2 Media Preparation

The prepared inoculums were added to the sterilized medium for pectinase production. The composition of media contains K_2HPO_4 0.5, casein 3.0, starch 10.0, peptone 1.0, yeast extract 1.0, pectin 1.0, malt extract 10.0 g L^{-1} in an Erlenmeyer flask and then incubated for 3 days under agitation of 100 rpm at 28 ± 2°C. The cultures (1 mL) were aliquot from the production medium and centrifuged at 10,000 × g for 10 min at 4°C. The *Streptomyces lydicus* MTCC 7505 strain was chosen as control for the study. The supernatant was taken for further cocktail pectinase enzymatic assays.

10.2.2.1 Extraction of Pectin

The fruit peels were rinsed with tap water and then followed by double dis.H_2O. The fruit peels were made into powder with the help of a blender. The 5 g of respective fruit peel powders were weighed and it was taken in 250 mL of dis.H_2O. The solution was acidified with 0.1 N (6.4 g L^{-1}) of citric acid and until it gets the homogeneous form and it was adjusted to 3.3 pH. Then, the solution was heated up to 70°C for 30 min. Then, the solution was kept at room temperature for 24 h. Then, it was allowed to centrifuge at 6000 rpm for 10 min to recover the precipitated pectin. The treated crude samples were filtered and added to a double volume of 95% ethanol (1:2 V/V) for pectin precipitation. Then, it was stored in dark condition at room temperature for 24 h. Then, it was separated by filtration using Whatman filter paper no. 1 and it was washed twice with 70% ethanol. The acetone was added in a dropwise manner and dried in an oven at 65°C (Liew et al., 2014).

The percentage yield of fruit peel pectin was determined as a gram of product obtained per 10 g of fruit peel powder used (Equation 3.1) (Patidar et al., 2018).

Pectin yield%) = product obtained (g)5g of fruit peel powder × 100 (3.3) 43

10.2.2.2 SEM Characterization of Fruit Pectin

The morphological characterization of fruit pectin was further characterized by a scanning electron microscope (ZEISS EVO 18 RESEARCH) which has been equipped with an Oxford X-ray probe employed with retro-disperse electrons under low vacuum conditions at 10 Kv.

10.2.2.3 FTIR Characterization of Fruit Pectin

The FTIR spectrum was utilized for the functional groups present in fruit peels. In our study, we have taken orange, mosambi, pomegranate, mango, and apple peels. The structural groups present in pectin were characterized using IR spectrum.

10.2.2.4 *In-vitro* Analysis of Fruit Peels

The pectin samples were dried in the oven at 105°C to evaluate the moisture content present in it. Ash, proteins, lipids, and crude fibers were analyzed according to AOAC methods (AOAC, 1995).

10.2.2.5 Postconstruction Cracks Repair by Spray Culture of Spore Bearing Nonpathogenic *S. fumigatiscleroticus* VIT-SP4

10.2.2.5.1 Culture Preparation

The prepared cultures were inoculated into the CSPYME media (100 mL) for pectinase yield and kept for incubation at 30°C, 100 rpm for 72 h. The culture was sprayed after the spore formation was observed.

10.2.2.6 Production of Biocement and Survivability of Actinomycete Spores

The viability of actinomycete endospores present in cement was investigated by inoculating endospores into cement simultaneously with the inclusion of water. Spore survivability in the cement paste was analyzed by total number of viable spores after 93 days. The isolate (1 g) was granulated utilizing motor and pestle, vortexing was carried out to improve separation of spores. The serial dilution of culture (1 mL) was carried out with sterilized Tris-HCl buffer (Wiktor and Jonkers, 2011).

10.2.2.7 Activity of Pectinase on a Pectin-coated Tablet in Controlled Drug Delivery System

The pectin-based drug delivery system was carried out. Indomethacin, an anti-inflammatory drug (NSAID), was selected as a model drug. The Ca-pectinate tablets were prepared with extracted pectin. The supernatant of 100 mL from *S. fumigatiscleroticus* VIT-SP4 was taken as a pectinase solution. The experiment was carried out for the observation of drug release at different pH and time intervals.

10.2.2.8 *In-vitro* Analysis of Fruit Peels

The *in-vitro* analysis for different fruit peels was carried out to estimate the proximate constituents such as moisture, protein, ash, carbohydrate, crude fiber, and fat (Garau et al., 2007).

10.2.2.9 Application of Pectinase as a Promoter for Plant Growth

The pectinase enzyme was added aseptically to the soil along with *Z. elegans* plant seeds that were sown in pots. The seeds sown without pectinase enzyme were taken as control and in the zinnia plants were grown in normal environmental condition; the root, shoot, and leaf length were measured after 2 weeks.

10.3 RESULTS AND DISCUSSION

10.3.1 Extraction of Pectin

In our study, we have done a comparative study on the pectinase production by utilizing different fruit peels such as mosambi, pomegranate, orange, and mango. The pectin was extracted from different fruit peels by using the citric acid extraction method. The pectin yield was calculated by using the above equation (Equation 3.5.1.1).

The pectin production of different fruit peels is given in Table 10.1. The pectin production was found to be maximum in orange peel of about 42.3% and it was

TABLE 10.1
Pectin% Yield from Different Fruit Peels

Fruit Peel	Pectin (%)
Mosambi	39.2
Pomegranate	29.1
Orange	42.3
Mango	41.1
Apple	34.5

higher compared to other substrates and it was also maximum compared to other methods of extraction 19.24% (Maran et al., 2013), 20.44 ±0.64 (Guo et al., 2003), 5.27% (Yeoh et al., 2008), as seen in earlier reports.

10.3.2 Morphological Characterization of Fruit Pectins by SEM

The characterization of different pectins were analyzed by using SEM. Morphological characteristics of mosambi pectin show cracks, cervices, and porous structure which are given in Figures 10.1 and 10.2. Pomegranate pectin shows rough and uneven structures with pores shown in Figures 10.3 and 10.4.

The orange pectin shows heterogeneous structure with pores shown in Figures 10.5 and 10.6; mango and other pectins show cuticles on cuticular wax ridges that are given in Figures 10.7 and 10.8 and these morphological characteristics were comparable to the previous studies (Arias and Ramon-Laca, 2005; Martínez-Avilaa et al., 2009).

10.3.3 FTIR Characterization of Fruit Pectins

The FTIR spectrum was utilized to investigate the functional groups present in fruit pectins. The structural groups present in different fruit pectins were characterized and compared with commercial pectins using IR spectrum. The FTIR spectrum of mosambi and pomegranate pectin is shown in Figure 10.9. The FTIR spectrum of orange and mango pectin is shown in Figure 10.10. The FTIR spectrum of commercial pectin is shown in Figure 10.11.

The characterization of functional compounds existing in pectin was analyzed by FTIR. The peaks 3331.07, 3323.35, 3334.92, 3342.64, and 3556.74 cm^{-1} of commercial pectin correspond to the OH groups due to the presence of alcohol and pectic components of pectin molecule. The peaks at 2916.37, 2922.16, 2978.09, 2916.37, and 2937.59 cm^{-1} of commercial pectin correspond to carbon–hydrogen asymmetrical vibrating stretch due to aliphatic formation caused by different mode of vibrations on carbohydrates and lignin. The peaks 1728.22, 1722.43, 1728.22,

FIGURE 10.1 Scanning electron microscopic images of cracks of mosambi (*Citrus limetta*) pectin.

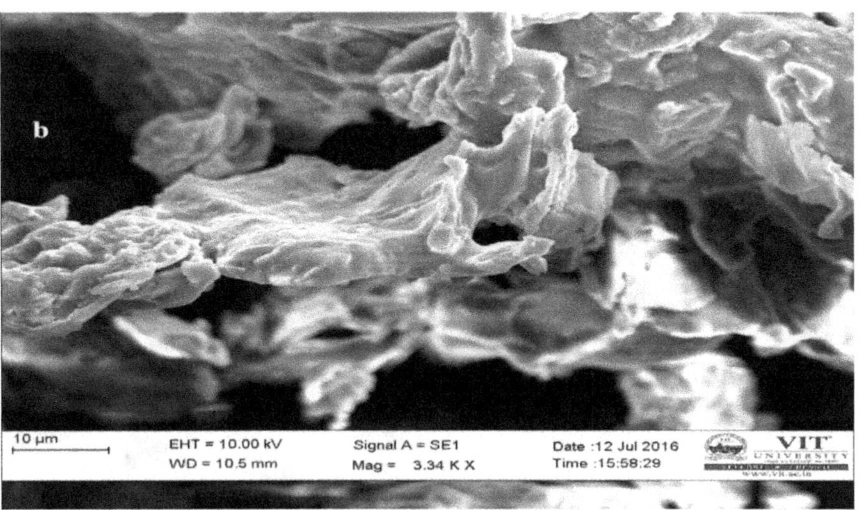

FIGURE 10.2 Cervical structure of mosambi (*Citrus limetta*) pectin.

2848.86, and 2937.59 cm^{-1} of pectin (commercial) correspond to a carbonyl group (C=O) stretching of methyl esterified $COCH_3$ in pectin that has been regarded as low methoxy pectin (Chatjigakis et al., 1998). The characteristic peaks at 1633.71, 1614.42, 1627.92, 1625.99, and 1629.85 cm^{-1} of commercially available pectin represent stretching vibration due to ionic carboxyl groups (Pappas et al., 2004). The peaks at 1238.30, 1055.06,1016.49, 1165.00, 1066.64, 1103.28, 1008.77, and 1031.92 cm^{-1} are caused by carbonyl group and alkynyl vibrations of glycoside

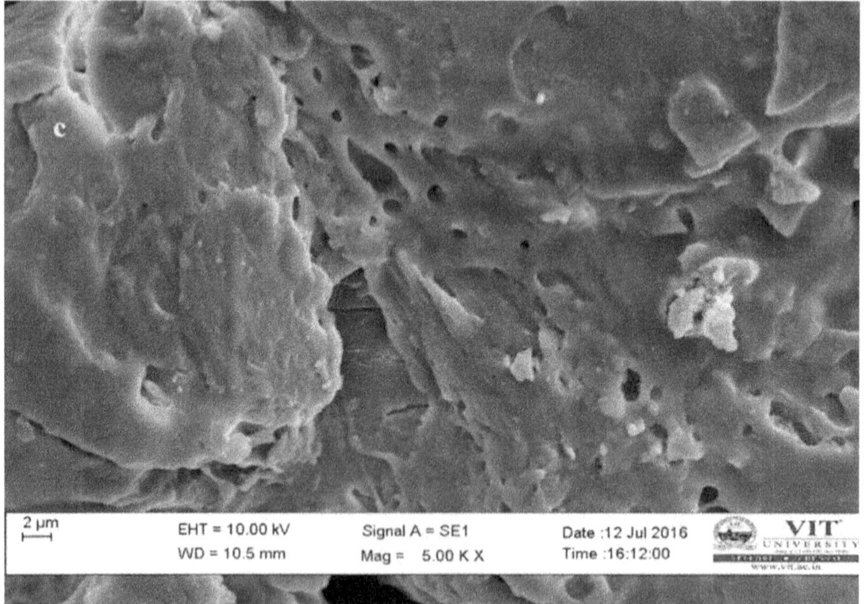

FIGURE 10.3 Scanning electron microscopic images of rough, uneven structures of pomegranate (*Punica granatum*) pectin.

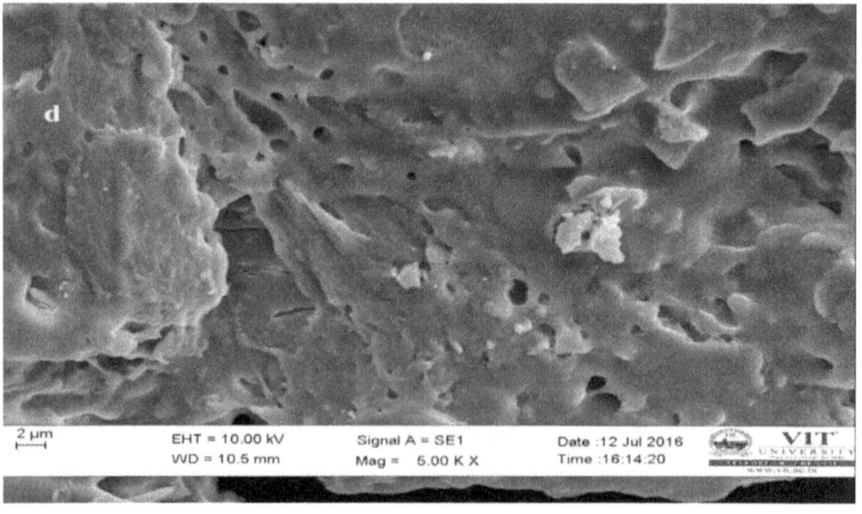

FIGURE 10.4 Porous structure of pomegranate (*Punica granatum*) pectin.

bond and pyrenoid rings. The FTIR spectra range around 800–1300 cm^{-1} cause due to OH bending and methyl plane distortion (Cerna et al., 2003; Gan et al., 2010). Finally, IR spectra at 3300 cm^{-1} confirm the presence of pectic acid in the peel; they were comparatively analyzed over commercially available pectin.

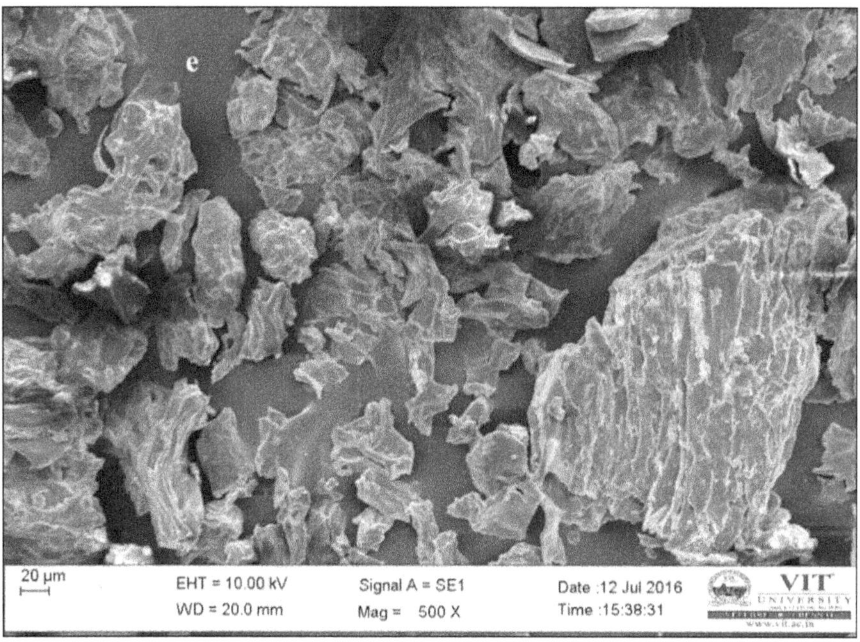

FIGURE 10.5 Scanning electron microscopic images of rough structure of orange (*Citrus aurantium*) pectin.

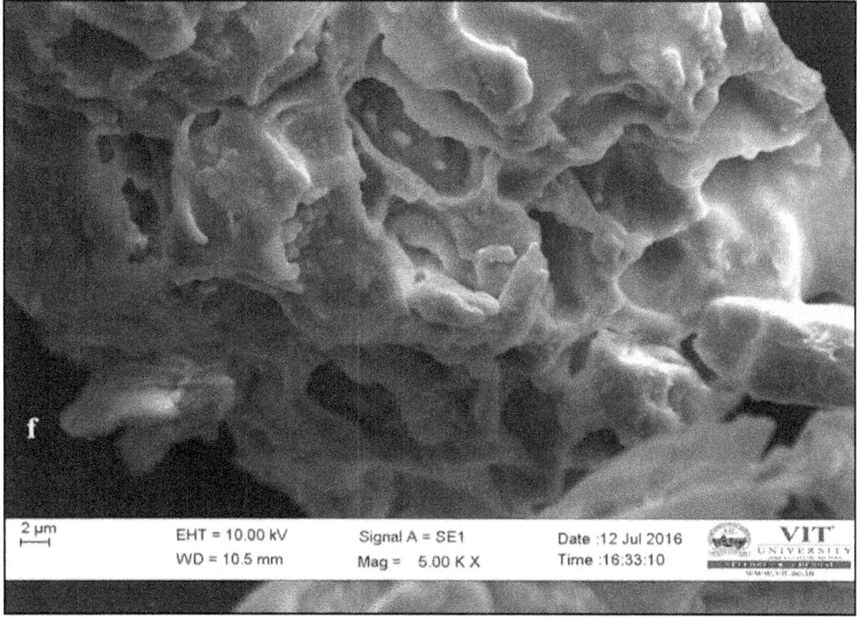

FIGURE 10.6 Heterogeneous structures of orange (*Citrus aurantium*) pectin.

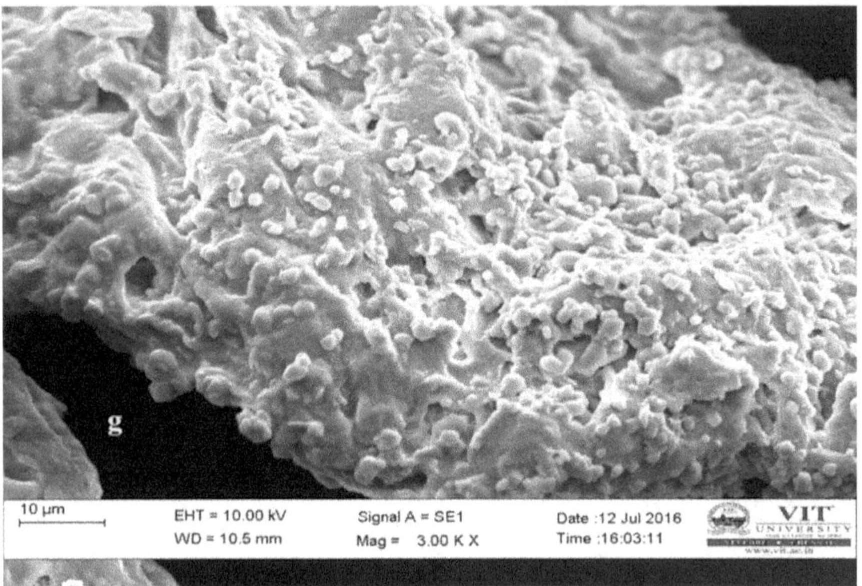

FIGURE 10.7 Scanning electron microscopic images of cuticles of mango (*Mangifera indica*) pectin.

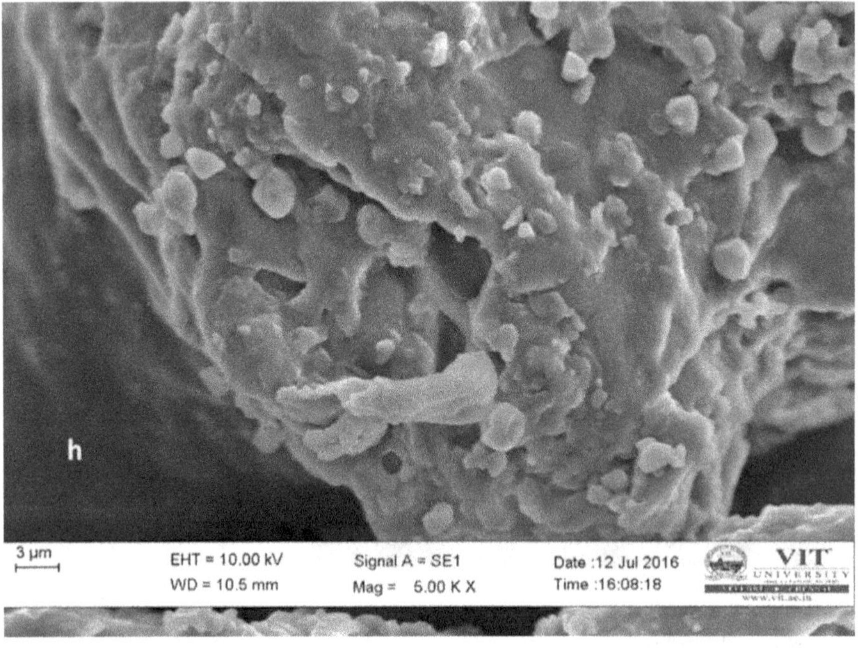

FIGURE 10.8 Cuticular porous structures of mango (*Mangifera indica*) pectin.

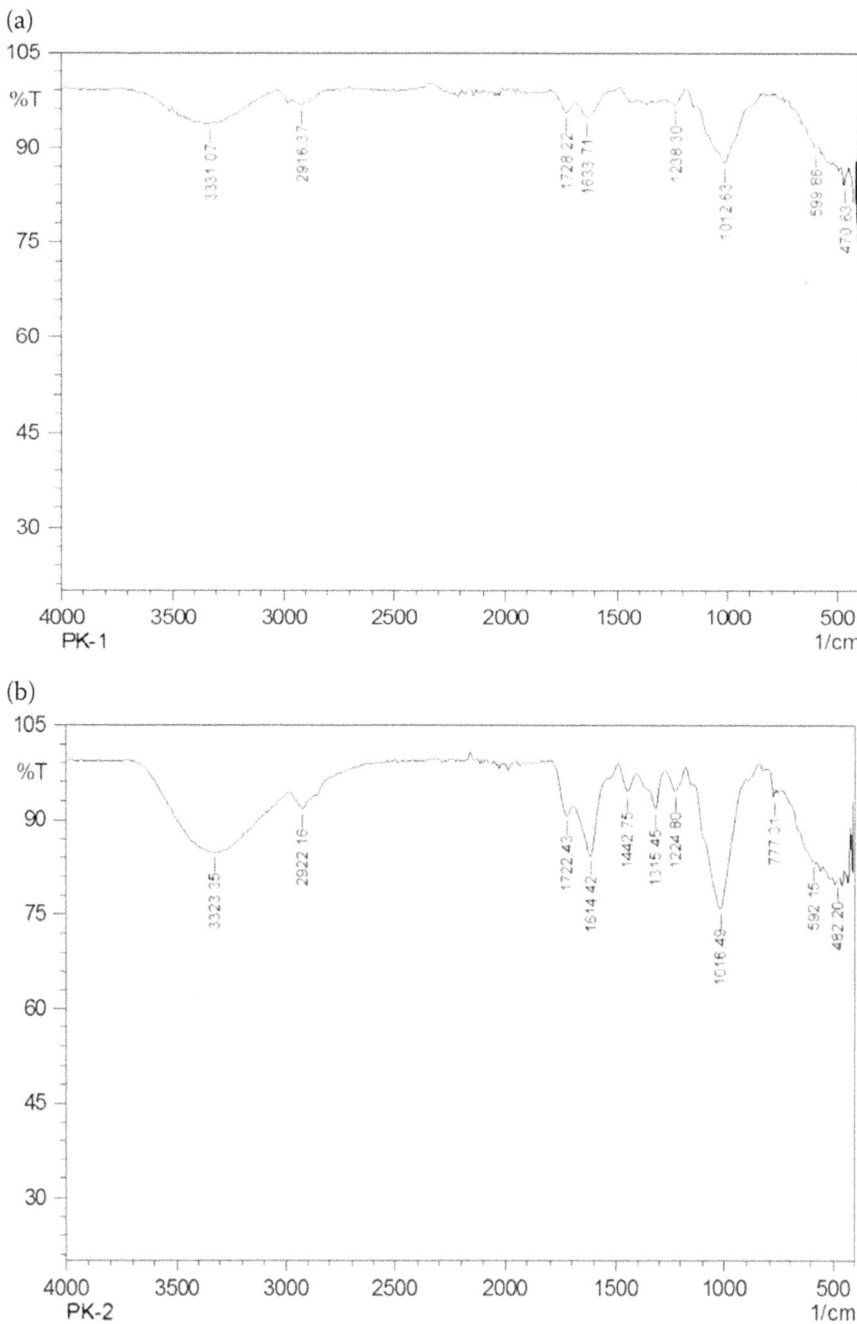

FIGURE 10.9 (a) FTIR spectra of mosambi (*Citrus limetta*) and (b) pomegranate (*Punica granatum*) pectin.

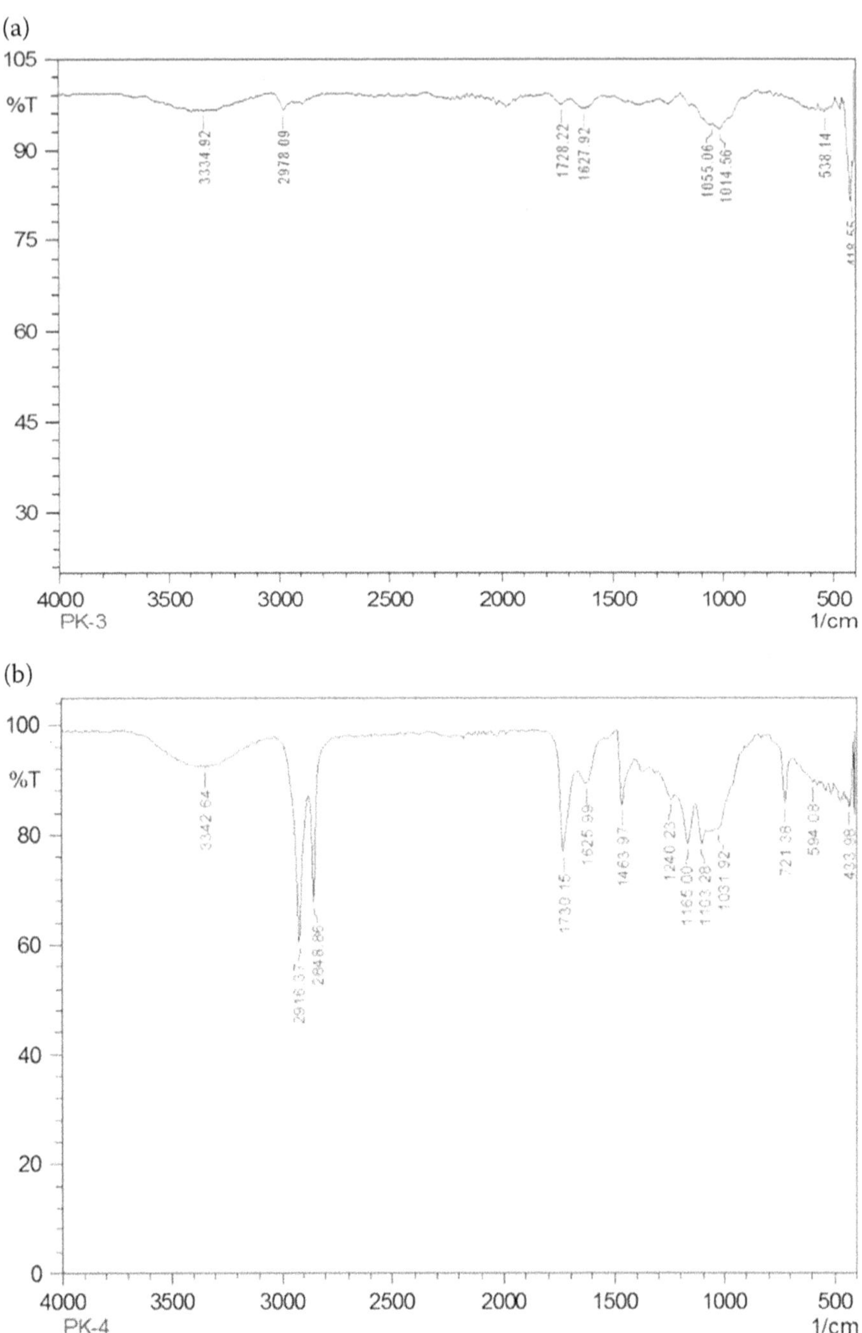

FIGURE 10.10 (a) FTIR spectra of orange (*Citrus aurantium*) and (b) mango (*Mangifera indica*).

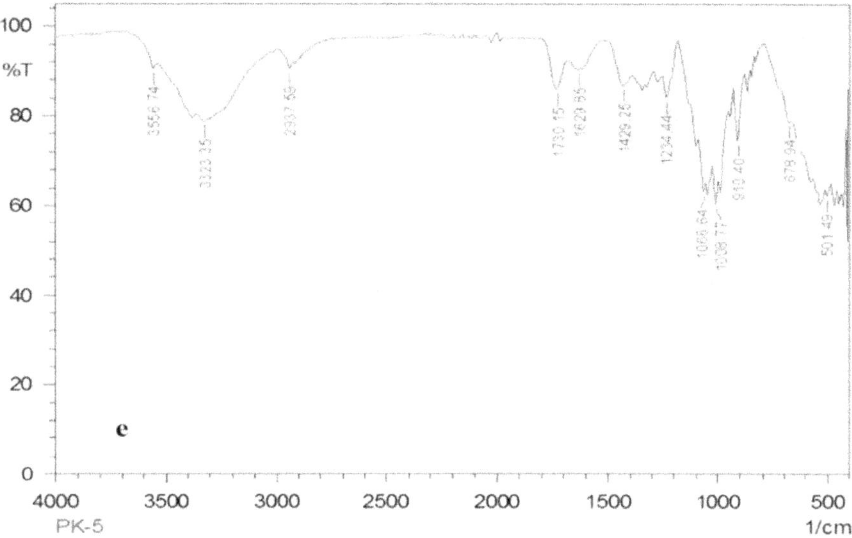

FIGURE 10.11 FTIR spectra of commercial pectin.

10.3.4 *In-vitro* Analysis of Pectins

The in-vitro analysis of different fruit peels was carried out by the AOAC method. The results revealed the proximate composition of lipid, ash, crude fiber, and carbohydrates, respectively, shown in Table 10.2 (AOAC, 1995) (Figure 10.12).

10.3.5 Postconstruction Crack Repair by Isolate VIT-SP4

The postconstruction crack repair of concrete block was carried out with potential isolate *S. fumigatisscleroticus* VIT-SP4. The inoculum of the potential isolate was sprayed on the test concrete material and observed for the self-crack healing mechanism. The results revealed that biomineralized calcite showed more resistance toward dissolution and least solubility compared to inorganic calcite. The viable spores of actinomycetes were visualized by a light microscope after spore staining shown in Figure 10.11. The sporulation % that was observed maximum after 30 h is shown in Figure 10.13. The presence of viable spores that were detected after 90 days is shown in Figure 10.11. The crack was made in test concrete and culture was sprayed on the cracks and it was observed for the healing process shown in Figure 10.13. The crack healing observed up to 0.55 mm wide was recorded as shown in Figure 10.13. However, concrete blocks were immersed in H_2O for 12 days that is comparative to earlier reports (Wiktor and Jonkers, 2011). The sporulation experiment revealed that spores have the capacity to survive in the cement blocks at higher pH by $CaCO_3$ precipitation. The ultrasonic measurement test was carried out for cracked and healed specimens. After, crack repair waves cannot pass through sealed crack leading to a decrease in the transmission time. The results revealed the transmission time of the crack specimen of about 31.05 μs and a

TABLE 10.2
Proximate Composition of Fruit Peels

Fruit	Yield of Fruit Peel (g/100g of Fruit Weight)	Lipid	Crude Fiber	Ash	Carbohydrates
	Proximate Composition (g/100g of dry peel)				
Mosambi	12.17 ± 0.05	8.60 ± 0.25	14.81 ± 0.01	4.29 ± 0.14	51.17 ± 0.10
Pomegranate	11.67 ± 0.03	3.35 ± 0.33	16.53 ± 0.05	6.03 ± 0.07	57.78 ± 1.22
Orange	13.27 ± 0.05	9.70 ± 0.55	15.20 ± 0.01	5.27 ± 0.98	53.80 ± 0.10
Mango	10.94 ± 0.03	4.81 ± 0.55	14.43 ± 0.13	3.34 ± 0.18	65.80 ± 0.16
Apple	9.21 ± 0.03	9.86 ± 0.52	12.85 ± 0.10	15.95 ± 0.10	69.36 ± 0.34

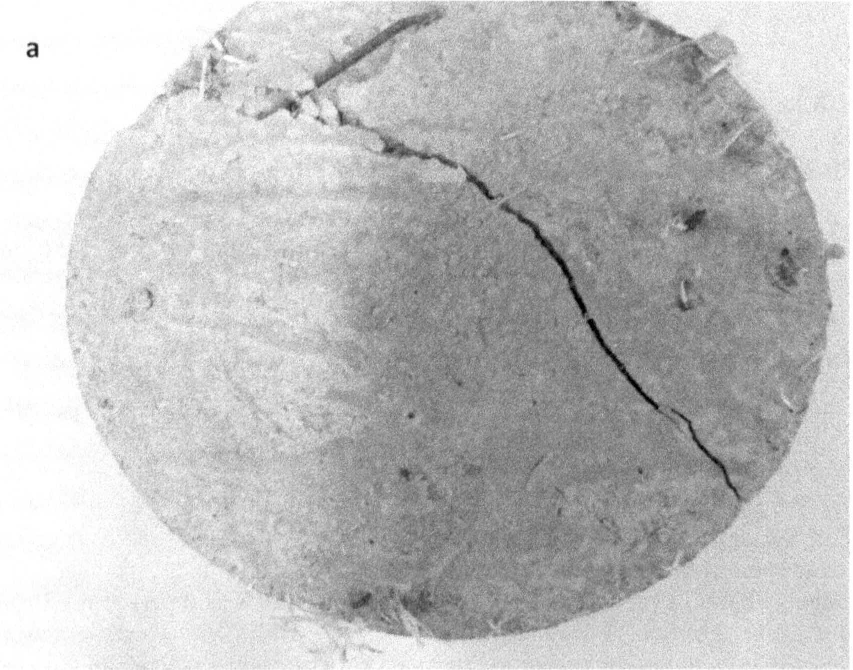

FIGURE 10.12 Cracks on the cement paste.

healed concrete specimen of about 30.07 μs. The ultrasonic treatment test was investigated for both cracked and healed specimens. The results confirm the self-healing of cracks by a decrease in the transmission time of a healed specimen (Van Tittelboom et al., 2010). *Streptomyces* are spore-forming actinomycetes that are resistant to high pH and remained viable after casting the concrete for several days and helps directly in self-healing system.

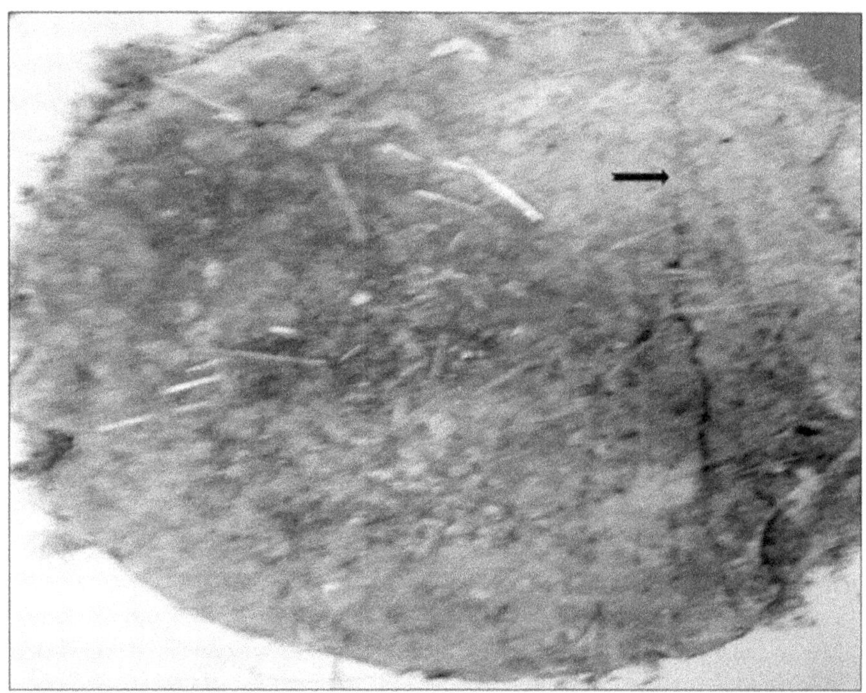

FIGURE 10.13 Healing of crack by spores of *Streptomyces fumigatiscleroticus* VIT-SP4.

10.3.6 Production of Biocement and Survivability of Actinomycete Spores

Actinomycetes germinate spores and they get colonized at the time of harsh conditions in the presence of H_2O and cause $CaCO_3$ precipitation to provide compressive strength and compatibility to the concrete. *S. fumigatiscleroticus* seems to be desirable isolate that can be utilized as crack-healing agent in concrete blocks. Thus, it helps in the production of biocement.

10.3.7 Drug Delivery Studies

The calcium pectinate pellets were prepared and pectinase from *S. fumigatiscleroticus* VIT-SP4 was investigated for controlled drug delivery studies. The calcium pectinate pellets were prepared with test drugs with orange pectin (Maestrelli et al., 2008). It was treated with pectinase from *S. fumigatiscleroticus* VIT-SP4 and it was observed for controlled drug delivery studies. The release of indomethacin drug from pectin by pectinase under different pH was studied. The result revealed that at initial pH (3.0) after some time of 2 h, drug release was absent. When pH was changed to pH 6 after the period of 6 h, rapid drug release was observed. At pH 3, the drug release was stopped. Once pH was changed to pH 6 after 17 h, drug release was observed. The maximum pectinase activity from *S. fumigatiscleroticus* VIT-SP4 was observed at pH 6 shown in Figure 10.14. The main cause for controlled

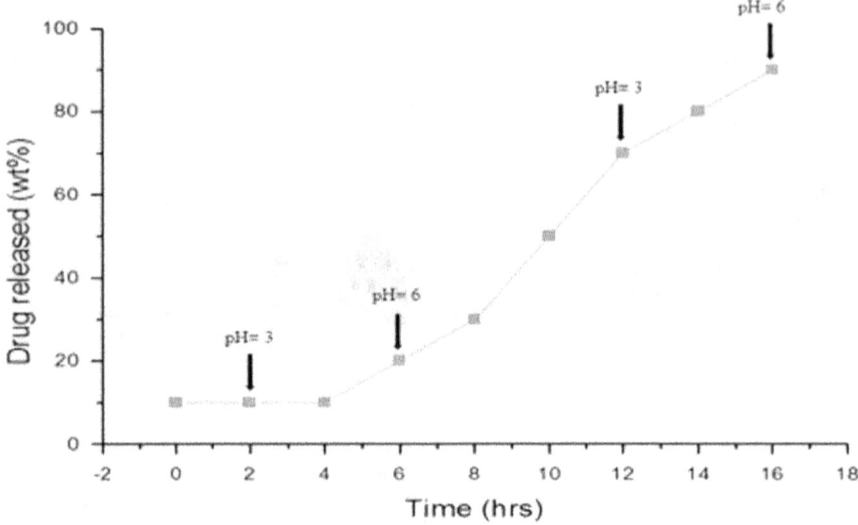

FIGURE 10.14 Controlled drug release from pectin by pectinase.

release of the drug is disintegration of pectin by pectinase enzyme at pH 6, whereas at pH 3 there was no disintegration, which confirms that the disintegration of pectin by pectinase enzyme at specific pH facilitates the controlled drug release as seen in earlier reports (Butte et al., 2014).

10.3.8 *In-vitro* Analysis of Pectin

The *in-vitro* analysis of pectin was carried out by the AOAC method after pectinase activity of isolate VIT-SP4 on different fruit peels. The results showed the presence of moisture protein, ash, carbohydrate, crude fiber, and fat in different fruit peels similar to previous studies. The proximate constituents present in the fruit peels were shown in Table 10.3. The moisture and fat content was observed maximal in mosambi peels followed by orange, pomegranate, and mango. The protein and ash

TABLE 10.3
Proximate Constituents of Fruit Peels after Pectinase Treatment

Proximate Constituents	Mosambi	Pomegranate	Orange	Mango
Moisture	10.65	4.10	9.80	3.2
Protein	5.27	8.52	16.20	3.4
Ash	3.28	5.13	5.41	2.7
Carbohydrate	38.25	31.26	40.37	87.10
Crude fiber	17.48	22.00	12.45	24.55
Fat	15.2	9.20	14.16	2.40

FIGURE 10.15 *Zinnia elegans* plant growth after a time period of 2 days.

FIGURE 10.16 Plant growth (pectinase enzyme) of seeds sown in pot after 1 week.

FIGURE 10.17 *Zinnia elegans* plant growth after 2 weeks time period.

content was found to be maximum in orange peel followed by pomegranate, mosambi, and mango. The crude fiber content was observed to be higher in mango peel followed by pomegranate, mosambi, and orange. The carbohydrate content was higher in mango followed by orange, mosambi, and pomegranate.

10.3.9 Pectinase as a Promoter for Plant Growth

The pectinase enzyme was added aseptically to the soil along with *Z. elegans* plant seeds that were sown in pots. The seeds sown without pectinase enzyme were taken as control and in the zinnia plants were grown in normal environmental condition; the root, shoot, and leaf length were measured after 2 weeks. The growth of the plant after 2 days was shown in Figure 10.15. The plant growth after 1 week was observed and it was shown in Figure 10.16. The plant growth was observed for both control and test after 2 weeks of the time period shown in Figure 10.17.

Then, both *Z. elegans* plants were measured for root, shoot, and leaf length. The result revealed that pectinase promotes the growth of root length of about 4.8 cm, shoot length of 9 cm, leaf length of the zinnia plant of about 3.8 cm, and it was maximum compared to control after a time period of 2 weeks.

ACKNOWLEDGEMENT

The authors want to express their gratitude to our honorable Chancellor, Dr. G. Viswanathan, Dr. Sekar Viswanathan, Mr. Sankar Viswanathan, and Mr. G.V. Selvam of VIT University for their constant encouragement and laboratory facilities from VIT University, Vellore, India to carry out this valuable work, DST, India for young scientist fund to carry out this research and University of Florida for helping in the analysis of value-added compounds. Special thanks to

My best brother Mr Raj Vuppu, Senior Vice President Citi Bank, USA for his constant support and encouragement.

REFERENCES

Abdi, M.R. and Mirzaeifar, H. (2017), 'Experimental and PIV evaluation of grain size and distribution on soil–geogrid interactions in pullout test', *Soils and Foundations* **57**(6), 1045–1058.

AOAC (1995), 'Official methods of analysis of AOAC International.

Arias, B. Á. and Ramón-Laca, L. (2005), 'Pharmacological properties of citrus and their ancient and medieval uses in the Mediterranean region', *Journal of Ethnopharmacology* **97**, 89–95.

Baker, R.A. and Wicker, L. (1996), 'Current and potential applications of enzyme infusion in the food industry', *Trends in Food Science and Technology* **7**(9), 279–284.

Blanco, P., Sieiro, C. and Villa, T.G. (1999), 'Production of pectic enzymes in yeasts', *FEMS Microbiology Letters* **175**(1), 1–9.

Blumich, B. (2005), '*Essential NMR: For scientists and engineers*', Springer publications.

Butte, K., Momin, M. and Deshmukh, H. (2014), 'Optimisation and in vivo evaluation of pectin based drug delivery system containing curcumin for colon', *International Journal of Biomaterials* **1**, 1–7.

Caffall, K.H. and Mohnen, D. (2009), 'The structure, function, and biosynthesis of plant cell wall pectic polysaccharides', *Carbohydrate Research* **344**, 1879–1900.

Casarotti, S.N., Borgonovi, T.F., Batista, C.L. and Penna, A.L.B. (2018), 'Guava, orange and passion fruit by-products: Characterization and its impacts on kinetics of acidification and properties of probiotic fermented products', *LWT* **98**, 69–76.

Chatjigakis, A.K, Pappas, C., Proxenia, N., Kalantzi, O., Rodis, P. and Polissiou, M (1998), 'FT-IR spectroscopic determination of the degree of esterification of cell wall pectins from stored peaches and correlation to textural changes', *Carbohydrate Polymers* **37**, 395–408.

Cerna, M., Barros, A.S., Nunes, A., Rocha, S.M., Delgadillo, I., Copıikova J. and Coimbra, M.A. (2003), 'Use of FT-IR spectroscopy as a tool for the analysis of polysaccharide food additives', *Carbohydrate Polymers* **51**(4), 383–389.

Das, S.K. and Varma, A. (2010), '*Soil enzymology*', Springer Publications.

Garg, G., Singh, A., Kaur, A., Singh, R., Kaur, J. and Mahajan, R. (2016), 'Microbial pectinases: An ecofriendly tool of nature for industries'. *3 Biotech* **6**, 1–13.

Garau, M.C., Simal, S., Rossello, C. and Femenia, A. (2007), 'Effect of air-drying temperature on physico-chemical properties of dietary fibre and antioxidant capacity of orange (*Citrus aurantium v. Canoneta*) by-products', *Food Chemistry* **104**(3), 1014–1024.

Govindaraji, P. K. and Vuppu, S. (December 2020). 'Characterisation of pectin and optimization of pectinase enzyme from novel *Streptomyces fumigatiscleroticus* VIT-SP4 for drug delivery and concrete crack-healing applications: An eco-friendly approach Saudi', *Journal of Biological Sciences* **27** (12), 3529–3540

Gummadi, S. N. and Panda, T. (2003), 'Purification and biochemical properties of microbial pectinases—a review'. *Process Biochemistry*, **38**, 987–996.

Guo, C., Yang, J., Wei, J., Li, Y., Xu, J. and Jiang, Y. (2003), 'Antioxidant activities of peel, pulp and seed fractions of common fruits as determined by FRAP assay'. *Nutrition Research* **23**, 1719–1726.

Jacob, N., Asha Poorna, C. and Prema, P. (2008), 'Purification and partial characterization of polygalacturonase from *Streptomyces lydicus*', *Bioresource Technology* **99**(14), 6697–6701.

Jayani, R. S., Saxena, S. and Gupta, R. (2005), 'Microbial pectinolytic enzymes: A review', *Process Biochemistry* **40** (9), 2931–2944.

Joshi, V. K., Parmar, M. and Rana, N. S. (2006), 'Pectin esterase production from apple pomace in solid-state and submerged fermentations', *Food Technology and Biotechnology* **44**(2), 253–256.

Kashyap, D.R., Vohra, P.K., Chopra, S. and Tewari, R. (2001). 'Applications of pectinases in the commercial sector: A review', *Bioresource Technology* **77**, 215–227.

Khaliq, S., Akhtar, K., Ghauri, M.A., Iqbal, R., Khalid, A.M. and Muddassar, M. (2009), 'Change in colony morphology and kinetics of tylosin production after UV and gamma irradiation mutagenesis of *Streptomyces fradiae* NRRL-2702', *Microbiological Research* **164**(4), 469–477.

Kumar, P. and Suneetha, V. (2017), 'Screening, biochemical and molecular identification of novel *Streptomyces fumigatiscleroticus* VIT-SP4 derived cocktail pectinase from border industrial pectin enriched places of Andhra Pradesh and Tamil Nadu', *Research Journal of Biotechnology* **12**, 22–29.

Latarullo, M. B. G., Tavares, E. Q. P., Padilla, G., Leite, D. C. C. and Buckeridge, M.S S. (2016), 'Pectins, endopolygalacturonases, and bioenergy' *Frontiers in Plant Science* **7**.

Liew, S. Q., Chin, N. L., and Yusof, Y. A. (2014), 'Extraction and characterization of pectin from passion fruit peels', *Agriculture and Agricultural Science Procedia* **2**, 231–236.

Luo, M., Qian, C.X. and Li, R.Y. (2015), 'Factors affecting crack repairing capacity of bacteria-based self-healing concrete', *Construction and Building Materials* **87**, 1–7.

Maran, J.P., Sivakumar, V., Thirugnanasambandham, K., and Sridhar, R. (2013). 'Optimization of microwave assisted extraction of pectin from orange peel', *Carbohydrate Polymers* **97**, 703–709.

Maestrelli, F., Cirri, M., Corti, G., Mennini, N. and Mura, P. (2008), 'Development of enteric-coated calcium pectinate microspheres intended for colonic drug delivery', *European Journal of Pharmaceutics and Biopharmaceutics* **69** (2), 508–518.

Martínez-Trujillo, A., Aranda, J. S., Gómez-Sánchez, C., Trejo-Aguilar, B. and Aguilar-Osorio, G. (2009), 'Constitutive and inducible pectinolytic enzymes from Aspergillus flavipes FP-500 and their modulation by pH and carbon source', *Brazilian Journal of Microbiology* **40**, 40–47.

Neill, O.M., Albersheim, P., and Darvill, A. (1990), 'The pectic polysaccharides of primary cell walls', *Methods in Plant Biochemistry, Carbohydrates*, **2**, 415–441.

Patidar, M. K., Nighojkar, S., Kumar, A., & Nighojkar, A. (2018), 'Pectinolytic enzymes-solid state fermentation, assay methods and applications in fruit juice industries: a review', *3 Biotech* **8**, 199–203.

Praveen, K.G. and Suneetha, V. (2014a), 'A cocktail enzyme – pectinase from fruit industrial dump sites: A review', *Research Journal of Pharmaceutical, Biological and Chemical Sciences* **5** (2), 1252–1258.

Praveen, K.G. and Suneetha, V. (2014b), 'Natural, culinary Fruit peels as a potential substrate for pectinolytic enzyme', *International Journal of Drug Development and Research* **6**(3), 109–118.

Praveen, K.G. and Suneetha, V. (2015a), 'Efficacy of Pectinase purified from Bacillus VIT sun-2 and in combination with xylanase and cellulase for the yield and clarification improvement of various culinary juices from South India for pharma and health Benefits', *International Journal of PharmTech Research* **7** (3), 448–452.

Praveen, K.G. and Suneetha, V. (2015b), 'Pectinases from actinomycetes: A through study', *International Journal of Chemtech Research* **8** (7), 445–450.

Ramírez-Tapias, Y.A., Lapasset Laumann, A.S., Britos, C.N., Rivero, C.W., and Trelles, J.A. (2017), 'Saccharification of citrus wastes by immobilized polygalacturonase in an improved alginate matrix', *3 Biotech* **7**, 1–16.

Ridley, B.L., O'Neill, M.A. and Mohnen, D. (2001), 'Pectins: structure, biosynthesis, and oligogalacturonide-related signaling', *Phytochemistry* **57**, 929–967.

Shankar, S. and Suneetha, V. (2019), '*In vitro* drug metabolism and pharmacokinetics of a novel thiazolidinedione derivative, a potential anticancer compound', *Journal of Pharmaceutical and Biomedical Analysis* **113000**, 1–17.

Shankar, S, Uppal, R. and Vuppu, S. (2019), 'Validation of a sensitive simultaneous LC-MS/MS method for the quantification of novel anti-cancer thiazolidinedione and quinazolin-4-one derivatives in rat plasma and its application in a rat pharmacokinetic study', *Journal of Chromatography. B, Analytical Technologies in the Biomedical and Life Sciences* **1121**, 18–27.

Suneetha, V (2010a) Screening, Characterisation and Optimization of Microbial Pectinase, Soil Enzymology, Soil Biology-22, G. Shukla and A. Varma (eds.), Springer-Verlag, Berlin, Heidelberg, 3, 329–337.

Suneetha, V (2010b). Actinomycetes: Sources for Soil Enzymes Soil Enzymology, Soil Biology-22, G. Shukla and A. Varma (eds.) Springer-Verlag, Berlin, Heidelberg, 3, 259–269.

Suneetha, V. (2014). 'High performance liquid chromatography analysis, production and brief comparative study of citric acid producing microorganisms from spoiled onions in and around Vellore district', *Iranian Journal of Science and Technology: Transaction A* **38**(A2), 193–197. TR impact factor-0.733

Suneetha, V. and Praveen Kumar, G. (2017), 'Kumkuma the social and religious marking of India is used as stain to improve the contrast of pectinolytic actinomycetes VIT S2', *Research Journal of Pharmacy and Technology* **10** (5), 1297–1300.

Van Tittelboom, K., De Belie, N., De Muynck, W., and Verstraete, W. (2010), 'Use of bacteria to repair cracks in concrete', *Cement and Concrete Research* **40**, 157–166.

Wiktor, V. and Jonkers, H.M. (2011), 'Quantification of crack-healing in novel bacteria-based self-healing concrete', *Cement and Concrete Composites* **33**, 763–770.

Yapo, B.M. (2011), 'Pectic substances: From simple pectic polysaccharides to complex pectins—A new hypothetical model', *Carbohydrate Polymers* **86**, 373–385.

Yeoh, S., Shi, J. and Langrish, T.A.G. (2008), 'Comparisons between different techniques for water-based extraction of pectin from orange peels', *Desalination* 218, 229–237.

11 Microbial Recycling of Feather Waste into Value-added Products

Suneetha Vuppu
School of Biosciences and Technology, VIT University, Vellore, Tamil Nadu, India

Arjun Chinamgari
Department of Industrial Engineering University of Florida, Gainesville, FL, USA

CONTENTS

DOI: 10.1201/9781003128977-11

11.1 INTRODUCTION

Enzymes have a wide work in industries for a variety of purposes. Microbial kerazyme offers a substantial advantage of degrading keratin waste as it is present in large amounts and easily available in the soil, hair, nails, feathers, and poultry waste (1). Kerazyme is an extracellular enzyme secreted by many micro-organisms. In our study, feather degrading bacteria were isolated from poultry waste soil and feathers (1,2). The keratinolytic activity was measured in units from culture filtrate of three isolates obtained. Strain C showed maximum keratinolytic activity of units/mL after 124 h of cultivation. Maximum growth of bacterial colonies in the feather medium was observed during the log phase.

The present modern world results the huge amount of poultry waste generated throughout the world due to industrial revolution, over population, and urbanization. The environmental burden due to such wastes is increasing at an alarming rate and thus there is a need for bioremediation (3–5). Previous treatment methods

FIGURE 11.1 Recycling of feather waste (Suneetha 2015).

include land filling. But land filling poses problems of secondary pollutants such as landfill leachate, greenhouse gases, and odour. The *Bacillus* spp. was screened from dumped poultry waste soils, sewage bed, river bed, and pond bed of different locations at Vellore and Chittoor and were identified on the basis of morphological, physiological, and biochemical characteristics. These bacilli species have the capability of degrading feathers by the release of an extracellular enzyme that is referred to as 'keratinase' (6–8) (Figure 11.1).

11.2 METHODOLOGICAL OVERVIEW

11.2.1 Collection and Processing of Enriched Soil

The waste feathers and soil sample required for isolation of bacteria were collected from a butcher shop near Chittor bus stand in Vellore. The feathers were washed properly three to four times with tap water then with distilled water, and allowed to dry in the sunlightfor 1 day. After drying the feathers were pretreated with chloroform:ethanol (3:1) and kept suspended for 2 days. Pretreated feathers were kept for sun drying for 2 days after which they were cut into very minute pieces. The media preparations were obtained from the labs of School of Biosciences and Technology, VIT University (Figure 11.2).

11.2.2 Feather Medium Preparation

The feather meal broth used for isolation, maintenance, and growth of bacteria was made by adding 0.5 g of NH4Cl, 0.5 g of NaCl, 0.3 g of K_2HPO_4, 0.4 g of KH_2PO_4,

FIGURE 11.2 Enriched soil sample source collection for potential micro-organisms.

FIGURE 11.3 Samples were collected in Vellore poultry and feathers are pretreated for screening of potential microbes.

0.1 g of $MgCl{\cdot}6H_2O$, 0.1 g of yeast extract, and 10 g of cut and milled chicken feathers. The pH was adjusted to 7.2 and volume was made up to 250 mL. Feather agar plate was prepared by adding 7 g of feather agar medium prepared in the VIT SBST lab and 2 g of agar–agar in 100 mL of distilled water. The broth and agar medium were sterilized at 121°C for 20 min (5,6,9) (Figure 11.3).

11.2.3 Production Medium Preparation

The production medium was prepared for different salts and added feather. From that petri plates colonies were inoculated to production medium (Figure 11.4).

11.2.4 Screening and Identification of Micro-organisms

The soil samples collected from the butcher shop in Vellore were diluted serially and 10^{-2} samples were used for plating. Plating was done using pour plate and spread plate techniques to obtain pure cultures. The plates were incubated at 35°C for 5 days. Distinct colonies were counted using colony counter. The colonies obtained were inoculated in fresh feather medium and subcultured to obtain sufficient quantity of individual colonies. Three different strains of bacteria were

FIGURE 11.4 Production medium.

obtained using gram staining. Biochemical tests were performed to test the presence of bacterial strains.

11.2.4.1 Micro-organism

Microbes were isolated from the soil where hairs and birds' feathers were dumped. These isolates were identified as actinomycetes species by doing different biochemical tests and further analyzed by analytical instruments (8,10,11).

11.2.4.2 Growth Conditions

The microbes were grown at an optimum condition using nutrient agar (HI-VEG MEDIA). The strain was cultivated at room temperature in salt medium. Birds' feathers and hairs were used as carbon and nitrogen source.

11.2.4.3 Feather Collection and Pretreatment of Feather

Feathers and hairs were collected from the poultry farm and tonsuring areas in and around Vellore shop, respectively. The feathers were rinsed with water thrice followed by drying in the sun light for 24 h. Thereafter, the feathers were cut into smaller pieces of 2–3 cm and treated with the methanol and chloroform in the ratio of 3:1 to make the keratin substrate more soluble and vulnerable that keratinase producing microbes can easily degrade. The obtained feathers were kept for 18 h of incubation period. Similar process was repeated several times to enrich the substrate used in the experiment.

11.2.4.4 Collection of Soil Sample

Two soil samples were collected from poultry and tonsuring areas in and around Vellore. These samples were tested for the presence of micro-organisms having the ability to grow and show keratinase activity. Tenfold serial dilution was made by mixing 1 g of sample in the first test tube and then transferred 1 mL to the second test tube and so on.

11.2.4.5 Enzyme Assays

The keratinolytic activity was measured using feather as a substrate (Anson 1938) and keratin as a substrate. The reaction mixture contains 5 mg of substrate and 0.8 mL of Tris-HCl buffer (pH = 7.5) and 0.2 mL of culture filtrates and it was incubated for 30 min at 50°C. The reaction was stopped by adding 1 mL, 0.1 M Tris-cetic acid (TCA). The mixture was filtered with the Whatman paper and 0.5 mL of filtrate added with 1 mL of ninhydrin. The filtrate was measured at 520 nm for liberation of amino acid. The quantity was determined from standard tyrosine solution (50–500 μg/mL).

11.2.4.6 Protein Determination

2 mL of reaction mixture was composed of 1 mL of 2% casein in 0.2 M of Tris-HCL, pH7.5 and 1 mL of appropriately diluted enzyme. The reaction mixture was incubated for 30 min in the water bath at 37°C. 2 mL of 20% TCA was then added and centrifuged at 5000 rpm for 15 min. Enzyme was added to process the control

after incubation and TCA was immediately added. The protein content of enzyme was determined by using Lowry et al. (1951).

11.2.4.7 Protease Activity

Protease activity was determined using casein as substrate according to the method reported by Gessesse et al. (2003). The reaction mixture in a total volume of 2 mL was composed of 1 mL of 2% casein in 0.2 M of Tris-HCl, pH7.5 and 1 mL of appropriately diluted enzyme. The reaction mixture was incubated at 40°C for 30 min in water bath. 20% TCA of 2 mL was added and the mixture was centrifuged at 5000 rpm for 15 min. A control was processed by adding enzyme to the mixture after incubation and TCA was immediately added.

11.2.5 Characterization of Physiological Properties of Isolated Micro-organisms for Recycling of Feather Waste

11.2.5.1 Effect of pH

Modified Bennett broth was sterilized, prepared, and pH adjusted to 5, 7, and 8.5 using 0.1 N HCL and NaOH. The tubes were incubated at 37°C for 7 and 14 days after incubation growth was recorded (Ivanko and Varvanets, 2004).

11.2.5.2 Effect of Temperature

Modified Bennett broth was prepared and sterilized and the actinomycetes culture were incubated in the broth. The tubes were incubated at 37 and 40°C for 7 to 14 days, then 4 and 10°C, respectively. After incubation growth was recorded. The pH and temperature were 7.9 and 43°C (Igantova, 1999).

11.2.6 Determination of Growth Characterstics

The culture innoculums obtained after 5 days of incubation were transferred in side arm flasks and kept in an orbital shaker at 100 rpm. Periodically O.D. values at 540 nm were calculated using calorimeter for a duration of 48 h.

11.2.7 Identification of Bacterial Strain

The potential bacterial isolate was identified using staining and biochemical characterization.

11.2.8 Staining

The refractive index of bacteria is nearly same as water, when the bacterial strain observed under microscope, it will be invisible to naked eye. Hence, stains were used to observe the bacterial morphology.

11.2.9 Gram Staining

In Gram staining, three major steps were followed such as smear preparation, staining, and decolourization.

Smear preparation: smear preparation was followed as described in simple staining.

Staining: in the smear crystal violet was added and it was kept undisturbed for 60 s. Then the slide was washed with distilled water. Gram's iodine was added to the smear and allowed to act for 60 s, then Gram's iodine was washed with distilled water. Decolorizer was added from the top of the slide until the dye color disappears. Then, counter stain was added to the smear and incubated for 30 s, then it was washed with tap water. The slide was dried with blotting paper and observed under microscope by high power oil immersion.

11.2.10 Motility Test

The fresh bacterial culture was taken and placed in the center of cover slip where all the corners of the cover slip have Vaseline. Depression microscope slide is placed on the cover slip and gentle press was made. Then the slide was inverted so the culture was hanging on the cover slip. The slide was observed under microscope.

11.2.11 Biochemical Characterization

Biochemical characterizations and identification of genus and species were achieved.

11.2.12 Catalase Test

Using a sterile loop, culture was taken and placed in the center of slide. Few drops of hydrogen peroxide were added to the culture using a fresh dropper. Result was observed for 1–2 min.

11.2.13 Oxidase Test

A filter paper is placed in sterile Petri plate, using a swab bacterial culture placed in the filter paper. To the culture few drops of tetra methyl-p-phenylenediamine are added with the help of sterile dropper. Results were observed within few seconds.

11.2.14 Citrate Test

In a sterile test tube, Simmons citrate agar was added and kept in a slant position until the agar got solidify. Using a cotton swab the pure bacterial culture was streak in the slant. The slant was incubated at 37°C for 24 h. Result was observed.

11.2.15 Starch Hydrolysis Test

Pure bacterial culture was inoculated in a starch plate and the plate was incubated at 39°C for 48 h. After, the incubation iodine was flooded in the plate. Result was observed in few minutes.

11.2.16 Indole Test

In the sterile test tube, nitrate broth was added; to this sterile broth, pure bacterial culture was inoculated and incubated for 48 h at 37°C. To the broth, few drops of sulfanilic acid and α-naphthylamine were added. Result was observed during the addition of α-naphthylamine.

11.2.17 Lactose Test

Phenol red lactose broth was transferred to sterile test tube and inoculated with the bacterial culture. The culture was incubated at 37°C for 24 h, then the results were observed. The yellow color shows the fermentation of lactose.

11.2.18 Methyl Red

Pure culture was inoculated in the MR-VP broth by using a sterile loop. The broth was incubated for 48 h at 37°C. To the broth, five drops of methyl red was added and the result was observed without mixing.

11.2.19 H_2S Formation

Pure bacterial culture was transformed aseptically to TSI agar slant. The culture was incubated for 24 h at 37°C. After incubation period the slant was observed for color change.

11.2.20 Optimization of Keratinase Enzyme

11.2.20.1 Effect of Various Incubation Time on Keratinase Production

Feather minimal medium was prepared and incubated at various time intervals (24, 48, 72, and 96 h). 0.0.5% inoculums from 24 h Luria broth of the culture was inoculated in 100 mL of basal feather medium and kept in shaker with 120 rpm at 37°C for 24 hrs.

11.2.20.2 Effect on pH on Keratinase Production

Feather minimal medium prepared at different PH (2, 4, 6, 8, and 10). 0.5% inoculums from 24 h Luria broth of the culture was inoculated in 100 mL of basal feather medium and kept in a shaker with 120 rpm for 24 h.

11.2.20.3 Effect of Temperature on Keratinase Production

Feather minimal medium was prepared at different temperatures (20, 30, 40, 50, and 60). 0.5% inoculums from 24 h nutrient broth of the culture was inoculated in 100 mL of basal feather medium at 80°C and kept in a shaker with 120 rpm for 24 h.

11.2.21 Immobilization of Keratinase Enzyme

11.2.21.1 Production of Beads Formation

4 g of calcium chloride was weighed and dissolved in 100 mL of distilled water and kept at 4°C for 2 h. To prepare 0.1 N NaCl sodium alginate for that weighed 0.685 g of sodium chloride and 3.5 g sodium alginate was dissolved in 100 mL of distilled water. In a beaker, 20 mL of sodium alginate is taken. To prepare 3 mL of enzyme filtrate is added in a beaker containing sodium alginate. Hence sodium alginate solutions are prepared. Now $CaCl_2$ is taken into a beaker. Then using a dropper, sodium alginate solution of crude enzyme is added drop by drop into beaker containing $CaCl_2$ so that beads are formed. Beads are dissolved in phosphate buffer and O.D. is taken at 24 h interval for 5 days at 280 nm.

11.2.21.2 Keratinolytic Activity Assay

The isolate of different strains was cultivated for 24 h in a feather medium. The culture medium was centrifuged at 10,000×*g* for 10 min, and the supernatant was used as enzyme preparation. The culture filtrate was used as enzyme source where 1 mL of the solution was added to 3.8 mL of 0.1 mol/L tris buffer (pH 7.7) containing feather. The solution was incubated for 30 min at 55°C. The reaction was stopped by adding 1 mL of TCA and placing the test tubes in ice. The amino acids liberated were measured at 280 nm in a UV spectrophotometer against a blank. The amount of enzyme was calculated using amino acid as standard which was prepared by ninhydrin method.

11.3 RESULTS AND DISCUSSION

11.3.1 Isolation of Feather Degrading Bacteria

Three distinct colonies of bacteria were obtained and were further subcultured for obtaining optimum quantity for further analysis. Strains A, B, and C degraded feathers and had 113, 105, and 127 units of keratinolytic activities, respectively. Strain 3 was found to have maximum keratino lytic activity at optimum temperature and pH conditions. Keratin degrading bacteria use keratin as a chief carbon source for growth maintenance. Keratin-degrading micro-organisms thrive under different ecological and environmental conditions and are known to have the capacity to solubilize keratinous substrates (12). Gram staining gave positive staining, thus confirming the presence of gram positive bacterial species. Bacteria were found to be positive for catalase test.

11.4 GROWTH CHARACTERISTICS AND KERATINOLYTIC ACTIVITY OF DIFFERENT BACTERIAL STRAINS

Strain C was shown to have maximum growth. Maximum enzyme activities were observed late in the logarithmic phase or at the beginning of the stationary phase for all three isolates. Enzyme activities remained with little changes in the late logarithmic growth phase or at the beginning of the stationary phase (Figures 11.5 and 11.6).

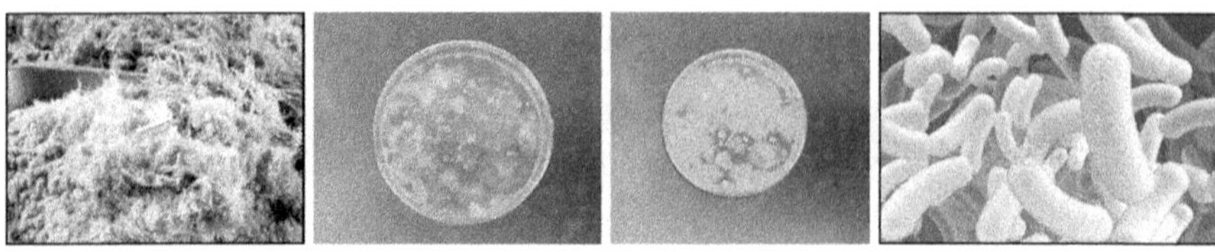

FIGURE 11.5 Degradation patterns and *Bacillus* sp.

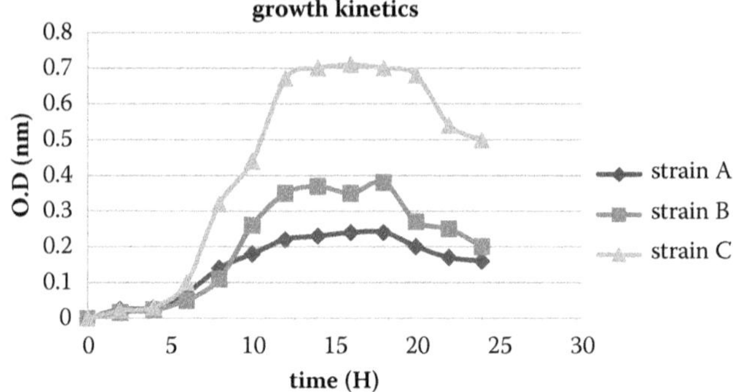

FIGURE 11.6 Physiological characterization.

11.4.1 Isolation of Microorganisms

Keratinase enzyme has been isolated from poultry soil and feather. The keratinase enzyme is isolated from the Baiting method. Prepared production medium in conical and inoculation samples was taken from streaking plates and kept in a shaker for 1 day. Soil organisms inoculated the production medium by degrading the large amount of enzymes which were present in that medium (Figures 11.7 and 11.8).

FIGURE 11.7 Inoculum with bacteria from broiler chicken feathers.

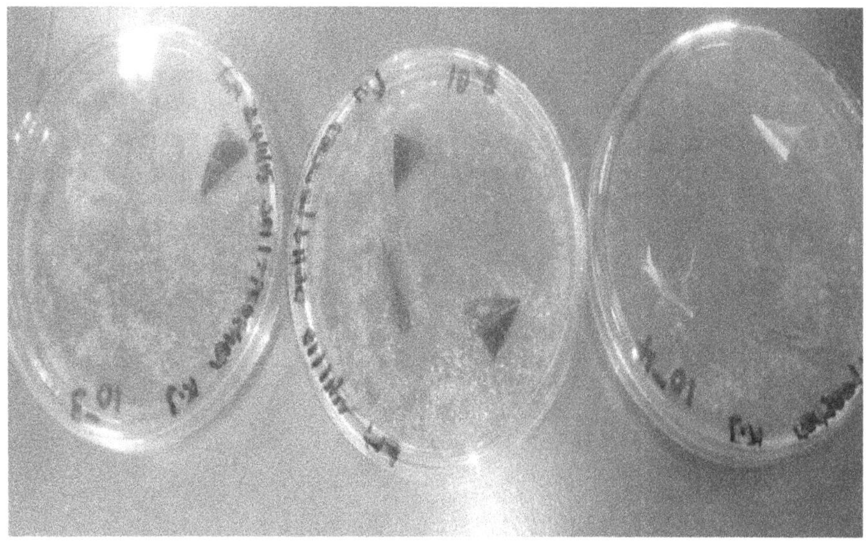

FIGURE 11.8 Baiting technique.

11.5 FEATHER DEGRADATION

The feathers were kept for feather degradation for a span of 7 days. It was viewed under the microscope before degradation, after 2 days and after 5 days.

11.5.1 STREAKING METHOD

Streaki plate technique is followed to isolate potential bacteria after screening the colonies and inoculated in production media for further analysis of keratinase enzyme (Figure 11.9).

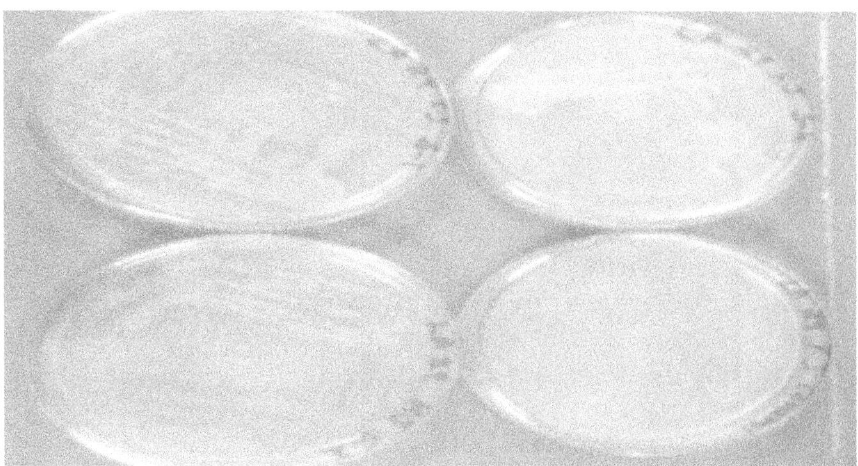

FIGURE 11.9 Streaking plates.

11.5.2 Screening of Keratinolytic Bacteria

Prepared production medium in conical and inoculation sample was taken from streaking plates and then they were kept in an orbital shaker system for 24 h. The production medium was inoculated by soil microorganism after degrading the large amount of enzymes which were present in that medium (Figures 11.10 and 11.11).

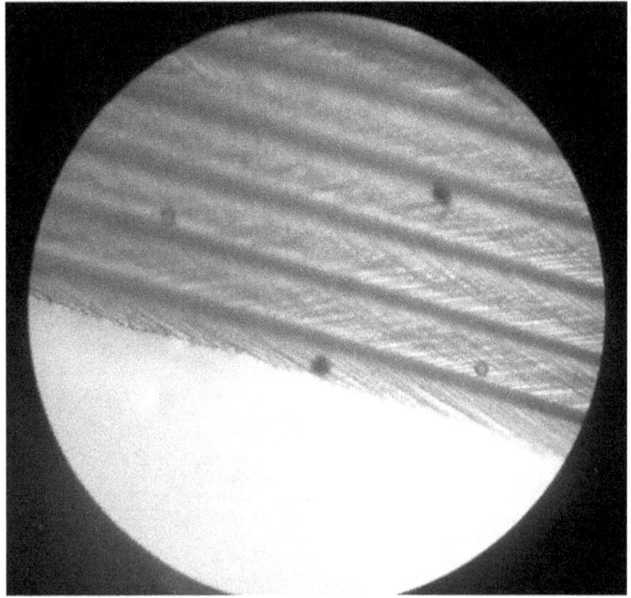

FIGURE 11.10 Feather viewed under the microscope before degradation.

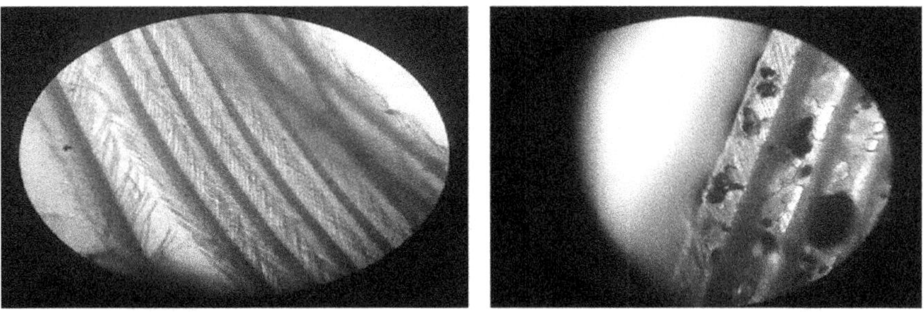

FIGURE 11.11 Degradation of feather after 2 days (broiler chicken).

11.5.3 Staining

Gram staining, spore staining, and hanging drop technique were performed to identify the keratinase producing bacteria (Figures 11.12–11.15) (Tables 11.1–11.3).

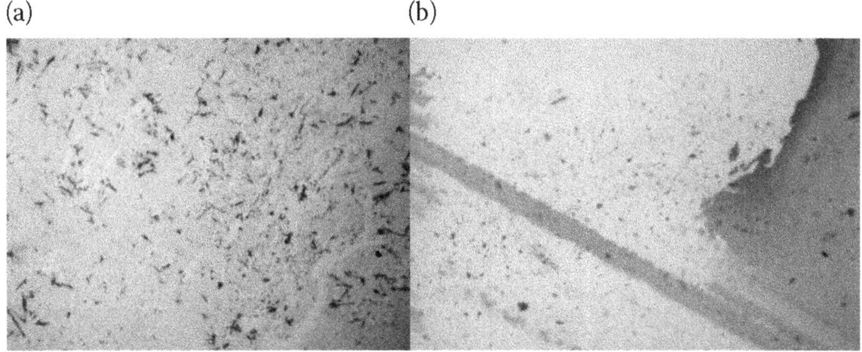

FIGURE 11.12 Gram staining and spore staining of micro-organism isolated from feather waste.

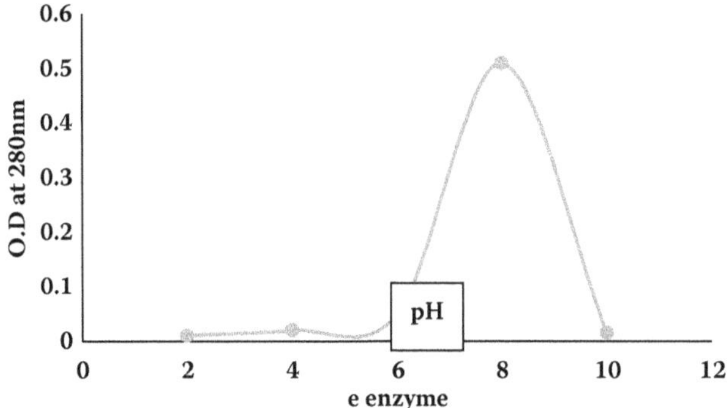

FIGURE 11.13 Effect of different pH on keratinase enzyme.

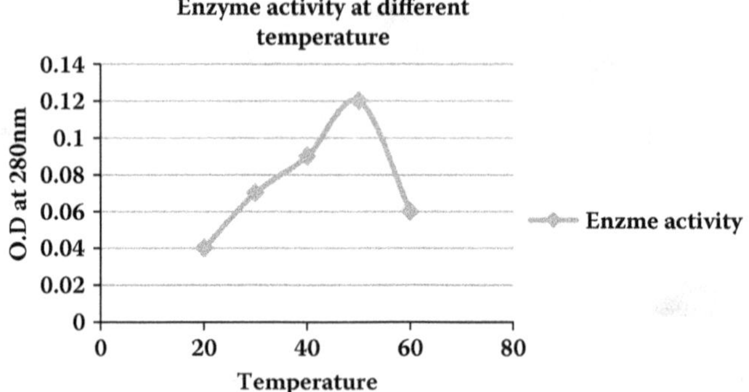

FIGURE 11.14 Effect of different temperatures on keratinase enzyme.

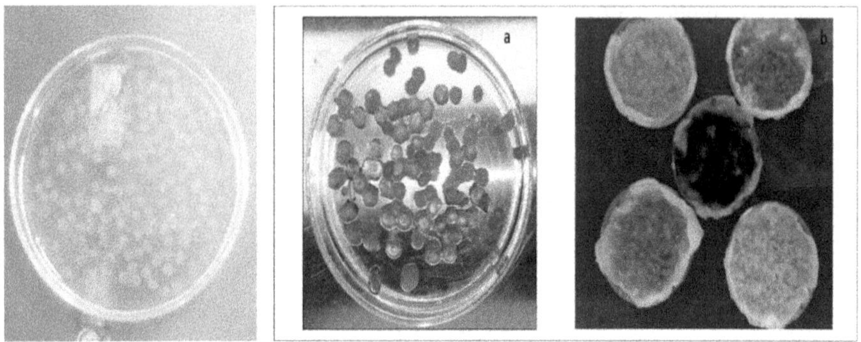

FIGURE 11.15 Sodium alginate beads formation of keratinase enzyme (Suneetha 2016, 2018).

TABLE 11.1
Biochemical Characterization of Keratinase Enzyme after Recycling of Feather

Biochemical test	VIT-JS1	VIT-JS2
1. Catalase test	-ve	-ve
2. Oxidase test	-ve	-ve
3. Citrate utilization test	+ve	+ve
4. Starch hydrolysis test	+ve	+ve
5. Indole test	-ve	-ve
6. Nitrate test	+ve	+ve
7. Lactose test	-ve	-ve
8. Methyl red test	+ve	+ve

TABLE 11.2
Effect of Different pH on Keratinase Enzyme

pH	Enzyme activity
2	0.01
4	0.02
6	0.05
8	0.51
10	0.015

TABLE 11.3
Effect of Temperature on Keratinase Enzyme

Temperature (°C)	Enzyme activity
20	0.04
30	0.07
40	0.09
50	0.12
60	0.06

11.5.4 Optimization of Keratinase Enzyme

11.5.4.1 Effect of pH on Keratinase Production

11.5.4.2 Immobilization of Keratinase Enzyme

Keratinase beads are formed by using production medium. They are preserved and maintained the enzyme activity.

11.6 DISCUSSION

Though feather keratin is highly regarded as resistant to degradation. These bacteria and half degraded feathers have an important role in being used as manure for the better growth of plants and flowers. The study aims to identify feather degrading bacteria and observe the effects of the bacteria if used as manure in flowering plants. There are some bacteria present in the environment containing mosquitocidal activity that has the ability to produce mosquitocidal toxins (5,13). Keratinolytic bacteria such as *Bacillus sphaericus (Bs)* and *Bacillus thuringiensis serovar israelensis (Bti)* are two main bacilli strains that degrade the chicken feather keratin (6) and are helpful in the production of endotoxins to kill the mosquito larvae (Figure 2). Chicken feather waste is generated in a huge quantity all over the world per year. About 18.5 million thousand tons of chicken feather waste are generated all over the world (9). This may be the cheapest bio-organic waste, which can be used as a substrate for culturing different mosquitocidal bacteria producing mosquitocidal toxins in the laboratories. Recent studies show that the degradation of feathers by potentially identified micro-organisms produces valuable amino acids such as cysteine, cysteine, methionine, arginine, etc. Thus, microbial recycling of feather waste acts as a source for value-added products.

ACKNOWLEDGEMENT

The authors want to express their gratitude to honorable Chancellor, Dr. G. Viswanathan, Dr. Sekar Viswanathan, Mr. Sankar Viswanathan, and Mr. G.V. Selvam of VIT University for their constant encouragement and laboratory facilities from VIT University, Vellore, India to carry out this valuable work, and Mr. Raj Vuppu, senior vice president, Citibank, USA and DST, India for constant help to carry out this research.

REFERENCES

1. Suneetha V. (2014). Statistical analysis and mathematical modeling of potential microbial enzymes keratinases and pectinases for pharmacological benefits: A review. *Der Pharmacia Lettre*, 6(2):169–172.
2. Suneetha V., Kumar R. and Nimesh N. (2014). *Nocardiopsis alba* DMS 43377: A noble potent feather degrading *Actinobacteria* isolated from feather waste in Tamilnadu, India. *Journal of Chemical and Pharmaceutical Sciences*, 7(2):165–168.
3. Sahoo S., Sonali R. and Suneetha V. (2014). Poultry waste is a source for kerazyme production. Der Pharmacia Lettre, 6(4):190–193.
4. Suneetha V. and Vuppu K.K. (2013). Partial purification of keratinase from Actinomycetes screened from surrounding places of VIT University for industrial applications. *Der Pharmacia Lettre*, 5 (5):7–11.
5. Suneetha V. and Raj V. (2012). Statistical analysis on optimization of microbial keratinase enzymes screened from Tirupati and Tirumala soil samples. *International Journal of Drug Development & Research*, 4:1–6.
6. Suneetha V. (2014). Kerazyme triggers the production of mosquitocidal toxins. *Research Journal of Pharmacy and Technology*, 7(11):1315–1318.
7. Srivastava A., Sharmaand A. and Suneetha V. (2011). Feather waste biodegradation as a source of Amino acids. *European Journal of Experimental Biology*, 1(2):56–63.
8. Suneetha V., Kumar S. and Ramalingam C. (2009). Bioremediation of poultry waste. *Advanced Biotech Journal of Biotechnology*, 10(01):28–32.
9. Ritika S., Abhishek G., Rahul G. and Suneetha V. (2012). An attempt and a brief research study to produce mosquitocidal toxin using *Bacillus* spp. (VITRARS), isolated from different soil samples (Vellore and Chittoor), by degradation of chicken feather waste. *Research Journal of Pharmaceutical, Biological and Chemical Sciences*, 3(4):40–48.
10. Suneetha V. (2012). A brief study on microscopic degradation pattern of chicken feather from Vellore industrial waste. *Cibtech Journal of Biotechnology*, 1(1):25–28.

11. Mohanapriya M., Parvathi L., Archana B. and Suneetha V. (2014). A potential beta-keratin degrading bacteria from Vellore emu feather dumped soil. *International Journal of Pharmaceutical Sciences, Review and Research*, 25(1): 224–228.
12. Revathi K., Singh S., Khan M.A. and Suneetha V. (2013). A potential strain of keratinolytic bacteria VIT RSAS2 from Katpadi and its pharmacological benefits. *International Journal of Pharmaceutical Sciences, Review and Research*, 20(2):89–92.

BOOK

13. Suneetha V. (2011). Screening, Characterization and Optimization of Keratinolytic Bacteria Isolated from Poultry Waste. Lambert Academic Publishing GmbH & Co. KG. ISBN978-3-8443-2787-8.

12 Soybean Meal Carbohydrates Wasted and the Microbial Role for Improved Digestibility

Kyle D. Teague and Lucas E. Graham
Department of Poultry Science, University of Arkansas, Fayetteville, AR, USA

X. Hernandez-Velasco
Departamento de Medicina y Zootecnia de Aves, FMVZ, Universidad Nacional Autonoma de Mexico, Ciudad de Mexico, Mexico

Guillermo Tellez-Isaias, Billy M. Hargis, and Samuel J. Rochell
Department of Poultry Science, University of Arkansas, Fayetteville, AR, USA

CONTENTS

DOI: 10.1201/9781003128977-12

12.1 INTRODUCTION

12.1.1 Indigestible Components of Poultry Feeds

Indigestible components in poultry feeds include the fractions that are either totally undigested or incompletely digested (Ravindran, 2013). These include substrates for which poultry do not produce suitable enzymes, or that are inaccessible to enzymes because they are encapsulated or bound to other compounds. Each vegetable cell contains nutrients such as starch, lipids, and proteins that poultry can easily digest using endogenously produced enzymes such as amylase, lipase, and protease (Meng et al., 2004). However, in cereal grains including corn, wheat, sorghum, barley, and rye, as well as other cereal byproducts like distiller's dried grains with solubles, these nutrients are often protected from enzymatic digestion by cell walls comprised of nonstarch polysaccharides (NSPs) (Knudsen, 1997). Fiber is one of the most quantitatively important indigestible fractions of feed (Choct et al., 2010). Insoluble fiber passes through the upper part of the intestinal tract intact (Knudsen, 1997), locking in the nutrients present within the cell wall making them unavailable to the animal. Insoluble fiber is commonly broken down further into analytical classification: neutral detergent fiber (NDF) and acid detergent fiber (ADF) (Raza et al., 2019). The NDF contains hemicellulose, cellulose, and lignin, while ADF consists of the more indigestible cellulose and lignin (Choct, 1997). The NSPs are mainly concentrated in the cell walls of the endosperm, but are also found in the bran (Knudsen, 1997). They can be divided into two categories, the water-soluble beta-glucans and pectins and water-insoluble cellulose fractions (Knudsen, 2014). The NSP content of cereal grains ranges from 7 to 19%. The outer cell wall components mainly comprise cellulose and lignin, while the cell walls of the endosperm are predominately arabinoxylans and beta-glucans. The complexity of arabinoxylans is variable among grains and highly influences the digestibility of feed (Knudsen, 1997).

Another indigestible component in poultry feeds is the phytate molecule, which is the storage form of phosphorus in cereal grains that consists of a fully phosphorylated myo-inositol ring. Within the plant, phytic acid readily forms complexes with K^+ and Mg^{2+}, and to a lesser extent Ca^{2+}, to form phytin (Bedford and Partridge, 2010). Phytic acid is poorly digested by monogastric animals and undigested phytate is known to have a negative impact on mineral and protein digestibility. Not only does phytate increase the cost of production by adding a need for additional sources of available phosphorus, but it also has detrimental effects on the environment and mineral and protein utilization (Bedford and Partridge, 2010).

Plant protein sources used in poultry feeds can also contain antinutrients that affect protein digestibility including, saponins, cyanogenic glycosides, tannins, gossypol, oxalates, goitrogens, lectins, protease inhibitors, and cholorogenic acids (Akande et al., 2010). These factors can be divided into heat-labile and heat-stable factors, and processing innovations to remove these ANF have expanded the availability of feedstuffs to the animal industry. The most common method for plant and animal processing is thermal treatment which decreases heat-labile ANF and increases the amino acid digestibility of product (Papadopoulos, 1989; Adeyemo and Longe, 2007).

NSPs are the predominant ANF found in both cereal grains and oilseed meals and can interact with other ANF introduced above. Thus, the role of NSP and other indigestible carbohydrates on nutrient utilization, particularly those found in soybean meal (SBM), will be the topic of this book chapter.

12.1.2 Effects of NSP on Intestinal Health

Increased levels of NSP in poultry diets, particularly the soluble fraction, lead to decreased nutrient digestion and absorption (Choct and Annison, 1990). It is accepted that the negative effects of NSP are linked to increased digesta viscosity, physiological and morphological changes to the digestive tract, and the interaction with the microflora of the gut (Choct, 1997). Elevated gut viscosity results from the water-soluble fraction as the arabinose branches provide soluble properties and the ability to bind water (Choct et al., 2010). This decreases the rate of diffusion of substrates and digestive enzymes and obstructs their interactions at the mucosal surface (Edwards et al., 1988). In addition to acting as a physical barrier to nutrient digestion and absorption, soluble NSP can also alter gut functions by changing endogenous secretions of electrolytes, water, lipids, and proteins (Angkanaporn et al., 1994). These changes within the gastro intestinal tract (GIT) are accompanied by increased digestive secretions, enlargement of digestive organs, and decreased nutrient digestion (Choct, 1997).

In addition, it has been suggested that the antinutritive effects of NSP may be exacerbated by some anaerobic, opportunistic intestinal microflora. For example, MacAuliffe and McGinnis (1971) reported that dietary supplementation with antibiotics moderately improved the nutritive value of rye in chicks at 28 days posthatch. Wagner and Thomas (1978) observed anaerobic bacteria counts in the ileum of birds fed rye or pectin-enriched diets were 2 to 3 logs higher compared to those fed a corn-soy diet. Furthermore, these counts were reduced by 5 logs with the addition of penicillin to the diet. Moreover, soluble NSP results in a slower digesta passage rate (Van der Klis and Van Voorst, 1993), which could decrease oxygen tension to benefit the overgrowth of anaerobic bacteria. Choct et al. (1996) reported an increase in fermentation in the small intestine of broilers by adding soluble NSP in the diet. While volatile fatty acid production was increased, the drastic modifications to the gut ecosystem resulted in decreased nutrient digestion accompanied by poor bird performance (Choct et al., 2010).

12.2 SOYBEAN MEAL AS A FEED INGREDIENT

12.2.1 Production and Nutritional Composition

United States SBM is included in 70% of the animal feed produced worldwide, making it the most frequently used source of protein for swine and poultry feeds (Dozier et al., 2011). It is highly valued by poultry feed producers due to its high protein content and amino acid profile that complements that of corn, the primary energy source in poultry diets. Soybean meal is produced from dehulled beans or beans having hulls (NRC, 1994). Dehulled SBM has a higher crude protein, amino acid, and metabolizable energy composition as opposed to SBM containing more hulls (NRC, 1994). In the US, soybean oil is removed from soybeans to generate SBM using one of two methods, mechanical or chemical extraction. While chemical processing is still the predominant method, recent increases in the production of biodiesel fuel have resulted in more plants using the extruder/expeller processing methods (Loeffler, 2012).

12.2.2 Solvent Extraction

Solvent extraction became the preferred method of oil extraction from soybeans around the 1930s because of its ability to remove all but about 0.5% of the soy oil. During solvent extraction, soybeans are dehulled and cracked using corrugated rollers. The hulls are removed because they contain very little oil and the cracked soybeans are conditioned at 70–75°C for 20–30 minutes. This steam conditioning hydrates the cracked beans to condition them for further processing. Next, the beans are flaked under pressure and flaking rollers. It is vital that the flakes are thin enough for solvent penetration, but strong enough so they do not crumble. Flaking ruptures the seed cellular structure, allowing the solvent to cover more surface area. The flaking process also exposes the oil cells, which allows the solvent to penetrate and increase oil extraction yield. The commercial solvent used is hexane at a ratio of 1:1 solvent-to-soybean on a per weight basis. After leaving the extractor, the beans can contain up to 40% solvent which is removed by air drying. The extracted flakes are toasted and steam heated to inactivate ANF and remove any remaining hexane. In the final step, the flakes are dried and ground to an acceptable size for feed ingredients. The beans used to make SBM are dehulled prior to flaking. The hulls can be added back to the meal to produce a product that will contain more NDF and less protein, thus explaining the crude protein percentage spread below. The resulting SBM has a particle size ranging from 700–1,000 microns, contains either 44 or 48% crude protein, 3.5 to 7% crude fiber, and 1% crude fat. The beans are dried to approximately 10% moisture content and tempered to allow a moisture equilibrium to be reached. For poultry, the dehulled, solvent-extracted SBM contains approximately 2,230–2,440 kcal/kg of metabolizable energy (NRC, 1994).

12.2.3 Expeller Extrusion

The dry extrusion process was first introduced in 1969, and as opposed to wet extrusion, the dry extruder does not require an external source of heat or steam. The process uses friction generated from the extrusion process as the sole source of energy to cook and dehydrate the soybeans. These are single screw extruders put together around a shaft. In between the screw a restriction plate of different diameters can be placed to increase the cook and shear. When material moves in the barrel and comes across these restriction plates, it is unable to pass through, and consequently, pressure builds up and a back flow is created. Usually these restriction plates are arranged in such a way that they increase in diameter toward the die end of the screw creating more pressure and shear as they reach the die. This buildup of pressure and temperature, together with shear stresses developed, tends to plasticize (gelatinize) the raw materials into viscous paste or puffed shape, depending upon the raw material. In dry extrusion, pressure and temperature should be at a maximum just before leaving the die. The optimum extrusion temperature is around 150–160°C. This temperature and pressure is enough to denature ANF and rupture the oil cells. As soon as the material leaves the extruder dies, pressure is instantaneously released from the products, which cause internal moisture to vaporize into steam, making the product expand (Extruding, 2007). The final product contains 38% crude protein, 5–10% crude fat, and a metabolizable energy value for poultry of 3,200 kcal/kg. (NRC, 1994; NRC, 2012). With a higher oil content in the final product, this method produces higher metabolizable energy than solvent extraction and may reduce or eliminate the need to add fat to the diet (Loeffler, 2012).

12.2.4 Overheating of Soybean Meal

Although soybeans must be heat-treated through one of these processes to remove ANF, they can be over-heated which reduces the nutritional value of the meal for poultry (Renner et al., 1953; Warnick and Anderson, 1968; Araba and Dale, 1990). The overcooking of SBM decreases the digestibility of amino acids in part due to Maillard reactions (Lee and Garlich, 1992; Parsons et al., 1992). Parsons et al. (1992) autoclaved dehulled, solvent extracted SBM at 121°C and 105 kPa for 0, 20, 40, and 60 minutes to examine the effects of over processing. Increasing autoclave time reduced total concentration of lysine, arginine, and cysteine, but did not influence other amino acids. A growth assay using broiler chicks found autoclaving at 121°C for 40 minutes reduced lysine availability by 15% compared to birds fed nonautoclaved SBM. The damage of lysine and arginine due to autoclaving and the browning of the meal indicate the presence of the Maillard reaction. During the Maillard reaction, the reactive carbonyl group of the sugar reacts with the nucleophilic amino group of the amino acid. This produces a Schiff base, which cyclizes to form a glycosylamine. The glycosylamine is then transformed into either an Amadori product (glucose) or a Heyns product (fructose). This reaction predominately affects the ε-amino group of lysine. The decreased absorption of lysine is due to the glycosylated lysine derivatives competing with lysine for absorption carriers. These derivatives are poorly utilized, with greater than 75% of the absorbed amounts being excreted (Dozier et al., 2011).

12.3 SOYBEAN ANTINUTRITIONAL FACTORS

Soybeans contain a number of ANFs that can result in a reduction in nutrient utilization, including oligosaccharides, NSPs, phytate, lectins, and trypsin inhibitors. These factors cause negative effects by different mechanisms, including binding to digestive enzymes and nutrients or increasing gut viscosity (Ravindran, 2013). The ANF can be split into two groups: (1) heat labile and (2) heat stable. The soybeans are heated to denature the native protein structure and inactivate the trypsin inhibitors and lectins. However, oligosaccharides, another carbohydrate ANF, are heat stable and the concentration of raffinose and stachyose cannot be reduced by heating (Loeffler, 2012).

12.3.1 Trypsin Inhibitors

Trypsin inhibitors are the primary ANF in SBM (Araba and Dale, 1990; Anderson-Hafermann et al., 1992; Mian and Garlich, 1995). It inhibits the conversion of zymogens to active proteases of trypsin and chymotrypsin by forming an irreversible compound with trypsinogen. The effects of trypsin inhibitors on chymotrypsin are less severe, as it forms a reversible dissociated compound that could potentially still become active (Loeffler, 2012). Chernick et al. (1948) fed chicks diets containing raw SBM or heat-treated SBM and observed a 43% increase in trypsinogen content per gram of pancreas nitrogen content and a 56% increase in pancreas weight as a percent of body weight in chicks fed the raw soybeans. This overstimulation of trypsinogen production in an attempt to compensate for the trypsin inhibitor leads to pancreatic hypertrophy. An enlarged pancreas not only results in increased enzyme production, but also in the increased secretion of nitrogenous products into the intestine, which could explain some of the growth-limiting effects of raw soybeans (Saini, 1989). Broiler growth has been reported to increase 140 to 150% by feeding heat treated raw, hexane extracted soybeans, or SBM compared to nonheat treated raw, hexane extracted soybeans, or SBM (Araba and Dale, 1990; Anderson-Hafermann et al., 1992).

12.3.2 Lectins

Another ANF found in soybeans is lectins. Lectins are glycoproteins that can bind to cell surfaces on specific oligosaccharides or glycopeptides. They also have a high binding affinity to the intestinal enterocytes which causes impairment of brush border continuity and ulceration of villi (Pusztai, 1991). Douglas et al. (1999) found that almost 15% of total growth reduction in chicks was associated with lectin presence in raw soybeans. Lectins, like trypsin inhibitors are heat labile and can be reduced with proper heating.

12.3.3 Oligosaccharides

The primary sugars present in SBM are sucrose and the galactooligosaccharides (GOSs) stachyose, raffinose, and verbascose. These GOSs contain a terminal sucrose that is linked to a chain of α-1,6 galactoses via an α-1,3 bond (Mul and Perry, 1994) and make up 7–8% of the DM within SBM as they are not removed during processing (Van Kempen et al., 2006). The GOS in SBM can only be enzymatically hydrolyzed by α-galactosidases that are not produced in the intestinal tract of nonruminants like poultry (Middelbos and Fahey Jr, 2008). Soy GOSs are considered ANFs because they are poorly digestible and can cause reduced transit time, which leads to lower fiber digestion and TME value (Coon et al., 1990). Leske et al. (1993) observed that the addition of raffinose and stachyose to a boiler diet significantly reduced its TME value compared to controls, but ethanol extraction removal of the oligosaccharides resulted in an increased TME for roosters and broiler chicks (Coon et al., 1990; Leske et al., 1991; Leske and Coon, 1999). The oligosaccharides can also cause wet feces, which can impact litter quality in poultry (Graham et al., 2002). This problem can be alleviated with low-oligosaccharide variety of soybeans or the addition of α-galactosidase enzyme. This also reduces the quantity of SBM needed due to greater nutritional value and the increased concentration of digestible amino acids (Perryman and Dozier, 2012). Graham et al. (2002) found that a lower concentration of raffinose and stachyose will reduce the viscosity of the gut, leading to a faster passage rate, greater access of digestive enzymes to substrates, and diffusion of absorbable nutrients to the intestinal mucosa.

12.3.4 Nonstarch Polysaccharides

Soybean meal contains between 20–30% of NSP on a dry matter basis, including 8 and 17% of insoluble and soluble NSP, respectively (Smits and Annison, 1996). Soy NSPs are classified into three main groups that include cellulosic, noncellulosic polymers, and pectic polysaccharides. The soy NSPs are predominately a mixture of pectic polysaccharides with rhamnogalacturonans as the most abundant carbohydrate and cellulose as the second most abundant in SBM (Choct et al., 2010). The solubility of NSP is the main factor affecting their digestibility with the soluble fraction being more digestible than the insoluble. Digestibility of a diet fed to cockerels containing 6.9% of NSPs from defatted, dehulled SBM had an NSP digestibility value of 13% (Carré et al., 1990). Carré et al. (1995) also reported higher NSP digestibility in adult cockerels compared to broilers fed a corn-soy diet. They hypothesized that the mature gut microflora of the adult birds adapts to more efficiently digest dietary NSP.

12.3.5 Phosphorus Digestibility

Limited phosphorus digestibility is also a factor in SBM as it contains 1 to 2% phytic acid, which results in approximately 2/3 of the P in soybeans being bound to phytate. Phytate bound P is nutritionally unavailable to pigs and poultry due to their insufficient endogenous phytase production. This increase in phytate bound P lowers the concentration of inorganic P, thus, decreasing absorption of P in the small intestine and decreasing its digestibility. Decreased absorption of P directly results in increased concentration of P in the feces. Phytate bound P can also bind minerals such as Zn, Ca, Mg, Mn, K, and Fe, reducing their availability to the animal and increasing their concentration in the feces as well. The addition of a phytase enzyme increases the amount of P available for absorption, decreases the concentration of fecal P, and reduces the amount of supplementary dietary P (Peper, 2015). Almost all commercial monogastric diets include supplemental phytase due to its environmental and health benefits and widespread availability.

12.4 LITTER QUALITY AND FOOTPAD DERMATITIS

Currently, a primary concern in the poultry industry in regard to litter quality is the condition known as footpad dermatitis (FPD). Footpad dermatitis was first described in broilers in the 1980s (McFerran et al., 1983; Greene et al., 1985). During this time period, the broiler paw market was beginning to develop, resulting in greater attention being given to paw quality. Due to the market value of this product along with increasing welfare concerns, the industry began to focus on how to reduce paw downgrades and condemnations (Shepherd and Fairchild, 2010). Footpad dermatitis is characterized by inflammation and necrotic lesions on the surface of the footpads and toes (Greene et al., 1985). Increased inclusion of SBM in all vegetable-based diets has been suggested as a possible predisposing factor that contributes to the incidence and severity of footpad dermatitis. This could be due to the high potassium content in SBM. It has been reported that dietary electrolyte balance is a key factor influencing litter moisture and high dietary inclusion of Na and K as well as high dietary electrolyte balance increases water intake and litter moisture (Borges et al., 2003; Ravindran et al., 2008; Koreleski et al., 2010). Cengiz et al. (2012) reported high dietary Na concentration enhanced water consumption and litter moisture. Fuhrmann et al. (2016) observed similar findings when broilers were fed a diet with high levels of K. Another factor is the indigestible NSP and GOS can increase gut viscosity resulting in sticky droppings that can adhere to the foot and over time deteriorate the epidermis and keratin layers (Hess et al., 2004).

12.5 STRATEGIES TO ENHANCE SBM UTILIZATION

12.5.1 Reduced ANF Varieties

There have been many attempts to reduce the negative effects of NSP and oligosaccharides in SBM. Soybean varieties have been genetically selected to have reduced raffinose and stachyose concentrations (Baker and Stein, 2009; Baker et al., 2011) which have demonstrated to yield higher ME_n values of the resulting SBM in poultry (Parsons et al., 2000). A 7 to 9% increase in TME_n was observed in roosters fed low oligosaccharide SBM compared to those fed conventional SBM (Parsons et al., 2000). In contrast, Baker et al. (2011) detected no difference in TME_n between low oligosaccharide SBM and conventional SBM. More recently, varieties of soybeans have been developed to ultra-low oligosaccharide content with over a 90% reduction in GOS compared to conventional SBM. Perryman and Dozier (2012) evaluated the use of a low oligosaccharide SBM and an ultra-low oligosaccharide SBM and reported increases in AME_n of 168 kcal/kg and 5.8% higher apparent ileal amino acid digestibility (AIAAD) for low oligosaccharide SBM compared to conventional SBM. They also observed an 8 and 17% increase in AIAAD for the first five limiting amino acids in broilers for low oligosaccharide and ultralow oligosaccharide SBM, respectively, compared to conventional SBM. Perryman et al. (2013) also observed similar performance and carcass characteristics from 1 to 40 days of age with birds fed low oligosaccharide SBM compared to conventional SBM. Similar results were observed from 1 to 42 days of age birds fed low oligosaccharide SBM, while those fed ultra-low oligosaccharide SBM also showed no differenfces in bodyweight gain and carcass characteristics, but there was a four-point reduction in FCR compared to those fed conventional SBM. The diets for the low and ultralow oligosaccharide SBM were formulated with 28 to 71% less supplemental oil compared to diets formulated with conventional SBM, translating to reductions in dietary costs.

12.5.2 Specialized Processing

There have also been reports of beneficial processing techniques to remove the oligosaccharide content of SBM. One of those methods is ethanol extraction (Coon et al., 1990; Leske and Coon, 1999). Ethanol extraction removes approximately 90% of water-soluble GOS and increases the TME_n of SBM for poultry (Coon et al., 1990). It has also shown to increase fiber digestibility, lengthen transit time, and increase cecal pH (Coon et al., 1990). Veldman et al. (1993) reported the addition of the ethanol extracted material back into the diet had a detrimental effect on ileal digestibility and resulted in fluid retention with increased microbial fermentation in the gut of piglets. However, some contradicting research with broilers has concluded little or no antinutritional effect of the SBM oligosaccharides (Irish et al., 1995).

Another specialized processing method is the fermentation of SBM by fungi and bacteria. The predominant organism used for SBM fermentation is *Aspergillus* due to its ability to produce enzymes such as hemicellulases, hydrolases, pectinases, protease, amylase, and lipases (Mathivanan et al., 2006). Bacterial fermentation is also accomplished using various *Lactobacillus* and *Bacillus* species (Yang et al., 2007). When compared to conventional SBM, fermented SBM was reported to have higher protein content and amino acid digestibility, lower ANF and allergenic compounds, and an overall improved nutritional value (Feng et al., 2007a; b; Frias et al., 2008; Song et al., 2008). Feng et al. (2007a) reported an increase in average daily gain and feed intake of broilers for 6 weeks with a significantly lower FCR from weeks 1 to 3. This is in agreement with Mathivanan et al. (2006) who reported increased body weight and lower FCR at 6 weeks for broilers fed fermented SBM at three different levels. Similarly, Hirabayashi et al. (1998) also showed increased body weight gain from weeks 1 to 5 in broilers fed fermented SBM compared to those fed conventional SBM. In addition to performance parameters, improvements in phosphorus digestibility (Hirabayashi et al., 1998; Feng et al., 2007a), blood parameters (Feng et al., 2007a), and histological characteristics (Mathivanan et al., 2006) were also reported. Fermented SBM has shown to provide performance benefits for poultry; however, under current production practices, those gains may not justify the increased cost of fermented SBM compared to conventional SBM.

12.6 *BACILLUS* AS DIRECT-FED MICROBIAL

12.6.1 Sporulation and Environmental Resiliency

Endospore formation by *Bacillus* and other Gram-positive bacteria, like *Clostridia*, is a strategy used by these organisms to survive environmental stress and inhabit harsh environments. When these rod-shaped bacteria are starved for carbon, nitrogen, or sometimes phosphorus, they produce an oval, dormant cell called a spore (Driks, 2002). This sporulation process takes approximately 7 hours at 37°C (Piggot and Hilbert, 2004). Sporulating *B. subtilis* cells are cannibalistic and feed on their siblings in order to delay committing to spore formation (González-Pastor et al., 2003). During initiation, the master transcription regulator Spo0A is activated by phosphorylation. The first morphological stage of sporulation from a vegetative cell (stage 0) is the formation of an axial filament of chromatin. During this stage, two copies of the chromosome condense and elongate to form a filament that reaches across the long axis of the cell (stage I). The cell then asymmetrically divides into two daughter cells (stage II). At the time of septation, only one-third of a chromosome is existent in the prespore. However, DNA translocase quickly transfers the remaining two-thirds yielding two cells with identical genomes, but unequal volumes. Subsequently, the prespore is engulfed by the mother cell and wrapped within the septal membrane, resulting in a free-floating protoplast encircled by two membranes (stage III). Following engulfment, two peptidoglycan layers, the primordial germ cell wall and the cortex are deposited between the membranes surrounding the prespore (stage IV) and are essential to spore dormancy (Driks, 2002). In stage V, the coat, a thick, complex structure of proteins on the outside surface of the prespore is assembled. The spore then matures, gaining resistance to high temperatures and UV radiation (stage VI). In the final stage (VII), the mother cell lyses, releasing the mature spore (Hilbert and Piggot, 2004). The spore coat is important in spore resistance to some chemicals, exogenous lytic enzymes that can degrade the spore cortex, and to predation by protozoa. However, the coat does not play a role in protection from chemicals, heat, or radiation (Setlow, 2006). The peptidoglycan-comprised cortex is essential for the formation of the dormant spore and in the reduction of water content in the core. While the germ cell wall under the cortex probably does not affect resistance, it does become the cell wall of the outgrowing spore during germination (Setlow, 2006). The inner membrane is a tough permeability barrier that aids in spore resistance to chemicals (Nicholson et al., 2000). The core also contains three small molecules whose concentrations are important for resistance. Water is the first molecule and it makes up only 27–50% of the core wet weight compared to 75–80% of a growing cell. This low amount of free water in the spore core restricts macromolecular movement, aids in enzymatic dormancy, and is the most important factor determining resistance to wet heat (Gerhardt and Marquis, 1989). The second molecule is dipicolinic acid (DPA) which makes up 5–15% of the dry weight of spores and is usually chelated to Ca^{2+} (Gerhardt and Marquis, 1989). The large amounts of DPA assist in reducing the core water content and UV photochemistry of the spore DNA (Setlow, 2006). The mature spore exhibits little to no metabolic activity and is considered dormant.

12.6.2 Germination

Once spores are exposed to a suitable stimulus (germinant), they promptly lose their dormancy and resistance properties. There are several germinant agents such as nutrients, calcium dipicolinic acid (CaDPA), and high hydrostatic pressure (HP) that can elicit spore germination. In nature, it is likely that the presence of specific nutrients that activates spore germination. Nutrient germinants bind to germinant receptors (GRs) in the inner membrane. For *B. subtilis*, L-alanine, L-valine, and L-asparagine have been shown to cause germination, while the D-amino acids are inactive (Atluri et al., 2006). When the nutrient germinants bind to receptors located in the spore's inner membrane a series of reactions is triggered that ultimately results in a metabolically active vegetative cell. In *B. subtilis*, the nutritional germinant will bind to the GerA, GerB, or GerK receptors encoded by gerA, gerB, and gerK operons, which initiates "commitment" after which germination will continue even if the germinant is removed (Paidhungat and Setlow, 2000). Around the same time as commitment, the release of monovalent cations, Na^+, K^+, and H^+ from the spore core result in an increase in spore core pH to approximately 8. It is unknown whether this cation release is causally related to the previous commitment step (Setlow, 2013). Shortly after these two steps, the spore core's huge CaDPA depot is completely released in about 2 minutes and replaced with water, thus increasing spore water content (Kong et al., 2010). With this, Stage I of germination is complete (Setlow et al., 2001). The time it takes for spores to complete stage I is widely variable with some spores taking less than 10 minutes, while others may take an hour or even days. The main reason for their differences seems to be in the time between germinant addition and initiation of rapid CaDPA release, termed "Tlag" (Yi and Setlow, 2010). This is predominately caused by variations in the spore's level of germinant receptors (Kong et al., 2010).

The final stage in spore germination, stage II, is triggered by events in stage I, particularly the release of CaDPA. In stage II, there is hydrolysis of the peptidoglycan (PG) cortex by cortex-lytic enzymes (CLEs). The CLEs specifically recognize cortical PG via the muramic acid-δ-lactam (MAL) in the polysaccharide backbone that is not present in growing cell or germ cell wall PG (Setlow, 2003). *Bacillus* spores have been shown to have two enzymes, CwlJ and SleB, which are involved in the degradation of the cortex PG (Chirakkal et al., 2002). The hydrolysis of the cortical PG allow for the expansion of the spore core and the inner membrane which increases 1.5- to 2-fold without new membrane synthesis (Cowan et al., 2004). Stage II of germination results in the spore core now containing approximately 80% wet weight as water and active enzymes within the core (Paidhungat and Setlow, 2001). Enzyme activity leads to the degradation of novel, acid-soluble proteins in the core and initiation of metabolism and macromolecular synthesis in the core (Setlow, 2013). Achievement of stage II of germination also leads to breakdown of the spore coat and escape of the outgrowing spore (Plomp et al., 2007).

12.6.3 Potential for *Bacillus* DFM to Improve Poultry Health and Performance

The increasing concern over multidrug resistant bacteria in recent decades has resulted in the Food and Drug Administration calling for companies to discontinue labeling antibiotics as growth promoters in agricultural animals (GFI #213). One promising option to replace some of the benefits conferred by antibiotics is the integration of probiotic bacteria into feed as DFMs. In this regard, *Bacillus* spp. have a distinct advantage over other microbes like *Lactobacillus,* which require more careful handling, storage, and administration by bird caretakers on site via drinking water. *Bacillus* spp. spores possess the ability to resist harsh environmental conditions such as extreme pH, high pressures, dehydration, and long storage periods all of which make them suitable for commercial use in the poultry industry (Cartman et al., 2007). These characteristics allow spores to survive pelletization during the feed milling process, which makes administration of *Bacillus* spp. spores as DFMs convenient to producers and helped gain traction for their commercial use (Hong et al., 2005). Latorre et al. (2014) reported that *Bacillus* spores can persist and possibly complete a full life cycle development within the GIT, indicating that these bacteria could be considered part of the metabolically active host microbiota. It is important to mention that not all *Bacillus* are created equal. Each isolate possesses distinctly different characteristics such as heat resistance, rate of growth and sporulation, enzyme production, and antimicrobial production (Larsen et al., 2014). Thus, they have been shown to have many possible modes of action for improving gut health and nutrient utilization. For example, Samanya and Yamauchi (2002) reported decreased blood ammonia concentration after supplementation with *B. subtilis natto*, which they hypothesize, resulted in increased cell mitosis in the intestines and greater villus heights, as ammonia has been reported to be toxic to enterocytes and reduce cell proliferation and gastric mucosal DNA synthesis. Similar results were observed by Yurong et al. (2005) who noted enhanced intestinal mucosal immunity following *Bacillus* supplementation. A second mode of action using *Bacillus* spp. as DFM could be competitive exclusion in limiting the colonization of pathogenic bacteria. Jin et al. (1996) reported decreased population counts of intestinal *E. coli* in broilers fed feed supplemented with *B. subtilis.* La Ragione and Woodward (2003) found that an oral inoculation of *B. subtilis* spores 24 hours prior to *C. perfringens* challenge suppressed colonization of the pathogen in the distal GIT of chickens. Similarly, Latorre et al. (2015b) showed antimicrobial activity against *C. perfringens* for several different species of *Bacillus.* Shivaramaiah et al. (2011) administered spores of different *Bacillus* spp. strains to *Salmonella* Typhimurium challenged chicks and poults and observed a reduction in pathogen recovery in the crop and ceca of birds fed *Bacillus* supplemented diets. Wolfenden et al. (2011) saw similar results with a decreased level of colonization of *Salmonella* in commercial turkeys supplemented with a *Bacillus*-based DFM and similar body weight at 23 days of age as turkeys consuming a diet medicated with Nitarsone. A third potential mode of action could be enzyme production. Latorre et al. (2015b) determined cellulase and xylanase production of several *Bacillus* isolates, as well as amylase, protease, lipase, and phytase activities (Latorre et al., 2016). These possible modes of action suggest that *Bacillus* isolates could be an effective tool in replacing antibiotic growth promotors and combatting ANF in feed components.

The addition of NSP-degrading enzymes and selected *Bacillus* DFMs have been shown to significantly reduce digesta viscosity in diets containing high NSP (Choct et al., 1995; Latorre et al., 2015b). Furthermore, research has evaluated the potential for synergistic effects when combining enzymes and DFMs. Dersjant-Li et al. (2015) evaluated the combination of a xylanase, amylase, and proteinase (XAP) with three strains of *Bacillus amyloliquefaciens* on welfare parameters in broilers reared under commercial conditions. They reported improved litter quality and reduced FPD lesion scores in treated birds compared to untreated controls. Similar results were observed by Flores et al. (2016) where a reduction in FPD lesion scores was observed in birds fed the combination of XAP and *Bacillus*-based DFM compared to the negative controls. In unpublished data from our lab, a *Bacillus*-based DFM selected for its ability to produce enzymes that hydrolyze SBM carbohydrates was administered alone to a low energy, high SBM diet in the absence of pathogen challenge and led to birds having significantly less FPD lesions than nonsupplemented control birds. This indicates *Bacillus* isolates may not only have probiotic effects for the host in the presence of pathogenic exposure, but also produce enzymes that could enhance nutrient utilization for healthy birds.

Work has also been done with *Bacillus*-based DFMs and alternative ingredients as complementary approaches to lowering poultry diet costs. Byproducts of biofuel production (distiller's dried grains with solubles) and cereals like wheat and barley are being included in poultry rations, effectively increasing the amount of poorly digestible NSP in the feed and raising concerns of digesta viscosity (Tellez et al., 2014, 2015). Latorre et al. (2015b) demonstrated that the inclusion of a selected *Bacillus*-based DFM to a high NSP rye-based diet significantly reduced digesta viscosity, increased body weight, and lowered FCR at 28 days posthatch compared to untreated control fed birds (Latorre et al., 2015a). Salim et al. (2013) reported birds fed a standard corn-soy diet supplemented with a *Bacillus*-based DFM increased body weight gain of broilers from 0 to 21 days posthatch and reduced FCR from 0 to 7 days posthatch when compared to untreated controls. These performance increases were similar to birds fed diets supplemented with an antibiotic growth promotor (AGP), virginiamycin. Additionally, Harrington et al. (2015) found that birds fed low energy diets supplemented with *B. subtilis* were able to achieve higher 42 days posthatch body weight gain and lower FCR than birds fed corresponding diets without *Bacillus* supplementation. In the same experiment, birds fed diets with a 2% reduction in metabolizable energy (ME) supplemented with *Bacillus* attained performance similar to that of birds fed nonsupplemented feed formulated without a reduction in ME. A regression analysis determined that supplementation of the *Bacillus*-based DFM had an overall ME contribution of +62 kcal/kg feed. Knap et al. (2011) had previously demonstrated birds fed diets with a 4% reduction in ME and *Bacillus* supplementation had improved FCR, but not body weight, compared to birds fed an untreated control energy diet. Additionally, recent work from our lab showed that birds fed a diet with a 4.3% reduction in ME improved BWG and FCR from 0 to 21 days posthatch in six trials compared to birds fed the reduced energy diet alone. This improvement was similar to performance of birds fed the industry relevant control with a standard energy level (unpublished data). In these trials, birds were fed high SBM diets to provide ample substrate for the *Bacillus* selected as superior digesters of GOS. By possibly increasing the ME of SBM, poultry producers can enhance performance of their animals or feed a lower cost feed while also maintaining bird performance and health.

12.7 CONCLUDING REMARKS AND FUTURE DIRECTIONS

With increasing demand for protein and outside pressure from consumers to eliminate antibiotics from use in livestock, producers are looking for alternatives that will allow for similar gains in performance and health seen with AGPs. *Bacillus* DFMs have proven to have the ability to increase health, immune status, and performance parameters of broilers, while providing convenient easy application methods. Moreover, there is great opportunity to leverage the ability of *Bacillus*-based DFM to produce enzymes in situ that can improve utilization of all vegetable diets that have a higher NSP and soy GOS content. Further research is needed to determine specific mechanisms, but there is potential for their use to possibly reduce the need for dietary fat supplementation, lower feed costs, and improve poultry health and performance.

REFERENCES

Adeyemo, G. O., and O. G. Longe. 2007. "Effects of graded levels of cottonseed cake on performance, haematological and carcass characteristics of broilers fed from day old to 8 weeks of age." *African Journal of Biotechnology* 6(8).

Akande, K. E., U. D. Doma, H. O. Agu, and H. M. Adamu. 2010. "Major antinutrients found in plant protein sources: their effect on nutrition." *Pakistan Journal of Nutrition* 9(8):827–832.

Anderson-Hafermann, J., Y. Zhang, C. Parsons, and T. Hymowitz. 1992. "Effect of heating on nutritional quality of conventional and Kunitz trypsin inhibitor-free soybeans." *Poultry Science* 71:1700–1709. 10.3382/ps.0711700

Angkanaporn, K., M. Choct, W. L. Bryden, E. F. Annison, and G. Annison. 1994. "Effects of wheat pentosans on endogenous amino acid losses in chickens." *Journal of the Science of Food and Agriculture* 66:399–404. 10.1002/jsfa.2740660319

Araba, M., and N. Dale. 1990. "Evaluation of protein solubility as an indicator of overprocessing soybean meal." *Poultry Science* 69:76–83. 10.3382/ps.0690076

Atluri, S., K. Ragkousi, D. E. Cortezzo, and P. Setlow. 2006. "Cooperativity between different nutrient receptors in germination of spores of *Bacillus subtilis* and reduction of this cooperativity by alterations in the GerB receptor." *Journal of Bacteriology* 188:28–36. 10.1128/JB.188.1.28-36.2006

Baker, K., and H. Stein. 2009. "Amino acid digestibility and concentration of digestible and metabolizable energy in soybean meal produced from conventional, high-protein, or low-oligosaccharide varieties of soybeans and fed to growing pigs." *Journal of Animal Science* 87:2282–2290. 10.2527/jas.2008-1414

Baker, K., P. Utterback, C. Parsons, and H. Stein. 2011. "Nutritional value of soybean meal produced from conventional, high-protein, or low-oligosaccharide varieties of soybeans and fed to broiler chicks." *Poultry Science* 90:390–395. 10.3382/ps.2010-00978

Bedford, M. R., and G. G. Partridge. 2010. *Enzymes in Farm Animal Nutrition*. UK: CABI.

Borges, S., A. F. Da Silva, J. Ariki, D. Hooge, and K. Cummings. 2003. "Dietary electrolyte balance for broiler chickens under moderately high ambient temperatures and relative humidities." *Poultry Science* 82:301–308. 10.1093/ps/82.2.301

Carré, B., L. Derouet, and B. Leclercq. 1990. "The digestibility of cell-wall polysaccharides from wheat (bran or whole grain), soybean meal, and white lupin meal in cockerels, muscovy ducks, and rats." *Poultry Science* 69:623–633. 10.3382/ps.0690623

Carré, B., J. Gomez, and A. Chagneau. 1995. "Contribution of oligosaccharide and polysaccharide digestion, and excreta losses of lactic acid and short chain fatty acids, to dietary metabolisable energy values in broiler chickens and adult cockerels." *British Poultry Science* 36:611–630. 10.1080/00071669508417807

Cartman, S. T., R. M. La Ragione, and M. J. Woodward. 2007. "Bacterial spore formers as probiotics for poultry." *Food Science and Technology Bulletin: Functional Foods* 4:21–30. 10.1616/1476-2137.14897

Cengiz, Ö., J. Hess, and S. Bilgili. 2012. "Influence of graded levels of dietary sodium on the development of footpad dermatitis in broiler chickens." *Journal of Applied Poultry Research* 21:770–775. 10.3382/japr.2011-00464

Chernick, S., S. Lepkovsky, and I. Chaikoff. 1948. "A dietary factor regulating the enzyme content of the pancreas: changes induced in size and proteolytic activity of the chick pancreas by the ingestion of raw soy-bean meal." *American Journal of Physiology* 155:33–41. 10.1152/ajplegacy.1948.155.1.33

Chirakkal, H., M. O'Rourke, A. Atrih, S. J. Foster, and A. Moir. 2002. "Analysis of spore cortex lytic enzymes and related proteins in *Bacillus subtilis* endospore germination." *Microbiology* 148:2383–2392. 10.1099/00221287-148-8-2383

Choct, M. 1997. "Feed non-starch polysaccharides: chemical structures and nutritional significance." *Feed Milling International* 191:13–26. 10.1071/AN15276

Choct, M., and G. Annison. 1990. "Anti-nutritive activity of wheat pentosans in broiler diets." *British Poultry Science* 31:811–821. 10.1080/00071669008417312

Choct, M., R. J. Hughes, R. P. Trimble, K. Angkanaporn, and G. Annison. 1995. "Non-starch polysaccharide-degrading enzymes increase the performance of broiler chickens fed wheat of low apparent metabolizable energy." *The Journal of Nutrition* 125:485–492. 10.1093/jn/125.3.485

Choct, M., R. J. Hughes, J. Wang, M. Bedford, A. Morgan, and G. Annison. 1996. "Increased small intestinal fermentation is partly responsible for the anti-nutritive activity of non-starch polysaccharides in chickens." *British Poultry Science* 37:609–621. 10.1080/00071669608417891

Choct, M., Y. Dersjant-Li, J. McLeish, and M. Peisker. 2010. "Soy oligosaccharides and soluble non-starch polysaccharides: a review of digestion, nutritive and anti-nutritive effects in pigs and poultry." *Asian-Australasian Journal of Animal Sciences* 23:1386–1398. 10.5713/ajas.2010.90222

Coon, C. N., K. Leske, O. Akavanichan, and T. Cheng. 1990. "Effect of oligosaccharide-free soybean meal on true metabolizable energy and fiber digestion in adult roosters." *Poultry Science* 69:787–793. 10.3382/ps.0690787

Cowan, A. E., E. M. Olivastro, D. E. Koppel, C. A. Loshon, B. Setlow, and P. Setlow. 2004. "Lipids in the inner membrane of dormant spores of *Bacillus* species are largely immobile." *Proccedings of the National Academy of Sciences of the United States of America.* 10.1073/pnas.0306859101

Dersjant-Li, Y., K. Van De Belt, J. Van Der Klis, H. Kettunen, T. Rinttilä, and A. Awati. 2015. "Effect of multi-enzymes in combination with a direct-fed microbial on performance and welfare parameters in broilers under commercial production settings." *Journal of Applied Poultry Research* 24:80–90. 10.3382/japr/pfv003

Douglas, M. W., C. M. Parsons, and T. Hymowitz. 1999. "Nutritional evaluation of lectin-free soybeans for poultry." *Poultry Science* 78:91–95. 10.1093/ps/78.1.91

Dozier, W., J. Hess, and H. El-Shemy. 2011. *Soybean Meal Quality and Analytical Techniques*. London, UK: IntechOpen. 10.5772/24161

Driks, A. 2002. "Overview: development in bacteria: spore formation in *Bacillus subtilis*." *Cellular and Molecular Life Sciences* 59:389–391. 10.1007/s00018-002-8430-x

Edwards, C., I. Johnson, and N. Read. 1988. "Do viscous polysaccharides slow absorption by inhibiting diffusion or convection?" *European Journal of Clinical Nutrition* 42:307.

Feng, J., X. Liu, Z. Xu, Y. Liu, and Y. Lu. 2007a. "Effects of *Aspergillus oryzae* 3.042 fermented soybean meal on growth performance and plasma biochemical parameters in broilers." *Animal Feed Science and Technology* 134:235–242. 10.1016/j.anifeedsci.2006.08.018

Feng, J., X. Liu, Z. Xu, Y. Lu, and Y. Liu. 2007b. "The effect of *Aspergillus oryzae* fermented soybean meal on growth performance, digestibility of dietary components and activities of intestinal enzymes in weaned piglets." *Animal Feed Science and Technology* 134:295–303. 10.1016/j.anifeedsci.2006.10.004

Flores, C., M. Williams, J. Pieniazek, Y. Dersjant-Li, A. Awati, and J. Lee. 2016. "Direct-fed microbial and its combination with xylanase, amylase, and protease enzymes in comparison with AGPs on broiler growth performance and foot-pad lesion development." *Journal of Applied Poultry Research* 25:328–337. 10.3382/japr/pfw016

Frias, J., Y. S. Song, C. Martínez-Villaluenga, E. González de Mejia, and C. Vidal-Valverde. 2008. "Immunoreactivity and amino acid content of fermented soybean products." *Journal of Agricultural and Food Chemistry* 56:99–105. 10.1021/jf072177j

Fuhrmann, R., and J. Kamphues. 2016. "Effects of fat content and source as well as of calcium and potassium content in the diet on fat excretion and saponification, litter quality and foot pad health in broilers." *European Poultry Science* 80:1–12. 10.1399/eps.2016.118

Gerhardt, P., and R. Marquis. 1989. *Regulation of Prokaryotic Development: Structural and Functional Analysis of Bacterial Sporulation and Germination.* Edited by I. Smith, R. A. Slepecky, P. Setlow, 43–63. Washington, D.C.: American Society for Microbiology.

González-Pastor, J. E., E. C. Hobbs, and R. Losick. 2003. "Cannibalism by sporulating bacteria." *Science* 301:510–513. 10.1126/science.1086462

Graham, K., M. Kerley, J. Firman, and G. Allee. 2002. "The effect of enzyme treatment of soybean meal on oligosaccharide disappearance and chick growth performance." *Poultry Science* 81:1014–1019. 10.1093/ps/81.7.1014

Greene, J. A., R. McCracken, and R. Evans. 1985. "A contact dermatitis of broilers-clinical and pathological findings." *Avian Pathology* 14:23–38. 10.1080/03079458508436205

Harrington, D., M. Sims, and A. Kehlet. 2015. "Effect of *Bacillus subtilis* supplementation in low energy diets on broiler performance." *Journal of Applied Poultry Research* 25:29–39. 10.3382/japr/pfv057

Hess, J., S. Bilgili, and K. Downs. 2004. "Paw Quality Issues." In *Proc. Deep South Poultry Conference*, Tifton, GA. University of Georgia, Athens.

Hilbert, D. W., and P. J. Piggot. 2004. "Compartmentalization of gene expression during *Bacillus subtilis* spore formation." *Microbiology and Molecular Biology Reviews* 68:234–262. 10.1128/MMBR.68.2.234-262.2004

Hirabayashi, M., T. Matsui, H. Yano, and T. Nakajima. 1998. "Fermentation of soybean meal with *Aspergillus usamii* reduces phosphorus excretion in chicks." *Poultry Science* 77:552–556. 10.1093/ps/77.4.552

Hong, H. A., L. H. Duc, and S. M. Cutting. 2005. "The use of bacterial spore formers as probiotics." *FEMS Microbiology Reviews* 29:813–835. 10.1016/j.femsre.2004.12.001

Irish, G., G. Barbour, H. Classen, R. Tyler, and M. Bedford. 1995. "Removal of the alpha-galactosides of sucrose from soybean meal using either ethanol extraction or exogenous alpha-galactosidase and broiler performance." *Poultry Science* 74:1484–1494. 10.3382/ps.0741484

Jin, L., Y. Ho, N. Abdullah, and S. Jalaudin. 1996. "Influence of dried *Bacillus substillis* and lactobacilli cultures on intestinal microflora and performance in broilers." *Asian-Australasian Journal of Animal Sciences* 9:397–404. 10.5713/ajas.1996.397

Knap, I., A. Kehlet, and T. Bente. 2011. *Bacillus subtilis* (DSM17299) - improved protein digestibility and equal performance in energy-reduced diets for broilers. In France: *Actes des 9èmes Journées de la Recherche Avicole*, Tours, France, 29 et 30 mars 2011:349–352.

Knudsen, K. E. B. 1997. "Carbohydrate and lignin contents of plant materials used in animal feeding." *Animal Feed Science and Technology* 67:319–338. 10.1016/S0377-8401(97)00009-6

Knudsen, K. E. B. 2014. "Fiber and nonstarch polysaccharide content and variation in common crops used in broiler diets." *Poultry Science* 93:2380–2393. 10.3382/ps.2014-03902

Kong, L., P. Zhang, P. Setlow, and Y. Li. 2010. "Characterization of bacterial spore germination using integrated phase contrast microscopy, Raman spectroscopy, and optical tweezers." *Analytical Chemistry* 82:3840–3847. 10.1021/ac1003322

Koreleski, J., S. Swiatkiewicz, and A. Arczewska. 2010. "The effect of dietary potassium and sodium on performance, carcass traits, and nitrogen balance and excreta moisture in broiler chicken." *Journal of Animal and Feed Sciences* 19:244–256. 10.22358/jafs/66285/2010

Larsen, N., L. Thorsen, E. N. Kpikpi, B. Stuer-Lauridsen, M. D. Cantor, B. Nielsen, E. Brockmann, P. M. Derkx, and L. Jespersen. 2014. "Characterization of *Bacillus* spp. strains for use as probiotic additives in pig feed." *Applied Microbiology and Biotechnology* 98:1105–1118. 10.1007/s00253-013-5343-6

Latorre, J., X. Hernandez-Velasco, G. Kallapura, A. Menconi, N. Pumford, M. Morgan, S. Layton, L. Bielke, B. Hargis, and G. Téllez. 2014. "Evaluation of germination, distribution, and persistence of *Bacillus subtilis* spores through the gastrointestinal tract of chickens." *Poultry Science* 93:1793–1800. 10.3382/ps.2013-03809

Latorre, J., X. Hernandez-Velasco, L. Bielke, J. Vicente, R. Wolfenden, A. Menconi, B. Hargis, and G. Tellez. 2015a. "Evaluation of a *Bacillus* direct-fed microbial candidate on digesta viscosity, bacterial translocation, microbiota composition and bone mineralisation in broiler chickens fed on a rye-based diet." *British Poultry Science* 56:723–732. 10.1080/00071668.2015.1101053

Latorre, J. D., X. Hernandez-Velasco, V. A. Kuttappan, R. E. Wolfenden, J. L. Vicente, A. D. Wolfenden, L. R. Bielke, O. F. Prado-Rebolledo, E. Morales, B. M. Hargis, and G. Tellez. 2015b. "Selection of *Bacillus* spp. for cellulase and xylanase production as direct-fed microbials to reduce digesta viscosity and *Clostridium perfringens* proliferation using an in vitro digestive model in different poultry diets." *Frontiers in Veterinary Science* 2:25. 10.3389/fvets.2015.00025

Latorre, J. D., X. Hernandez-Velasco, R. E. Wolfenden, J. L. Vicente, A. D. Wolfenden, A. Menconi, L. R. Bielke, B. M. Hargis, and G. Tellez. 2016. "Evaluation and selection of *Bacillus* species based on enzyme production, antimicrobial activity, and biofilm synthesis as direct-fed microbial candidates for poultry." *Frontiers in Veterinary Science* 3:95. 10.3389/fvets.2016.00095

Lee, H., and J. Garlich. 1992. "Effect of overcooked soybean meal on chicken performance and amino acid availability." *Poultry Science 71*:499–508. 10.3382/ps.0710499

Leske, K. L., O. Akavanichan, T. Cheng, and C. Coon. 1991. "Effect of ethanol extract on nitrogen-corrected true metabolizable energy for soybean meal with broilers and roosters." *Poultry Science* 70:892–895. 10.3382/ps.0700892

Leske, K. L., C. J. Jevne, and C. N. Coon. 1993. "Effect of oligosaccharide additions on nitrogen-corrected true metabolizable energy of soy protein concentrate." *Poultry Science* 72:664–668. 10.3382/ps.0720664

Leske, K. L., and C. N. Coon. 1999. "Nutrient content and protein and energy digestibilities of ethanol-extracted, low alpha-galactoside soybean meal as compared to intact soybean meal." *Poultry Science* 78:1177–1183. 10.1093/ps/78.8.1177

Loeffler, T. 2012. "The Effect of Trypsin Inhibitors on the Nutritional Value of Various Soy Products and Broiler Performance." PhD dissertation, University of Georgia.

La Ragione, R. M., and M. J. Woodward. 2003. "Competitive exclusion by *Bacillus subtilis* spores of *Salmonella enterica* serotype Enteritidis and *Clostridium perfringens* in young chickens." *Veterinary Microbiology* 94:245–256. 10.1016/s0378-1135(03)00077-4

MacAuliffe, T., and J. McGinnis. 1971. "Effect of antibiotic supplements to diets containing rye on chick growth." *Poultry Science* 50:1130–1134. 10.3382/ps.0501130

Mathivanan, R., P. Selvaraj, and K. Nanjappan. 2006. "Feeding of fermented soybean meal on broiler performance." *International Journal of Poultry Science* 5:868–872.

McFerran, J., M. McNulty, R. McCracken, and J. Greene. 1983. "Enteritis and associated problems." In: *Disease Prevention and Control in Poultry Production*, edited by T. G. Hungerford, 129–138. Sydney: University Press.

Meng, X., B. Slominski, and W. Guenter. 2004. "The effect of fat type, carbohydrase, and lipase addition on growth performance and nutrient utilization of young broilers fed wheat-based diets." *Poultry Science* 83:1718–1727. 10.1093/ps/83.10.1718

Mian, M., and J. Garlich. 1995. "Tolerance of turkeys to diets high in trypsin inhibitor activity from undertoasted soybean meals." *Poultry Science* 74:1126–1133. 10.3382/ps.0741126

Middelbos, I. S., and G. C. Fahey Jr. 2008. "Soybean Carbohydrates." In *Soybeans: Chemistry, Production, Processing, and Utilization*, Oilseed Monograph Series, Vol. 2, edited by P. White, L. Johnson, and R. Galloway, 269–296. Urbana, IL: AOCS Press.

Mul, A., and F. Perry. 1994. "Role of fructo-oligosaccharides in animal nutrition." *Recent Advances in Animal Nutrition* 1994:57–79.

Nicholson, W. L., N. Munakata, G. Horneck, H. J. Melosh, and P. Setlow. 2000. "Resistance of *Bacillus* endospores to extreme terrestrial and extraterrestrial environments." *Microbiology and Molecular Biology Reviews* 64:548–572. 10.1128/mmbr.64.3.548-572.2000

NRC. 1994. *Nutrient Requirements of Poultry*. Washington, D.C.: National Academies Press. 10.17226/2114

Paidhungat, M., and P. Setlow. 2000. "Role of Ger proteins in nutrient and nonnutrient triggering of spore germination in *Bacillus subtilis*." *Journal of Bacteriology* 182:2513–2519. 10.1128/JB.182.9.2513-2519.2000

Paidhungat, M., and P. Setlow. 2001. "Spore Germination and Outgrowth *Bacillus subtilis* and its closest relatives: from genes to cells." In *Spore Germination and Outgrowth*, edited by A. L. Sonenshein, J. A. Hoch, R. Losick, 537–548. Wiley Online Library, Inc. 10.1128/9781555817992.ch37

Papadopoulos, M. C. 1989. "Effect of processing on high-protein feedstuffs: a review." *Biological Wastes* 29(2), 123–138.

Parsons, C., K. Hashimoto, K. Wedekind, Y. Han, and D. Baker. 1992. "Effect of overprocessing on availability of amino acids and energy in soybean meal." *Poultry Science* 71:133–140. 10.3382/ps.0710133

Parsons, C. M., Y. Zhang, and M. Araba. 2000. "Nutritional evaluation of soybean meals varying in oligosaccharide content." *Poultry Science* 79:1127–1131. doi: 10.1093/ps/79.8.1127

Peper, K. M. 2015. "Effects of Origin on the Nutritional Value of Soybean Meal." PhD diss., University of Illinois.

Perryman, K., and W. Dozier III. 2012. "Apparent metabolizable energy and apparent ileal amino acid digestibility of low and ultra-low oligosaccharide soybean meals fed to broiler chickens." *Poultry Science* 91:2556–2563.

Perryman, K., H. Olanrewaju, and W. Dozier III. 2013. "Growth performance and meat yields of broiler chickens fed diets containing low and ultra-low oligosaccharide soybean meals during a 6-week production period." *Poultry Science* 92:1292–1304.

Piggot, P. J., and D. W. Hilbert. 2004. "Sporulation of *Bacillus subtilis*." *Current Opinion in Microbiology* 7:579–586. 10.1016/j.mib.2004.10.001

Plomp, M., T. J. Leighton, K. E. Wheeler, H. D. Hill, and A. J. Malkin. 2007. "*In vitro* high-resolution structural dynamics of single germinating bacterial spores." *Proceedings of the National Academy of Sciences of The United States of America* 104:9644–9649. 10.1073/pnas.0610626104

Pusztai, A. 1991. *Plant Lectins*. NY., USA.: Cambridge University Press.

Ravindran, V. 2013. "Feed enzymes: The science, practice, and metabolic realities." *The Journal of Applied Poultry Research* 22:628–636. 10.3382/japr.2013-00739

Ravindran, V., A. Cowieson, and P. Selle. 2008. "Influence of dietary electrolyte balance and microbial phytase on growth performance, nutrient utilization, and excreta quality of broiler chickens." *Poultry Science* 87:677–688. 10.3382/ps.2007-00247

Raza, A., S. Bashir, and R. Tabassum. 2019. "An update on carbohydrases: growth performance and intestinal health of poultry." *Heliyon* 5:e01437. 10.1016/j.heliyon.2019.e01437

Renner, R., D. Clandinin, and A. Robblee. 1953. "Action of moisture on damage done during over-heating of soybean oil meal." *Poultry Science* 32:582–585.

Riaz M. 2007. *Extruding Full Fat Soy for Maximum Quality*. https://www.allaboutfeed.net/Nutrition/Raw-Materials/2007/12/Extruding-full-fat-soy-for-maximum-quality-AAF011248W/

Saini, H. 1989. "Legume Seed Oligosaccharides." In *Recent Advances of Research in Antinutritional Factors in Legume Seeds*, edited by T. F. B. Van Der Poel, J. Huisman, and I. E. Lienr, 329–341. The Netherlands: Pudoc, Wageningen. http://edepot.wur.nl/313366

Salim, H., H. Kang, N. Akter, D. Kim, J. Kim, M. Kim, J. Na, H. Jong, H. Choi, O. Suh, and W. K. Kim. 2013. "Supplementation of direct-fed microbials as an alternative to antibiotic on growth performance, immune response, cecal microbial population, and ileal morphology of broiler chickens." *Poultry Science* 92:2084–2090. 10.3382/ps.2012-02947

Samanya, M., and K. Yamauchi. 2002. "Histological alterations of intestinal villi in chickens fed dried *Bacillus subtilis* var. natto." *Comparative Biochemistry and Physiology Part A: Molecular & Integrative Physiology* 133:95–104. doi: 10.1016/s1095-6433(02)00121-6

Setlow, B., E. Melly, and P. Setlow. 2001. "Properties of spores of *Bacillus subtilis* blocked at an intermediate stage in spore germination." *Journal of Bacteriology* 183:4894–4899. 10.1128/JB.183.16.4894-4899.2001

Setlow, P. 2003. "Spore germination." *Current Opinion in Microbiology* 6:550–556. 10.1016/j.mib.2003.10.001

Setlow, P. 2006. "Spores of *Bacillus subtilis*: their resistance to and killing by radiation, heat and chemicals." *Journal of Applied Microbiology* 101:514–525. 10.1111/j.1365-2672.2005.02736.x

Setlow, P. 2013. "Summer meeting 201-when the sleepers wake: the germination of spores of *Bacillus* species." *Journal of Applied Microbiology* 115:1251–1268. 10.1111/jam.12343

Shepherd, E., and B. Fairchild. 2010. "Footpad dermatitis in poultry." *Poultry Science* 89:2043–2051. 10.3382/ps.2010-00770

Shivaramaiah, S., R. Wolfenden, J. Barta, M. Morgan, A. Wolfenden, B. Hargis, and G. Téllez. 2011. "The role of an early *Salmonella* Typhimurium infection as a predisposing factor for necrotic enteritis in a laboratory challenge model." *Avian Diseases* 55:319–323. 10.1637/9604-112910-ResNote.1

Smits, C. H., and G. Annison. 1996. "Non-starch plant polysaccharides in broiler nutrition-towards a physiologically valid approach to their determination." *World's Poultry Science Journal* 52:203–221. 10.1079/WPS19960016

Song, Y.-S., J. Frias, C. Martinez-Villaluenga, C. Vidal-Valdeverde, and E. G. de Mejia. 2008. "Immunoreactivity reduction of soybean meal by fermentation, effect on amino acid composition and antigenicity of commercial soy products." *Food Chemistry* 108:571–581. 10.1016/j.foodchem.2007.11.013

Tellez, G., J. D. Latorre, V. A. Kuttappan, M. H. Kogut, A. Wolfenden, X. Hernandez-Velasco, B. M. Hargis, W. G. Bottje, L. R. Bielke, and O. B. Faulkner. 2014. "Utilization of rye as energy source affects bacterial translocation, intestinal viscosity, microbiota composition, and bone mineralization in broiler chickens." *Frontiers in Genetics* 5:339. 10.3389/fgene.2014.00339

Tellez, G., J. D. Latorre, V. A. Kuttappan, B. M. Hargis, and X. Hernandez-Velasco. 2015. "Rye affects bacterial translocation, intestinal viscosity, microbiota composition and bone mineralization in turkey poults." *PLoS One* 10:e0122390. 10.1371/journal.pone.0122390

Veldman, A., W. Veen, D. Barug, and P. Van Paridon. 1993. "Effect of alpha-galactosides and alpha-galactosidase in feed on ileal piglet digestive physiology." *Journal of Animal Physiology and Animal Nutrition* 69:57–65.

Van der Klis, J., and A. Van Voorst. 1993. "The effect of carboxy methyl cellulose (a soluble polysaccharide) on the rate of marker excretion from the gastrointestinal tract of broilers." *Poultry Science* 72:503–512. 10.1080/0007166930841765

Van Kempen, T., E. Van Heugten, A. Moeser, N. Muley, and V. Sewalt. 2006. "Selecting soybean meal characteristics preferred for swine nutrition." *Journal of Animal Science* 84:1387–1395. 10.2527/2006.8461387x

Wagner, D., and O. Thomas. 1978. "Influence of diets containing rye or pectin on the intestinal flora of chicks." *Poultry Science* 57:971–975. 10.3382/ps.0570971

Warnick, R., and J. Anderson. 1968. "Limiting essential amino acids in soybean meal for growing chickens and the effects of heat upon availability of the essential amino acids." *Poultry Science* 47:281–287. 10.3382/ps.0470281

Wolfenden, R., N. Pumford, M. Morgan, S. Shivaramaiah, A. Wolfenden, C. Pixley, J. Green, G. Tellez, and B. Hargis. 2011. "Evaluation of selected direct-fed microbial candidates on live performance and *Salmonella* reduction in commercial turkey brooding houses." *Poultry Science* 90:2627–2631. 10.3382/ps.2011-01360

Yang, Y., Y. Kim, J. Lohakare, J. Yun, J. Lee, M. Kwon, J. Park, J. Choi, and B. Chae. 2007. "Comparative efficacy of different soy protein sources on growth performance, nutrient digestibility, and intestinal morphology in weaned pigs." *Asian-Australasian Journal of Animal Sciences* 20:775–783. 10.5713/ajas.2007.775

Yi, X., and P. Setlow. 2010. "Studies of the commitment step in the germination of spores of bacillus species." *Journal of Bacteriology* 192:3424–3433. 10.1128/JB.00326-10

Yurong, Y., S. Ruiping, Z. ShiMin, and J. Yibao. 2005. "Effect of probiotics on intestinal mucosal immunity and ultrastructure of cecal tonsils of chickens." *Archives of Animal Nutrition* 59:237–246. 10.1080/17450390500216928

13 Microbial Valorization of Beverage Crop Residues

E Varun and Pushpa S Murthy
Department of Spice and Flavor Science, CSIR-Central Food Technological Research Institute, Mysuru, India

CONTENTS

13.1 INTRODUCTION

The development and modernization of agro-food industrialization has resulted into production of enormous quantities of diverse wastes. These wastes can cause serious environmental pollution if not managed properly and also significantly

DOI: 10.1201/9781003128977-13

impact on water quality and a general loss of aesthetics. The much-used natural beverages such as coffee, tea, and cocoa require major processing steps before entering into global market. Additionally, huge quantity of residues is produced in almost every processing stage till it reached the final form, and these byproducts especially obtained from beverage crops are utilized as a renewable resource. The accumulation of waste increases with increasing worldwide populace (Ravindran and Jaiswal 2016). The rational conduct of byproducts has a vital part in the sustainability of beverage industries (Waldron 2009).

The recycling and reuse of agri-food residues for commercial utilization is highly growing fast in the scientific community as a sustainable concept. Many biotechnological approaches including anaerobic digestion, fermentation, and composting technologies are potentially helpful for conversion of litters into biofertilizers, biofuels, biomass, and bioactive chemicals (Schieber, Stintzing, and Carle 2001). The sustainable management of residues from agro-processing industries includes environmentally sound management of waste, in which micro-organisms play a huge role in bioconversion and composting technologies. The micro-organisms have a variety of essential functions that are underutilized in resolving several issues that manhood needs to manage the environment. Biological methods include application of microbes such as bacteria, fungi, algae, virus, and protozoa in techniques such as composting, activated sludge, trickling filters, and oxidation ponds. Scientific management of microbes in environment composes hybrid approaches such as bioreactor, anaerobic digestion, and vermiculture technology. The chapter summarizes the sustainable management of beverage crops residues, i.e., coffee, cocoa, and tea wastes with notably on microbial bioconversion.

13.2 GENERATION OF BEVERAGE CROP WASTE

The accumulation of agri-food wastes and ill impact on the environment and economy are increasing. Yearly, around 998 million tons of agro-food wastes are produced all over the world. Each country contributes almost 15% of total world waste generation. Studies from Baiano (2014) estimates that around 26% of food wastes are obtained from the beverage industry, followed by manufacture of dairy products (21%), fruit/vegetable (14.8%), cereals (12.9%), meat (8%), vegetable and animal oils (3.9%), fish (0.4%), and others (12.7%) (Baiano 2014). These beverage crops are having the high potential for generation of waste/byproducts, thus require attention for eco-friendly management. The generation of waste from various beverage crops (coffee, cocoa, and tea) is described as follows.

13.2.1 Coffee

About 11 million hectares of farmland worldwide is coffee cultivation with the production of 84.3 and 59.1 million bags of Arabica and Robusta coffee, respectively. The Europe continent possesses highest per capita coffee consumption among the world coffee segments. Coffee undergoes various postharvest processing stages such as dry process, wet process, and semi-dry process. Coffee residues/ byproducts such as cherry pulp, parchment husk, husk, silverskin, and spent coffee

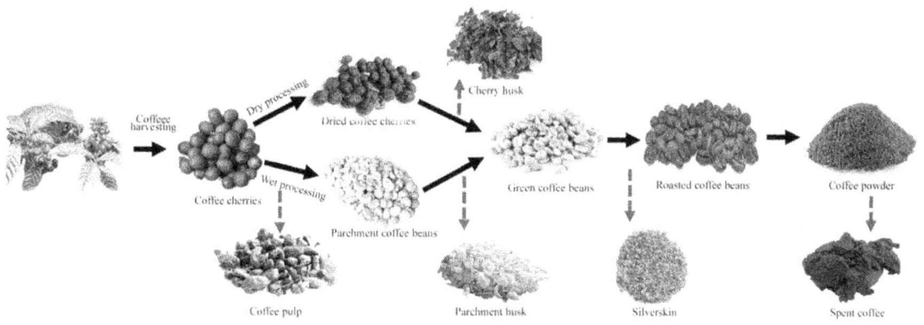

FIGURE 13.1 Coffee processing and production of various byproducts.

are generated through various processing stages of coffee (Figure 13.1). Coffee pulp (CP) is the skin of the coffee fruit and accounts to 29% of dry weight of the berry which is approximately 1 ton for every 2 tons of coffee produced. (Roussos et al. 1995). During the dry processing, coffee cherry husks (CCHs) are obtained, which is almost 12% of the berry on dry weight basis. The coffee cherries generate about 0.18 tons of husks for every 1 ton. Water is required in large quantities for coffee processing especially for wet processed coffee. Huge quantity of effluents is released during wet processing, estimated around 187 billion liters in 2019 in Brazil, which are toxic and deplete the oxygen in the water (Caldarelli, Gilio, and Zilberman 2019). All the waters used in various stages contribute to the total coffee wastewater (CWW) produced. The 50% of the coffee generated globally is utilized for soluble coffee manufacture. (Ramalakshmi et al. 2009). One-ton green coffee, on an average, produces around 650 kg of spent coffee (SC) and about 2 kg of wet SC are released to each 1 kg of soluble coffee produced (Murthy and Naidu 2012b).

13.2.2 Cocoa

Cocoa is a chief crop of the tropical world with generation of 3.9 million tons has an export value of 47 billion dollars (Vásquez et al. 2019). The cocoa postharvest practices generate number of residues or byproducts that are generally discarded (Figure 13.2). The generation of cocoa pod waste covers almost 60–70% of pods and cocoa shell accounts for 17% of cocoa bean weight. Thus, by comparing with

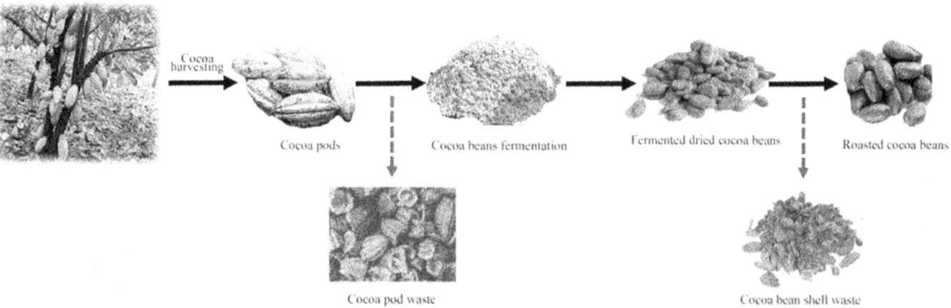

FIGURE 13.2 Cocoa processing and generation of waste.

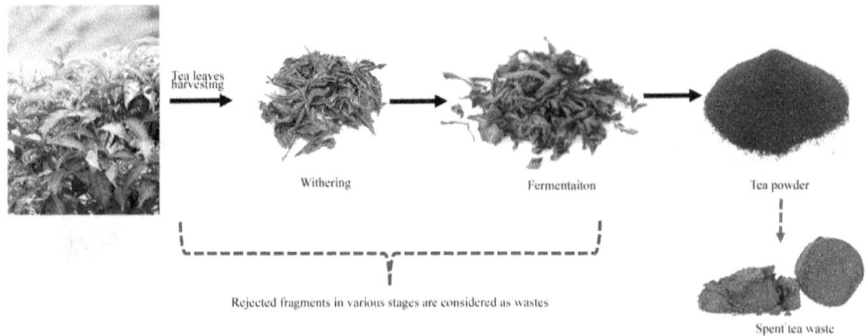

FIGURE 13.3 Tea processing and generation of waste.

the world cocoa production, the annual generation of this waste can be calculated at approximately 7 lakh tons worldwide (Handojo, Triharyogi, and Indarto 2019).

13.2.3 Tea

In 2013, world tea production (black, green, and instant) has spiked eloquently by 6% having an overall production of 5.07 million tons, in which China accounts for 38% of world's produce, being the largest tea producing country with an output of 1.9 million tons followed by India with 1.2 million tons in 2013 (Soni et al. 2015). Additionally, processing of tea is marginally diverse than coffee and cocoa. In which rather than byproducts, waste is generated after preparation of final cup, i.e., spent tea waste (Figure 13.3). Out of 860,000 metric tons of production, 190,000 metric tons constitute leaf and fragment wastes that are excluded from the final packaged product (Menzie et al. 2010). In the tea processing industry, fiber portion of the leaves is removed and casted off as waste, which is called factory tea waste (FTW). These FTWs are rich in caffeine and subjected to extraction process and the remains after caffeine removal is called decaffeinated tea waste (DCTW). The tea waste produced also contaminates the environment (Chowdhury et al. 2016).

13.3 CONVENTIONAL MANAGEMENT OF CROP WASTE/RESIDUES

Around 30% of global agriculture produce are generated as residual wastes and refuses. Enormous quantity of solid and liquid waste produced from various agro-/food industries are still not disposed in a sustainable manner. Given below are the prevalent waste disposal methods followed in most countries.

13.3.1 Landfill and Open Dumping

Landfill is the primeval and common mode of disposal. Generally, in many countries, 60–80% of the wastes are disposed in swamps and in low lying regions (Ngoc and Schnitzer 2009). These methods of waste disposal have become one of the major sources of environmental pollution.

13.3.2 Incineration or Thermal Treatment

Incineration is another method of treating the waste by combustion. All the thermal treatment systems including incineration require high temperature. Though burning of these wastes seems to be simple and easiest way, the adaptability depends on the components of the waste. These applications need huge economical support and operating capital compulsions.

13.3.3 Composting or Biodigesting

Composting is a biological method of treatment, where micro-organisms involve in conversion of organic waste. Micro-organisms decompose the organic waste leading to highly nutritive organic manure (biofertilizer for plants). Even though composting involves lesser technologies, the operational and maintenance costs are high. The composting produces low cost of organic manure/biofertilizer compared to the commercial chemical fertilizers.

13.3.4 Recycling or Recovery

Many developed countries have been effectively carried out recycling of the solid wastes. In the developed countries, nearly 44% of solid wastes are recycled, where as it is 12% in the underdeveloped countries and 8–11% in other low-income countries (Ngoc and Schnitzer 2009). However, recycling of various wastes is primarily composed of paper, rubber, glass, plastic, ferrous, etc., which are raw materials for new products.

13.4 MICROBIAL BIOCONVERSION OF BEVERAGE CROP WASTE

13.4.1 Bioconversion of Coffee Wastes

The coffee processing generated various coffee byproducts during dry and wet processing such as coffee cherry husk, pulp, silverskin, and spent coffee, which are rich in cellulose, hemicellulose, and lignin content. Additionally, coffee residues comprise substantial quantities of nutrients rich in fermentable sugars. These constitute suitable substrates for micro-organism growth and in turn producing its metabolites or valuable products such as organic acids, enzymes, flavor, and volatile compounds or pigments (Pandey et al. 2000; Zheng and Shetty 2000; Soares et al. 2000; Silveira, Daroit, and Brandelli 2008).

13.4.1.1 Enzyme Production

The enzymes regulate the majority of the chemical changes that occur in living tissues. Over a decade, much interest has been renewed for enzymes' production through microbial fermentation. Agricultural byproducts and wastes can often be used as substrates for microbial growth because of their carbon-rich composition containing nitrogen and minerals. Murthy and Naidu (2010) evaluated coffee byproducts with the application of *Penicellium* Sp CFR 303 mediated solid-state fermentation for production of xylanase. The coffee cherry husk was found to be highest xylanase activity (9,475 U/g) compared to other residues (Murthy and Naidu 2012a). These xylanase enzymes have high demand in the food and paper-making industries, and pharmaceutical industries. Additionally, the coffee husk also worked excellent for the production of commercial enzymes exoglucanase and β-glucosidase with the application of *Rhizopus stolonifera* (Navya, Bhoite, and Murthy 2012). The *A. oryzae* significantly performed against coffee cherry husk for the production of protease (Murthy and Naidu 2010). Another industrially important enzyme, tannase, was studied in coffee husk with the application of *P. variotii*. The solid-state fermentation of husk using *P. variotii* for 5 days by maintaining temperature of 29–34°C with the supplement of tannic acid (8.5–14%) and coffee husk:wheat bran (50:50) observed to the best optimum condition for the production of tannase (Battestin and Macedo 2007). Similarly, Bhoite and team (2013) obtained the tannase from a new fungal strain *Penicillium verrucosum* using coffee pulp as a substrate by solid-state fermentation (Bhoite, Navya, and Murthy 2013a). And the same team reported Gallic acid (3, 4, 5-trihydroxybenzoic acid) production by microbial biotransformation of coffee pulp tannins by P. verrucosum (Bhoite, Navya, and Murthy 2013b). Amylase is another important commercial enzyme that is also obtained with steam-pretreated coffee pulp demonstrating maximum α-amylase activity of 7084 U g^{-1} ds (Pushpa and Manonmani 2008). Hence, these coffee wastes are proved to be excellent carbon source for synthesis of various enzymes through micro-organisms.

13.4.1.2 Flavor Compound Production

Majority of the flavoring additives are produced from extraction from natural materials or chemical synthesis. Presently, there has been a growing interest in food flavor production and other aroma additives of plant/microbial origin. Many micro-organisms, including bacteria and fungi, are exploited for synthesis of different aroma compounds. Coffee wastes are also having potential for producing flavor compounds and demonstrated by many researches. Soares and co-workers (2000) researched on the application of Ceratocystis fimbriata for aroma compound production: fruity aroma and strong pineapple flavor (Soares et al. 2000). Similarly, Medeiros and team (2006) also obtained fruity aroma compounds, such as ethanol, ethyl acetate, ethyl propionate, acetaldehyde, and isoamyl acetate by subjecting coffee husk with the *C. fimbriata* in bioreactors (Medeiros et al. 2006)

13.4.1.3 Alcohol Production

Fermentation technology can be employed in bioethanol production from byproducts of coffee processing such as husk, pulp, and spent coffee (Choi et al. 2012; Gouvea et al. 2009; Shenoy et al., 2011: Woldesenbet, Woldeyes, and Chandravanshi 2016). Gouvea and co-workers (2009) evaluated the feasibility of coffee husks for the production of ethanol by fermentation using *S. cerevisiae*. The optimum results were observed when coffee husks were inoculated with 3 g of yeast per liter of the substrate and incubated at 30°C temperature. The ethanol production was found to be 8.49 ± 0.29g/100g (dry basis) (13.6 ± 0.5 g ethanol/L), while residues such as corn stalks, barley straw, and hydrolyzed wheat stillage observed 5–11 g ethanol/L

(Gouvea et al. 2009). Sampaio and team (2013) utilized spent coffee for ethanol production by subjecting it into acid hydrolysis process followed by fermentation using yeast (*S. cerevisiae*) and observed the ethanol yield of 50.1% (Sampaio et al. 2013). Dadi and co-workers (2018) utilized the coffee waste for the production of bioethanol in an intergraded approach involving cellulolytic enzymes and ligno-cellulosic yeast. The coffee wastes (spent coffee and husk) were subjected to hydrolysis expending cellulose complex and b-glucosidase enzymes, fermenting with lignocellulosic yeast, purification and the bioethanol yielded 51.7 ± 7.4 and 132.2 ± 40g/L (Dadi et al. 2018).

13.4.1.4 Organic Acid Production

Vandenberghe et al. (1999) studied coffee husk for manufacture of organic acids using *A. niger* mediated solid-state fermentation and found coffee husk as a suitable substrate for citric acid production (Vandenberghe et al. 1999). Similarly, coffee husk was also utilized by Maria and team (2008) for the gibberellins (GA) production in solid-state fermentation and submerged fermentation. The five strains of *Gibberella fujikuroi* with comparison using *Fusarium moniliforme* were used against coffee husk and observed that GA production reached 1100 mg/kg of dry coffee husk. Also concluded by comparing all the fermentation, solid-state fermentation appeared superior to submerged fermentation (Maria, Machado, and Soccol 2008).

13.4.1.5 Protein Production

The coffee pulp composed of 10.5% protein content and transformed by application of *S. chattanoogensis*, S UAH 47 and S UAH Nic-C. The noticeable improvement has been observed in protein compostion of the residues (30, 21, and 28% for S. chattanoogensis, *S.* UAH 47, and *S.* UAH Nic-C, respectively). The obtained protein enhancement in coffee pulp is higher compared to other lignocellulosic residues (cassava or banana) on fermentation with Aspergillus or Rhizopus strains. The coffee pulp protein composition can be enriched by the action of *Streptomyces* strains, which can be utilized as essential feed (Pandey et al. 2000).

13.4.1.6 CWW Treatment

The washed coffee processing generates enormous amount of wastewater, which requires high attention of systemic treatment prior to disposal. The CWW is highly potent in causing environment pollution due to intense values of chemical oxygen demand (COD) (50 g/L) and biochemical oxygen demand (BOD) (6–20 g/L), a low pH (<4.0), high ammoniacal nitrogen (40–60 mg/L), total nitrogen (180–250 mg/L), phosphorus (60–800 mg/L), and total solids (1–7.5 g/L) (Campos, Prado, and Pereira 2010; Villanueva-Rodríguez et al. 2014; Pires et al. 2017). Pires and team (2020) demonstrated by application of CWW with a bacterial consortium of *S. marcescens* CCMA 1010 and CCMA 1012, *C. flavescens* CCMA 1006 and *A. indonesiensis* CCMA 1002 resulted in 85% decrease in BOD and 60% decrease in COD. The application of consortium (11.18 log CFU/mL) represented 80% reduction in phosphorus and nitrogen with an increase in final pH (6.0–7.5). Also, ~59% of eco-toxicity was reduced with application of Daphnia similis (Pires et al. 2020).

13.4.1.7 Other Bioconversions of Coffee Waste

Green coffee is another wing of coffee processing getting popular in recent years. The processing and preparation of green coffee also generates large amount of green coffee spent (70%). As these residual wastes composed of high quantity of polyphenols such as trigonelline, caffeine, theobromine, and theophylline, various value-added products can be developed. There is less information available on the reuse/bioconversion of silverskin. This silverskin composed of antioxidants, dietary fiber can be a source of substrate for *A. japonicas* for fructo-oligosaccharides and β-furctofuranosidase production through solid-state fermentation (Mussatto et al. 2009; Pourfarzad, Mahdavian-Mehr, and Sedaghat 2013).

13.4.2 Bioconversion of Cocoa Wastes

Cocoa industry is one of the comprehensive food supply chains with $47 billion globally. During the cocoa process about 80% of the byproducts in form of cocoa pod, husk, bean shell, and sweatings are released. The transformation of these products for value addition is vital.

13.4.2.1 Enzyme Production

The theobromine and caffeine present in cocoa meal, which are byproducts of cocoa processing, are considered as anti-nutritional for animals. The researchers Amorim et al. 2017 explored application of *A. awamori* IOC-3914 with cocoa meal and reported reduction of methyxanthines followed by xylanse production. Laccase is an enzyme that can be used in various processes in the field of pharmaceutical, biotechnology, and food. The laccase enzyme can be produced by white-weathered fungi, one of them is *Marasmius* sp. by utilizing the lignin components of ligno-cellulose of agricultural waste and bark of cocoa beans (Ramadiyanti et al. 2020). Other research shows that the larger the mesh of bark of cocoa beans, the higher the enzyme activity.

13.4.2.2 Alcohol Production

Cocoa sweating (CSW) is a liquid generated/leach-out during fermentation of beans. These CSWs are rich in sugars and can be subjected to yeast (*S. cerevisiae*) for alcohol production and vinegar. The pH of CSW ranges 3–4 with suitable content of alcohol (11%). CSW with high alcoholic content would be more preferable for people that cherish high alcohol content. Similarly, coco powder (CPW) for wine production with the good amount of alcohol (8 %) can be obtained. The CPW with low alcohol content is sweet and would target consumers who prefer sweet taste and could be served as table wine. Developing wines from cocoa would increase value addition opportunities available to farmers and further augment their incomes (Jayeola and Lawal 2008). Cocoa shell is another byproduct generated during the roasting process. These shells are rich in lignin content and most suitable for ethanol production. Awolu & Oyeyemi (2015) studied the cocoa shells for ethanol production with the application of acid hydrolysis and yeast (*S. cerevisiae*) and found effective. The pH highly influences the ethanol productivity along with the additional factors such as fermentation time and concentration of yeasts (Awolu and Oyeyemi 2015).

13.4.2.3 Other Bioconversions of Cocoa Waste

The fermentation of the cocoa beans releases a mucilaginous liquid (cocoa honey), which has a sweet-sour flavor and rich in sugars (glucose and fructose) and bioactive compounds. Thus, a good substrate for the development of micro-organisms. These microbes are essential in fermentation and responsible for development of the aromas and flavors of the end product (Ouattara et al. 2014). The lignocellulosic cocoa shells generated during roasting exhibited fair potential in biogas production with cumulative methane yields (Awolu and Oyeyemi 2015).

13.4.3 Bioconversion of Tea Wastes

Maximal methane was produced by using spent tea (25%) and cow manure (75%) in anaerobic digesters (Ozbayram 2020). *A. niger* (ARNU 4) was studied for the production of gluconic acid using tea wastes as substrate along with molasses-based fermentation medium (Sharma, Vivekanand, and Singh 2008). Gluconic acid is an oxidative product of glucose and has extensive usage in textile, cement, leather, food, and pharmaceutical industries (Roehr, Kubicek, and Komínek 1996) (Table 13.1).

TABLE 13.1
Production of Industrially Important Products Through Microbial Bioconversion

Residues/byproducts	Micro-organisms	Production	References
Coffee cherry husk	*Penicillium sp. CFR 303*	Xylanase	Murthy and Naidu (2012a)
Coffee cherry husk	*Rhizopus stolonifera*	*Exoglucanase and β-glucosidase*	Navya, Bhoite, and Murthy (2012)
Coffee cherry husk	*A. oryzae*	Protease	Murthy and Naidu (2010)
Coffee cherry husk	*P. variotii*	*Tannase*	Battestin and Macedo (2007)
Coffee pulp	*Penicillium verrucosum*	*Tannase*	Bhoite, Navya, and Murthy (2013a)
Coffee husk		*Amylase*	Pushpa and Manonmani (2008)
Coffee husk	*Ceratocystis fimbriata*	*Fruity aroma and Strong pine apple flavor*	Soares et al. (2000); Medeiros et al. (2006)
Coffee husk	*S. cerevisiae*	*Ethanol*	Gouvea et al. (2009); Dadi et al. (2018)
Coffee husk	*A. niger*	*Citric acid*	Vandenberghe et al. (1999)
Coffee husk	*Gibberella fujikuroi*	*Gibberellic acid*	Maria, Machado, and Soccol (2008)
Coffee pulp	*S. chattanoogensis, Streptomyces strains*	Protein	Pandey et al. (2000)

(Continued)

TABLE 13.1 (Continued)
Production of Industrially Important Products Through Microbial Bioconversion

Residues/ byproducts	Micro-organisms	Production	References
Silverskin	*Aspergillus japonicas*	*Fructo-oligosaccharides and β-furctofuranosidase*	Mussatto et al. (2009)
Cocoa waste	*Aspergillus awamori*	Xylanase	Amorim et al. (2017)
Bark of cocoa beans	*Marasmius* sp.	Laccase	Ramadiyanti et al. (2020)
Cocoa shell	*S. cerevisiae*	*Ethanol, Biogas*	Awolu and Oyeyemi (2015)
Cocoa sweatings	*S. cerevisiae*	Ethanol, Vinegar	Adams et al. (1982); Jayeola and Lawal (2008)
Tea waste	*Aspergillus niger*	Gluconic acid	Sharma, Vivekanand, and Singh (2008)

Demand for eco-friendly/sustainable way of pollution prevention and waste management is growing worldwide. The crop residues and byproducts are regarded as a new raw material source for industries and for the production of value-added products. Effective management and utilization of wastes generated from agriculture and food sector can become an asset to the agro-food industries. The agro-byproducts, which are rich in bioenergy can enhance the productivity of agriculture and increase the efficiency of resource utilization by deploying appropriate technologies for the processing and reuse of the same. Various micro-organisms and variants are gaining much exploration with emphasis on composting or bio-conversion. More microbial cultures with various formulations or encapsulations for wide applications are the highly demanding insights.

13.5 CONCLUSION

The utilization of agro-food wastes to develop economically important products such as ethanol, mushroom, enzymes, organic acids, and compost are in high demand. Particularly, enzymes are the most evident biocatalysts for industries and pharmaceutical sector in a various preparative stages of production. The worldwide requirement of enzymes is rising 6.3% annually with a strong demand in the specialty enzymes. Majority of the agro-food wastes are abundant and cost-effective. The application of these bioconversion micro-organisms could be a potential solution for agro-food waste management in a sustainable manner, which can generate industrially important products, bioplastics, biofertilizers, biosurfactants, biofuels, etc. The drawback is the lack of infrastructure and technology implementation to deal with this large supply. Hence, the current issue requires critical attention to maintain the ecological balance.

REFERENCES

Adams, M. R., J. Dougan, E. J. Glossop, and D. R. Twiddy. 1982. "Cocoa Sweatings—An Effluent of Potential Value." *Agricultural Wastes* 4 (3): 225–229.

Amorim, G. M., A. C. Oliveira, M. L. E. Gutarra, M. G. Godoy, and D. M. G. Freire. 2017. "Solid-State Fermentation as a Tool for Methylxanthine Reduction and Simultaneous Xylanase Production in Cocoa Meal." *Biocatalysis and Agricultural Biotechnology* 11: 34–41.

Awolu, O. O., and S. O. Oyeyemi. 2015. "Optimization of Bioethanol Production from Cocoa (*Theobroma cacao*) Bean Shell." *International Journal of Current Microbiology and Applied Science* 4 (4): 506–514.

Baiano, A. 2014. "Recovery of Biomolecules from Food Wastes—A Review." *Molecules* 19 (9): 14821–14842.

Battestin, V., and G. A. Macedo. 2007. "Tannase Production by *Paecilomyces variotii*." *Bioresource Technology* 98 (9): 1832–1837.

Bhoite, R. N., P. N. Navya, and P. S. Murthy. 2013a. "Purification and Characterisation of a Coffee Pulp Tannase Produced by *Penicillium verrucosum*." *Journal of Food Science and Engineering* 3 (6): 323.

Bhoite, R. N., P. N. Navya, and P. S. Murthy. 2013b. "Statistical Optimization of Bioprocess Parameters for Enhanced Gallic Acid Production from Coffee Pulp Tannins by *Penicillium verrucosum*." *Preparative Biochemistry and Biotechnology* 43 (4): 350–363.

Caldarelli, C. E., L. Gilio, and D. Zilberman. 2019. "The Coffee Market in Brazil: Challenges and Policy Guidelines." *Revista de Economia* 39 (69).

Campos, C. M. M., M. A. C. Prado, and E. L. Pereira. 2010. "Physical-Chemical, Biochemical and Energetic Characterization of Wastewater Originated from Wet Coffee Processing." *Bioscience Journal* 26 (4): 514–524.

Choi, I. S., S. G. Wi, S. B. Kim, and H.-J. Bae. 2012. "Conversion of Coffee Residue Waste into Bioethanol with Using Popping Pretreatment." *Bioresource Technology* 125: 132–137.

Chowdhury, A., S. Sarkar, A. Chowdhury, S. Bardhan, P. Mandal, and M. Chowdhury. 2016. "Tea Waste Management: A Case Study from West Bengal, India." *Indian Journal of Science and Technology* 9 (42): 1–6.

Dadi, D., A. Beyene, K. Simoens, J. Soares, M. M. Demeke, J. M. Thevelein, K. Bernaerts, P. Luis, and B. Van der Bruggen. 2018. "Valorization of Coffee Byproducts for Bioethanol Production Using Lignocellulosic Yeast Fermentation and Pervaporation." *International Journal of Environmental Science and Technology* 15 (4): 821–832.

Gouvea, B. M., C. Torres, A. S. Franca, L. S. Oliveira, and E. S. Oliveira. 2009. "Feasibility of Ethanol Production from Coffee Husks." *Biotechnology Letters* 31 (9): 1315–1319.

Handojo, L., H. Triharyogi, and A. Indarto. 2019. "Cocoa Bean Shell Waste as Potential Raw Material for Dietary Fiber Powder." *International Journal of Recycling of Organic Waste in Agriculture* 8 (1): 485–491.

Jayeola, C. O., and J. O. Lawal. 2008. "Comparative Study of Wine Produced from Cocoa Sweatings and Cocoa Powder." *Journal of Applied Biosciences* 8 (2): 331–333.

Maria, C., M. Machado, and C. R. Soccol. 2008. "Gibberellic Acid Production." In *Current Developments in Solid-State Fermentation*, 277–301. Springer.

Medeiros, A. B. P., A. Pandey, L. P. S. Vandenberghe, G. M. Pastore, and C. R. Soccol. 2006. "Production and Recovery of Aroma Compounds Produced by Solid-State Fermentation Using Different Adsorbents." *Food Technology and Biotechnology* 44 (1): 47–51.

Menzie, W. D., J. J. Barry, D. I. Bleiwas, E. L. Bray, T. G. Goonan, and G. Matos. 2010. *The Global Flow of Aluminum from 2006 through 2025*. US Department of the Interior, US Geological Survey.

Murthy, P. S., and M. M. Naidu. 2010. "Protease Production by Aspergillus Oryzae in Solid-State Fermentation Utilizing Coffee by-Products." *World Applied Sciences Journal* 8 (2): 199–205.

Murthy, P. S., and M. M. Naidu. 2012a. "Production and Application of Xylanase from Penicillium Sp. Utilizing Coffee by-Products." *Food and Bioprocess Technology* 5 (2): 657–664.

Murthy, P. S., and M. M. Naidu. 2012b. "Sustainable Management of Coffee Industry By-Products and Value Addition—A Review." *Resources, Conservation and Recycling* 66: 45–58.

Mussatto, S. I., C. N. Aguilar, L. R. Rodrigues, and J. A. Teixeira. 2009. "Fructooligosaccharides and β-Fructofuranosidase Production by *Aspergillus japonicus* Immobilized on Lignocellulosic Materials." *Journal of Molecular Catalysis B: Enzymatic* 59 (1–3): 76–81.

Navya, P. N., R. N. Bhoite, and P. S. Murthy. 2012. "Improved β-Glucosidase Production from *Rhizopus stolonifer* Utilizing Coffee Husk." *International Journal of Current Research* 4 (8): 123–129.

Ngoc, U. N., and H. Schnitzer. 2009. "Sustainable Solutions for Solid Waste Management in Southeast Asian Countries." *Waste Management* 29 (6): 1982–1995.

Ouattara, D. H., H. G. Ouattara, B. G. Goualie, L. M. Kouame, and S. L. Niamke. 2014. "Biochemical and Functional Properties of Lactic Acid Bacteria Isolated from Ivorian Cocoa Fermenting Beans." *Journal of Applied Biosciences* 77: 6489–6499.

Ozbayram, E. G.. 2020. "Waste to Energy: Valorization of Spent Tea Waste by Anaerobic Digestion." *Environmental Technology*, 14: 1–24.

Pandey, A., C. R. Soccol, P. Nigam, D. Brand, R. Mohan, and S. Roussos. 2000. "Biotechnological Potential of Coffee Pulp and Coffee Husk for Bioprocesses." *Biochemical Engineering Journal* 6 (2): 153–162.

Pires, J. F., L. de Souza Cardoso, R. F. Schwan, and C. F. Silva. 2017. "Diversity of Microbiota Found in Coffee Processing Wastewater Treatment Plant." *World Journal of Microbiology and Biotechnology* 33 (12): 211.

Pires, J. F., D. C. Viana, R. A. Braga Jr, R. F. Schwan, and C. F. Silva. 2020. "Protocol to Select Efficient Microorganisms to Treat Coffee Wastewater." *Journal of Environmental Management* 278: 111541.

Pourfarzad, A., H. Mahdavian-Mehr, and N. Sedaghat. 2013. "Coffee Silverskin as a Source of Dietary Fiber in Bread-Making: Optimization of Chemical Treatment Using Response Surface Methodology." *LWT-Food Science and Technology* 50 (2): 599–606.

Pushpa, S. M., and H. K. Manonmani. 2008. "Bioconversion of Coffee Industry Wastes with White Rot Fungus Pleurotus Florida." *Research Journal of Environmental Sciences* 2 (2): 145–150.

Ramadiyanti, Mita, M. Djali, E. Mardawati, and R. Andoyo. 2020. "Production of Laccase Enzyme by Marasmius Sp. from the Bark of Cocoa Beans." *Systematic Reviews in Pharmacy* 11 (3): 405–409.

Ramalakshmi, K., L. Jagan Mohan Rao, Y. Takano-Ishikawa, and M. Goto. 2009. "Bioactivities of Low-Grade Green Coffee and Spent Coffee in Different in Vitro Model Systems." *Food Chemistry* 115 (1): 79–85.

Ravindran, R., and A. K. Jaiswal. 2016. "Exploitation of Food Industry Waste for High-Value Products." *Trends in Biotechnology* 34 (1): 58–69.

Roehr, M., C. P. Kubicek, and J. Komínek. 1996. "Gluconic Acid." *Biotechnology: Products of Primary Metabolism*, 6: 347–362.

Roussos, S., M. De los Angeles Aquiahuatl, M. del Refugio Trejo-Hernández, I. Gaime Perraud, E. Favela, M. Ramakrishna, M. Raimbault, and G. Viniegra-González. 1995. "Biotechnological Management of Coffee Pulp—Isolation, Screening, Characterization, Selection of Caffeine-Degrading Fungi and Natural Microflora Present in Coffee Pulp and Husk." *Applied Microbiology and Biotechnology* 42 (5): 756–762.

Sampaio, A., G. Dragone, M. Vilanova, J. M. Oliveira, J. A. Teixeira, and S. I. Mussatto. 2013. "Production, Chemical Characterization, and Sensory Profile of a Novel Spirit Elaborated from Spent Coffee Ground." *LWT-Food Science and Technology* 54 (2): 557–563.

Schieber, A., F. C. Stintzing, and R. Carle. 2001. "By-Products of Plant Food Processing as a Source of Functional Compounds—Recent Developments." *Trends in Food Science & Technology* 12 (11): 401–413.

Sharma, A., V. Vivekanand, and R. P. Singh. 2008. "Solid-State Fermentation for Gluconic Acid Production from Sugarcane Molasses by Aspergillus Niger ARNU-4 Employing Tea Waste as the Novel Solid Support." *Bioresource Technology* 99 (9): 3444–3450.

Shenoy, D., A. Pai, R. K. Vikas, H. S. Neeraja, J. S. Deeksha, C. Nayak, and C. Vaman Rao. 2011. "A Study on Bioethanol Production from Cashew Apple Pulp and Coffee Pulp Waste." *Biomass and Bioenergy* 35 (10): 4107–4111.

Silveira, S. T., D. J. Daroit, and A. Brandelli. 2008. "Pigment Production by *Monascus purpureus* in Grape Waste Using Factorial Design." *LWT-Food Science and Technology* 41 (1): 170–174.

Soares, M., P. Christen, A. Pandey, M. Raimbault, and C. R. Soccol. 2000a. "A Novel Approach for the Production of Natural Aroma Compounds Using Agro-Industrial Residue." *Bioprocess Engineering* 23 (6): 695–699.

Soares, M., P. Christen, A. Pandey, and C. R. Soccol. 2000b. "Fruity Flavour Production by *Ceratocystis fimbriata* Grown on Coffee Husk in Solid-State Fermentation." *Process Biochemistry* 35 (8): 857–861.

Soni, R. P., M. Katoch, A. Kumar, R. Ladohiya, and P. Verma. 2015. "Tea: Production, Composition, Consumption and Its Potential as an Antioxidant and Antimicrobial Agent." *International Journal of Food and Fermentation Technology* 5 (2): 95–106.

Stadler, R. H., I. Blank, N. Varga, F. Robert, J. Hau, P. A. Guy, M. C. Robert, and S. Riediker. 2002. "Acrylamide from Maillard Reaction Products." *Nature* 419 (6906): 449–450.

Vandenberghe, L. P. S., C. R. Soccol, A. Pandey, and J. M. Lebeault. 1999. "Microbial Production of Citric Acid." *Brazilian Archives of Biology and Technology* 42 (3): 263–276.

Vásquez, Z. S., D. P. de Carvalho Neto, G. V. M. Pereira, L. P. S. Vandenberghe, P. Z. de Oliveira, P. B. Tiburcio, H. L. G. Rogez, A. G. Neto, and C. R. Soccol. 2019. "Biotechnological Approaches for Cocoa Waste Management: A Review." *Waste Management* 90: 72–83.

Villanueva-Rodríguez, M., R. Bello-Mendoza, D. G. Wareham, E. J. Ruiz-Ruiz, and M. de Lourdes Maya-Treviño. 2014. "Discoloration and Organic Matter Removal from Coffee Wastewater by Electrochemical Advanced Oxidation Processes." *Water, Air, & Soil Pollution* 225 (12): 2204.

Waldron, K. W. 2009. *Handbook of Waste Management and Co-Product Recovery in Food Processing*. Elsevier.

Woldesenbet, A. G., B. Woldeyes, and B. S. Chandravanshi. 2016. "Bio-Ethanol Production from Wet Coffee Processing Waste in Ethiopia." *SpringerPlus* 5 (1): 1903.

Zheng, Z., and K. Shetty. 2000. "Solid State Production of Polygalacturonase by Lentinus Edodes Using Fruit Processing Wastes." *Process Biochemistry* 35 (8): 825–830.

14 Microbial Role in Nanonutrient Production and Its Potential in Agriculture

J. C. Tarafdar
Former UGC-Emeritus Professor & ICAR Emeritus Scientist, ICAR-CAZRI, Jodhpur, Rajasthan, India

CONTENTS

14.1 INTRODUCTION

Fertilizers are accountable for higher agricultural production and it has been estimated that it can contribute on an average 35–40% of the productivity of any crop (Tarafdar, 2021b). Generally, fertilizers are applied by spraying or broadcasting but very less concentration reaches the desired sites due to leaching, runoff, evaporation, hydrolysis by soil moisture, or photolytic and microbial degradation. It has been estimated that about 65–70% N, 80–85% P, and 60% K of the applied fertilizers are lost in the environment, resulting in major economic losses (Kumar et al., 2014; Tarafdar, 2020). Therefore, it is essential to increase the nutrient use efficiency and reduce the risk of the environmental pollution. All essential plant nutrients can assemble as nano form from the respective salts through the physical, chemical, aerosol, and biological techniques (Tarafdar and Raliya, 2011). The synthesis of nanoparticles through microbial means has emerged very fruitful input alternative to conventional methods (Tarafdar and Rathore, 2016). The reduced nanoscale materials exhibit some unusual

DOI: 10.1201/9781003128977-14

properties enabling various systematic applications. The production of nanoparticles through the physical, chemical, or aerosol techniques is energy and capital intensive. Moreover, there is a need for toxic chemicals and nonpolar solvents in the synthesis procedure in most of the cases. Therefore, production of nanonutrient through microbial means is very important especially for agricultural use. Bacteria, fungi, actinomycetes, yeast, and viruses are found to have potential to develop nanoparticles intra- or extracellularly. They are also considered as potential biofactories for different nanonutrient synthesis (Tarafdar et al., 2018). The potential of living organisms is well known with desirable shapes and sizes of nanoparticles for different applications (Tarafdar et al., 2012a; Jeevanadam et al., 2016). Nanonutrient is made by nanoparticles and it can be defined as fertilizers contained within nanostructure formulations that can be dispatched to the targeted sites to permit the release of active ingredients keeping the plant nutrient demands. The advantages of microbial synthesized nanonutrients are: cheap, naturally encapsulated by mother protein, slow and controlled release, can maintain soil health, can trigger plant enzyme systems when applied through foliage, low requirement, more stress tolerant, enhance plant physiological activities, more soil carbon build up and nutrient use efficiency (Tarafdar, 2021a). Microbial polysaccharides produced after induced by nanoparticles have also found to be commercially important for sand dune stabilization under deserts after trapping and retaining the moisture and humidity in the sandy soils which also help in protecting the upper layer of desert soils from soil erosion (Raliya et al., 2014).

14.2 MICROBIAL SYNTHESIS OF NANONUTRIENTS

Oxidation and reduction are the major reaction intricates in biosynthesis of nanonutrients. The microbial enzymes are accountable for reduction of respective salts to their nanoparticles (Kaul et al., 2012). The nanoparticles produced by microbial synthesis process are generally more stable as they are naturally encapsulated by mother protein. In case of fungi, the fungal protein may go for biomimetic mineralization, whereas in bacteria the microbial cell reduces metal ions of specific reducing enzymes like NADH-dependent reductase or nitrate-dependent reductase. In case of yeast, the possible reasons for nanonutrients' production are membrane bound as well as cytosolic oxido-reductases reaction. Many organisms have so far identified which are capable of 100% nanonutrients' production from the respective salt solutions. Micro-organisms can be isolated from the native soils and their spore can be used for the preparation of microbial balls. Then the balls were allowed to release the enzymes to breakdown the desired salts into nanosize (Tarafdar and Raliya, 2019). Important factors identified so far for the efficient nanonutrients' production are: pH, salt concentration, temperature, microbial protein:salt ratio, and stirring speed. In general, 0.1 mM salt concentration, pH 5.5–7.0, a temperature of 28°C, and protein to salt ratio of 1:1 with stirring of 60–80 rpm are ideal for the production of nanonutrients (Tarafdar et al, 2018). A list of selected organisms developed in our laboratory for the production of different nanonutrients is shown in Table 14.1.

TABLE 14.1
Some Selected Organisms Fabricating Nanonutrients from the Respective Salts

Organism Name	NCBI Accession No.	Salt Used	Type of Nanonutrients	Average Size (nm)
Pantoeatarafdar JCT14	KC 806057	$NH_4(NO_3)_2$	N	2.5
Aspergillus tubingensis TFR5	JQ 675292	$Ca_3(PO_4)_2$	P	28.2
Aspergillus orchraceus TFR23	KC 806053	KNO_3	K	2.2
Aspergillus terreus CZR1	JF 681300	$MgNO_3$	Mg	5.1
Penicillium limosum TFR26	KF 729585	$CuSO_4$	S	23.4
Staphylotrichumcoccosporum TFR27	KF 729586	H_3BO_3	B	20.9
Aspergillus oryzae TFR9	JQ675292	Fe_2O_3	Fe	17.3
Aspergillus terreus TFR2	JN 194186	$(NH4)_2\ MoO_4$	Mo	15.7
Rhizoctonia bataticola TFR6	JQ675307	$ZnNO_3$	Zn	18.5

In general, extracellular protein is mainly responsible for the production of various monodispersed nanonutrients. It has been noticed that 32 kDa protein weight group was largely responsible for breakdown of salts into nanoform. The breakdown of salts depends on released protein group's efficiency and the concentration of the protein of the organism. The major benefit of this process is that nanoparticles produced are naturally encapsulated by mother protein, which keeps them under stable condition till protein layer breaks.

14.3 CHARACTERIZATION OF MICROBIAL SYNTHESIZED NANONUTRIENTS

Microbial synthesized nanonutrients can be characterized after using particle size analyzer(PSA)/dynamic light scattering (DLS), scanning electron microscopy (SEM), transmission electron spectroscopy (TEM), UV-VIS absorption spectroscopy, atomic force microscopy (AFM), X-ray diffraction, energy dispersive X-ray spectroscopy, Fourier transform infrared spectroscopy (FTIR), inductively coupled plasma mass spectroscopy (ICP-MS), and inductively coupled plasma atomic emission spectroscopy (ICP-OES). The properties of nanonutrients mainly depend on the diversity of parameters such as particle size, dispersity index, surface area, porosity, solubility, aggregation, zeta potential, etc. The most important instrument is particle size analyzer or dynamic light scattering which can measure the average nanoparticle size with distribution range, polydispersity index of the particle that will indicate whether the particles are monodispersed or not, as well as can provide the zeta potential value that indicates the stability of the particle. It can also give the protein and colloid size. In this instrument, it is possible to conduct experiments on a wide range of sample buffer and wide temperature range.

SEM fabricates images of the object and furnishes the information of the surface structure of nanoparticles indicating their actual shapes. Moreover, the sample requires minimum preparation actions. TEM can exactly measure the length, breadth, and diameter of the nanoparticles. It can also determine the position of the particle with high resolution and magnification. UV-VIS may provide the information about the molecule or ion and the dimensions of absorption are proportional to the amount of the particle species absorbing the light. AFM may provide size information such as length, width, and height of the particle as well as other physical properties such as morphology and surface structure of the particle. It has high resolution than SEM and may furnish beautiful three-dimensional structure of the particle. X-ray diffraction can determine the arrangement of atoms as well as provide information about the mean position of the atoms, their elemental bonds, and their disorders beside other information. FTIR provides information of vibrational states of the molecule. The functional group involved in breakdown or absorption of the particle may also be educated by FTIR. It is a nondestructive and ecofriendly technique and provides better speed as well as reliable output. EDS will provide the chemical composition of the nanonutrients by peak height ratio relative to a standard. It may also provide the information of the purity percentage of the synthesized nanonutrients. ICP-MS and ICP-OES can be used for elemental analysis of the particle produced. Both the instruments can trace the nanonutrients in ppb level while ICP-MS may detect as low as 0.6 ppb concentration level and ICP-OES can detect up to 20 ppb concentration of the particle. They have the advantages of multielemental analysis at a time with minimum interference and maintain highest sensitivity.

14.4 APPLICATION OF MICROBIAL SYNTHESIZED NANONUTRIENTS

Mainly liquid nanonutrients are prepared through microbial technique. It may apply both foliar as well as soils. It can also be applied through drip, sprinkler, hydroponic, aqua, and aeroponic. But mainly foliar application is advisable where it can activate the plant co-enzyme systems and additional benefit from the plant activity can be obtained. The nitrile gloves and musk use is advisable to use during application of nanonutrients. Nebulizer or very fine nozzle sprayer should be used during spray to minimize the environmental losses (Tarafdar et al., 2012b). It was noticed that 33% loss of the material during spray with the use of normal sprayer can be reduced to 14% if nebulizer is used. The fertilizer effect period under nanosize has been increased to 40–50 days as compared to 5–10 days for chemical fertilizers. Countless experiments were conducted with different microbial synthesized nutrients to find out the optimum doses for foliar application. The results suggested that the doses differ with different nanonutrients as well as cereal and legume groups of plants. Moreover, the horticultural crops need 5–15% more concentration than the field crops. The optimum concentration standardized for foliar application is presented in Table 14.2.

TABLE 14.2
Concentration of Microbial Synthesized Nanonutrients Standardized for Foliar Application (modified after Tarafdar et al., 2018)

Type of Nanonutrients	Cereals	Legumes
N (ppm)	80–120	80–100
P (ppm)	40–60	40–50
K (ppm)	40–75	20–35
Mg (ppm)	20–30	20–25
S (ppm)	15–28	20–35
B (ppm)	04–06	10–12
Fe (ppm)	30–45	30–40
Mo (ppm)	06–09	02–04
Zn (ppm)	10–20	10–18

In general, nanonutrients' size less than 20 nm is more effective and less than 5 nm is most effective. The cube shaped particles are found to be better than plate, wire, or cage shaped particles for agricultural use due to their better and faster movement inside the cell. The rate of plant uptake of nanonutrients depends on the size and surface properties of nanoparticles. Nanoparticles can enter through the cuticle, stomata, hydathods, stigma, and other parts of the shoot as well as through the tips, rhizodermis/cortex, lateral root junctions, and wounding of roots (Tarafdar and Rathore, 2016). The most effect of microbial synthesized nanoparticles was observed when sprayed in the recommended doses on two weeks old plants. The nanonutrients may again apply to plants at critical growth stages for better results. The stability of foliar applied nanonutrients depends on the zeta potential value of the particle. Normally, beyond +30 mV and −30 mV zeta potentials are found to be stable for more than 2 years under room temperature.

14.5 EFFECT OF MICROBIAL SYNTHESIZED NANONUTRIENTS ON SOIL AND PLANTS

The nanonutrients after spraying may be adsorbed on the plant surfaces and enter through the plant openings as stated above. The uptake rate by plants generally depends on the size and shape of the particles. For example, smaller size nanonutrients can penetrate through the cuticle-free areas, such as hydathodes, the stigma of flowers or the stomata. After entering the nanonutrients particle possibly will travel through cell sap and gradually become the large particles due to agglomeration and some of them may accumulate in the vacuole (Tarafdar and Rathore, 2016). During travel they may trigger the plant co-enzyme system to make plants more active. Due to the triggering of plant co-enzyme system by nanonutrients during traveling, an enhancement of different beneficial enzyme activities in the rhizosphere was observed. In general, an increase in dehydrogenase activities

(25–68%), esterase activities (23–90%), acid phosphatase activities (21–72%), alkaline phosphatase activities (18–136%), phytase activities (23–83%), nitrate reductase activities (12–47%), and aryl sulphatase activities (19–68%) were noticed under different crops (Tarafdar, 2021b). Microbial synthesized nanonutrients also significantly enhance the water stress tolerance by enhancing root hydraulic conductance and water uptake of plants. It also helps to overcome the plant biotic and abiotic stresses such as salinity, drought, flooding, chilling, freezing, ultraviolet radiation, pathogenic attack, insect attack, herbivore attack, etc. The abiotic stresses can be controlled due to the production of primary metabolites with the help of nanonutrients while the biotic stress can be controlled by nanonutrients induced phytohormones production resulted rebuild the strong plant defense. The different microbial synthesized nanonutrients helping to overcome the plant stress conditions are shown in Table 14.3.

It is undoubtedly proved that use of nanonutrients is an emerging solution to overcome plant biotic and abiotic stresses.

Nanonutrients can make various morphological and physiological changes in plants. It has great positive impact on overall plant development when applied as recommended doses. It starts helping plant from seed germination, crop growth to the crop yield. Improved photosynthesis and biomass production noticed almost all the crops where microbial synthesized nanonutrients are applied. Magnetic nanonutrients enhanced the activity of different antioxidant enzymes, such as catalase (CAT), superoxide dismutase (SOD), and peroxides (POD) in many plants. These antioxidant enzymes can make strong defense system of plants to quench ROS generating free radicals. Nanonutrients have shown to positively affect the production of callus, shoots, roots including adventitious roots and hairy roots. Nanonutrients also found to influence different plant growth hormones such as indolebutyric acid (IBA), giberellic acid (GA3), abscisic acid (s-ABA), salicylic acid (SA), indoleacetic acid (IAA), etc. It may also enhance photosynthetic

TABLE 14.3
Nanonutrients Proved to be Efficient Against Plant Stress Conditions

Plant stress condition	Effective nanonutrients
Moisture stress	Zn, Fe
Salinity stress	Se, cerium oxide
Temperature stress	Se, Ag
Viral attack	Ag, Au
Insect attack	ZnO, Al, Ag, SiO_2, chitosan
Pathogenic attack	S, Zn, Cu, Ag, Au, TiO_2, Cu-chitosan
Nematode attack	ZnO_2, Ag, Al_2O_3, TiO_2, SiO_2
Weed attack	Ag, Cu, Fe, Zn, Mn, Ag-chitosan
Oxidizing agents	TiO_2, ZnO, ZnS, SnO_2

pigments, total protein, and total amino acids in the plant (Tarafdar, 2021b). Microbial synthesized nanonutrients were found to be much superior than chemical fertilizers for crop yield improvement, soil and environmental health, as well as long-term crop benefit.

Nanonutrients may help in soil aggregation, carbon build up, and soil microbial build with recommended doses of application (Tarafdar and Adhikari, 2015). It has been noticed that nanonutrients are very good for maintenance of soil health and far better than chemical fertilizer and gluconate-based product (Tarafdar, 2021a). In general, 33–81% improvement in soil aggregation, 10–14% more soil moisture retention, and 3–5% enhancement in organic carbon build up in the soil were noticed with the application of microbial based nanonutrients (Tarafdar, 2012; Tarafdar et al., 2012b). The nanonutrient particles possess low surface energy; therefore, they are more stable in soil. Some of the microbial nanonutrients like Zn and Fe have the property to activate polysaccharide production from the polysaccharide producing organisms. It has immense commercial importance. The types of polysaccharides generally produced are Dextron, Curdlan, Xanthan, and Pollulan.

14.6 NUTRIENT USE EFFICIENCY

The fertilizer industry is facing the problem with improving the nutrients' use efficiency of the existing fertilizers in the market. The use efficiency of nutrients applied as chemical or organic fertilizers continues to remain very low as 30–50% for N and K, 15–20% for P, and 2–5% for different micronutrients such as Zn, Fe, Mo, B, and Cu. As the fertilizer nutrients are costly and used in large quantities in global level, any increase in use efficiency may lead to a sizeable cut in nutrient requirement by the crops and massive economic benefit in the global level. Nanonutrients can enter through plant pores well, triggering the plant co-enzyme systems before absorbing, resulted more nutrient use efficiency. Due to their smaller size (between 1 and 100 nm), more surface area, and slow rate of release, plants can take up most of the nutrients with negligible waste. The important properties of nanofertilizer which help in more nutrient use efficiency are: higher surface area to facilitate the various metabolic process in the plant system, their high solubility in water, due to the smaller size it facilitates more penetration of the nanonutrients in the plant system, increased penetration of nanonutrients in the pores of root and leaves, more number of nutrient particles per unit area than other fertilizer group, more uptake of the plant nutrients to the crop. A comparison of some of the plant nutrients to their use efficiency between nanonutrients with chemical fertilizers is presented in Figure 14.1.

On an average 2–20 times increase in nutrient use efficiency was noticed when nutrients are applied as nanoform. The higher nutrient use efficiency is an essential qualification for expansion of crop production in marginal lands with very low nutrient availability. It may reduce the crop cultivation cost as well as biosphere pollution. It is very clear that nanonutrients can rectify the limited nutrient use efficiency and environmental constraints faced by chemical fertilizer.

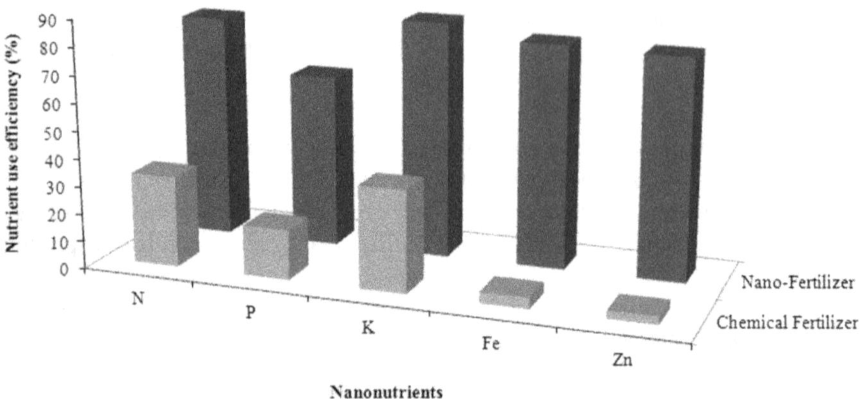

FIGURE 14.1 A comparison of average nutrient use efficiency between chemical fertilizers and nanonutrients (modified from Tarafdar, 2021b).

14.7 MICROBIAL SYNTHESIZED NANONUTRIENTS ON CROP YIELD

Microbial synthesized nanonutrients such as N, P, K, Mg, S, Fe, Zn, B, Cu, Mn, Mo, etc., can deliver exact nutrient requirements to the crops. Multiplication trails showed an increase in yield between 17–54% of different crops. On an average 24–32% increase in yield was noticed which was much higher than the 12–18% expected average increase in yield for chemical fertilizer application. The more effect of nanonutrients was observed on cauliflower (38–54%) followed by maize (25–38%) and potato (24–37%). Yield improvement of some of the nanonutrients on arid crops is shown in Figure 14.2.

In general, most effective size of nanonutrients is 5–40 nm to influence the maximum crop yield. N, P, and Zn are most potent nanonutrients influencing higher crop production. Substantial increase in root length, root area, dry biomass as well as nodulation under legumes was noticed with nanonutrients application as

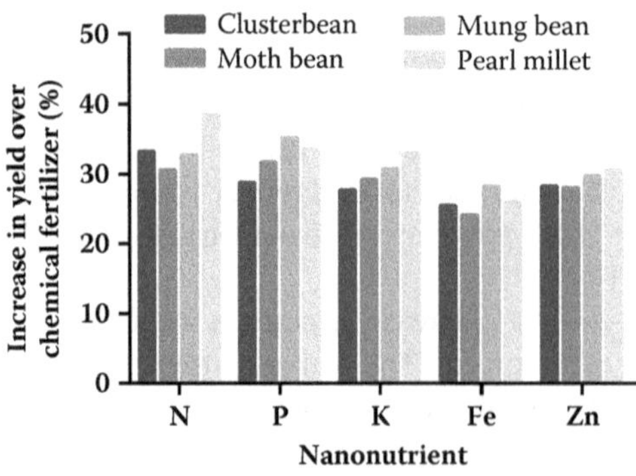

FIGURE 14.2 Yield improvement due to nanonutrients over chemical fertilizer (modified from Tarafdar, 2021a).

TABLE 14.4
Upgradation (%) of Microbial Population with Different Nanonutrients' Application (modified from Tarafdar, 2021a)

Biosynthsized Nanonutrients	Fungi	Bacteria	Actinomycetes
N	15–22	32–35	25–29
P	20–25	40–44	32–37
K	10–15	20–23	17–20
Mg	9–11	30–39	17–19
Fe	28–33	40–47	15–19
Mo	5–8	10–13	17–20
Zn	45–49	37–41	28–33

Concentration applied: N – 80 ppm, P – 40 ppm, K – 40 ppm, Mg – 20 ppm, Fe – 30 ppm, Mo – 6 ppm, and Zn – 10 ppm.

recommended doses irrespective of soil and climatic conditions. The improvement of root length varies between 2–32%, the root area has been improved between 4–21%, dry biomass has advance between 2–10%, and the nodulation has been enhanced between 5–68% of the different crops reported so far with the application of nanonutrients as compared to the chemical fertilizer application. The rhizosphere of the crops showed an upward improvement in microbial population in nanonutrient treated plants. The increase in population of beneficial micro-organisms may vary among the crops, environmental conditions, and soil types. The average increase in population under different crops with the recommended doses of nanonutrients application is shown in Table 14.4.

14.8 SAFETY ASSESSMENT OF MICROBIAL SYNTHESIZED NANONUTRIENTS

The risk linked to the application of biosynthesized nanonutrients is yet to be completely evaluated but the results obtained so far clearly indicated that biosynthesized nanonutrients application with recommended doses to the crops is completely safer. A number of experiments showed it has no adverse effect on seed germination, soluble seed protein content, rhizosphere microbial population, body weight, grain consumption, and blood pH of mice with the feeding of nanonutrients treated plants (Tarafdar, 2021b). The preclinical safety evaluation results on arid crops showed recommended doses of nanonutrients applied plants has manifest no pre-terminal death of the animal after taking the nanotreated food material. No abnormal clinical signs, behavioral activity etc. were noticed in animals that accept nanonutrients sprayed test material. As well as no significant effect on feed intake, body weight gain was perceived between the control and nanonutrients treated material intake (Tarafdar et al., 2018). No changes in gross necropsy and any organ weight was observed. Moreover, the plants grown under recommended doses of nanonutrients treatments did not persuade any adverse effect in rats even after feeding more than two and half times of the limit dose. Histopathology analysis of animal liver, kidney and spleen tissues disclose that oral exposure of test substances build no adverse effect as evidence by the normal tissue architecture. In general, gross architecture was intact with no perceptible necrosis or fibrosis within the analyzed tissue of the tested animals. Bioinformatics study with recommended doses of nanonutrients application showed some changes in the unigenes (Tarafdar et al., 2018); however, the changes are rather positive and helping in plant embryogenesis, seed germination, shoot growth, leaf formation, flower development, fruit ripening, carbohydrate metabolism, lipid metabolism, nucleotide metabolism, amino acid metabolism, and biosynthesis of secondary metabolism. Nanonutrient treated material may exhibit certain physicochemical properties, biokinetic behavior, and biological interactions that are different from the chemical or organic treated material.

Nanonutrients particle absorbed into the body by various roots may lead to immunological effects. But their effect mainly depends on their size, shape, composition, surface properties, protein binding and administration routes. Effects may happen from induction of reactive oxygen species, apoptosis, cell cycle inhibition, complement activation, enhanced secretion of cytokines and chemokines, interaction through toll-like receptors, etc. However, a systematic and quantitative analysis regarding the potential health impacts of nanonutrients as well as environmental clearance can lead to further application of nanonutrients. So far, it was noticed that nanonutrients coming out from microbial sources are very safe when applied with recommended doses.

14.9 FUTURE PROSPECTS

Biogenic nanonutrients' application has the potential to attain sustainability toward global food production. It can also reduce transportation and application cost to minimum level. Moreover, due to the smaller requirements soil does not get loaded with salts which is very common in case of conventional fertilizer application. Nanonutrients also can be administered according to the nutrient requirements of the intended crops. Biosensor may be attached to the nanonutrients that control the delivery of nutrients according to the soil nutrient status, growth period of a crop, crop demand as well as environmental conditions. Nanonutrients also may be used as nanobioformulations after blending with one or more beneficial micro-organisms helping plant growth. The formulations can be done through electrospinning, electrospray, or nanospray. The formulations may help in plant photosynthesis rate, improvement in plant growth and biomass production, increase in plant protein, reducing the stress conditions, more siderophore production, accumulation of more protein as well as more plant hormone synthesis besides higher crop production.

Microbial synthesized nanonutrients can very well use as seed treatment, resulting to increase in moisture content of seeds as well as give protection against soil borne diseases. Nanonutrients can effectively use in precision farming that help in minimizing inputs to get maximum crop yield or output. Nanonutrients can easily be used as nanosensor to monitor crop health and removal of soil contaminants. It can be used as herbicides in commercial vegetable crops. It may also operate against plant pathogen. It can help in minimizing agricultural run-off as well as water management. Bio-synthesized nanocoated pesticides may also reduce environmental and health risks. Nanonutrients have a character in symbiotic exchange between soil and plant system. Due to their slow and efficient release and native nutrient mobilization ability, all the required nutrients are boosted by the plants and reimpose the required and efficient energy in it that results in the drastic increase in yields. The most important benefits of biogenic nanonutrients over conventional fertilizers are: nanonutrient control the delivery of nutrients in crops through the slow and control release mechanism, required small amounts, negligible accumulation of salts in soil, it helps in more bioavailability of nutrients, helping plant from biotic and abiotic stresses as well as providing balance nutrition to the plants.

14.10 CONCLUSIONS

Microbial synthesized nanonutrients have a cost effective technology and has great future for global agriculture. It is an eco-friendly low-cost protocol for synthesis and commercial production. It has positive effect on plant growth and development, stress management, native nutrient mobilization, nutrient use efficiency as well as crop yield and soil health management. The recommended doses of application clearly showed no adverse effect on soil, plant and environment but have multiple benefits. It can be used globally to enhance seed germination, triggering of plant beneficial enzymes, rhizosphere microbial build up, overcome the plant stress conditions, nutrient use efficiency, soil carbon build up, higher physiological activities, more soil moisture retention and enhancement of plant photosynthesis. The major benefits over chemical fertilizers are: low requirement, more use efficiency, more enzyme release and plant physiological activities, low cost, higher crop yield and maintenance of soil health.

REFERENCES

Jeevanadam J, Chan YS, Danquah MK (2016). Biosynthesis of metal oxide nanoparticles. *Chem Bio Eng Rev*, 3(2): 55–67.

Kaul RK, Kumar P, Burman U, Joshi P, Agrawal A, Raliya R, Tarafdar JC (2012). Magnesium and iron nanoparticles production using microorganisms and various salts. *Material Sci – Poland*, 30: 254–258.

Kumar R, Pandey DS, Vijay P, Singh I, Singh P (2014). *Nanotechnology for better fertilizer use*. Reasearch Bulletin No. 201, G. B. Pant University of Agriculture and Technology, Pantnagar, Uttarakhand, India.

Raliya R, Tarafdar JC, Mahawar H, Kumar R, Gupta P, Mathur T, Kaul RK (2014). ZnO nanoparticles induced exopolysaccharide production by *B. subtilis* strain JCT 1 for arid soil applications. *Int J Biol Macromol* 65: 362–368.

Tarafdar JC (2012). Perspectives of nano technological applications for crop production. *NAAS News*, 12: 8–11.

Tarafdar JC (2020). Novel bioformulations for nano-phosphorus synthesis and its use efficiency. *Ind J Fert*, 16: 1278–1282.

Tarafdar JC (2021a). Bionanofertilizer: A new invention for global farming. In: *Biofertilizers*, Vol. 1, Chapter 26. Rakshit A, Meena VS, Parihar M, Singh HB and Singh AK (Eds.), Elsevier Inc., pp. 347–358.

Tarafdar JC (2021b). *Nanofertilizers: Challenges and prospects*. Scientific Publisher (India), pp. 363.

Tarafdar JC, Raliya R (2011). *The nanotechnology*. Scientific Publisher (India), pp. 215.

Tarafdar JC, Adhikari T (2015). Nanotechnology in soil science. In: *Soil science: An introduction*. Rattan RK, Katyal JC, Dwivedi BS, Sarkar AK, Bhattacharyya T, Tarafdar, JC and Kukal, SS (Eds.), Indian Society of Soil Science, New Delhi, pp. 775–807.

Tarafdar JC, Rathore (2016). Microbial synthesis of nanoparticles for use in agriculture ecosystem. In: *Microbes for plant stress management*. Bagyaraj DJ and Jamaluddin (Eds.), New India Publishing Agency, India, pp. 105–118.

Tarafdar JC, Raliya R (2019). Biosynthesis of metal nanoparticle from fungi. Indian Patent No. 311785.

Tarafdar JC, Raliya R, Rathore I (2012a). Microbial synthesis of phosphorus nanoparticles from tri-calcium phosphate using *Aspergillus tubingensis* TFR-5. *J Bionanosci*, 6: 84–89.

Tarafdar JC, Xiang Y, Wang WN, Dong Q, Biswas P (2012b). Standardization of size, shape and concentration of nanoparticle for plant application. *Appl Biol Res*, 14: 138–144.

Tarafdar JC, Rathore I, Kaur R, Jain A (2018). Biosynthesis of nanonutrients for agricultural applications. In: *NanoAgroceuticals and NanoPhytoChemicals*. Singh B (Ed.), CRC Press: Boca Raton, FL, USA, pp. 15–30.

Index

Page numbes in **bold** indicate tables; page numbers in *italics* indicate figures.

For Product Safety Concerns and Information please contact our EU
representative GPSR@taylorandfrancis.com
Taylor & Francis Verlag GmbH, Kaufingerstraße 24, 80331 München, Germany

www.ingramcontent.com/pod-product-compliance
Lightning Source LLC
Chambersburg PA
CBHW060327220326
41598CB00023B/2631
9780367653361